INTRODUCTORY MODERN BIOLOGY

INTRODUCTORY MODERN BIOLOGY

DR. S. SUNDARA RAJAN
Prof. and Head
Department of Botany
St. Joseph's College, Bangalore-560 025

ANMOL PUBLICATIONS PVT. LTD.
NEW DELHI-110 002 (INDIA)

ANMOL PUBLICATIONS PVT. LTD.
4374/4B, Ansari Road, Daryaganj
New Delhi-110 002

Introductory Modern Biology

First Edition 1999
ISBN 81-261-0259-4

PRINTED IN INDIA

Published by J.L. Kumar for Anmol Publications Pvt. Ltd., New Delhi-110 002 and Printed at Mehra Offset Printers, Delhi.

Contents

BOTANY

Preface

Introductory Modern Biology is a text cum reference book primarily meant for plus two level students. The contents of the book are compiled together keeping in view the syllabi of plus two level of many state boards as well as CBSE.

The subject matter has been dealt exhaustively, but keeping the narration in a straight forward and simple manner. The author hopes that this will help the students to easily grasp the intricate details of topics.

The author is grateful to Mr. J.L. Kumar and Mr. Negi, the M.D and General Manager respectively of Anmol Publications for undertaking the publication of the book.

Dr. S. Sundara Rajan

PART ONE
BOTANY

One

Chromosmes

The coloured thread like structures that appear during cell division have hidden within them all the secrets of genetic characters that are passed on from one generation to another generation. Housing the particles of inheritance, the genes, these vehicles of heredity are so important for all organisms; because it is these that delimit one individual from another. These are the chromosomes (Gr.*Chromo* = coloured, *soma* = body) which appear like coloured threads in the cell during divsion. In this chapter we will learn about the structure and functions of these unique threads of life.

History: It was Waldeyer (1888) who first coined the term chromosome to the darkly staining bodies of the nucleus. Fleming (1879) actually had described the splitting of the chromosomes, but he had coined the term chromatin to refer to them. Sutton and Boveri (1902) suggested that the chromosomes are the carriers of hereditary particles. Morgan (1933) discovered the role of chromosomes in the transmission of hereditary characters.

Chromosomes occur as a network of fibrils called choromonema in the interphase nucleus. At the time of cell division the chromonema breaks and condensation takes place forming specific number of chromosomes.

Number of Chromosomes: The number of chromosomes is constant for a given species. The number of chromosome in various species is as varied as the organisms themselves. They range from four to 500.

Ploidy of chromosomes: The total number of chromosomes present in the cell of an organism is called ploidy. In all sexually reproducing higher organisms the ploidy is of two types - haploid and diploid. The somatic cells contain diploid or two sets of chromosomes (2x). In these organisms during reproduction, the chromosome number would be halved in meiosis. As a result, the gametes have haploid or one set (x) of chromosomes. When two gametes fuse, the resultant zygote has two sets of chromosomes (diploid). It is this zygote which developes into the adult individual.

In lower organisms like many algae and fungi, the somatic cells have haploid chromosome numbers. In these organisms only the zygote will have diploid chromosomes. The meiotic division which halves the number of chromosomes takes place in zygote rather than during gamete formation. In some unusual cases the cells have many sets (polyploid) of chromosomes. In which case it is a multiplication of the base number called x.

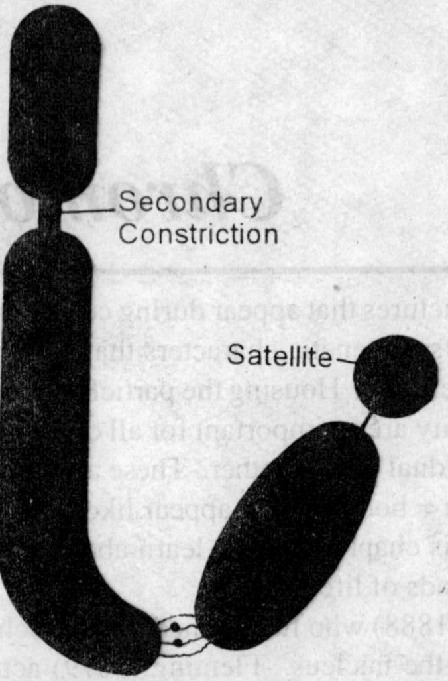

Fig. 1.1 External view of a Metacentric Chromosome

Group	Name of the organism	Chromosome number
Protozoa	*Paramecium*	30-40
Coelenterata	*Hydra vulgaris*	32
Nematoda	*Ascaris lumbricoides*	24
Chordata	*Rana esculenta*	26
	Homo sapiens	46
Fungi	*Mucor heimalis*	2
Pteridophyte	*Ophioglossum spp*	500
Gymnosperm	*Pinus ponderosa*	24
Angiosperm	*Allium cepa*	14

Size of the chromosomes: There is a great variety of chromosome size and shape. In size they range from 0.1 m to about 30 m in length. In exceptional cases however (polytene chromosomes) the chromosomes may be as long as 2mm. All the chromosomes in a species may be of similar size or they may vary. In *Yucca arakansana,* there are two distinct size groups. In some human beings there is a gradation in size. In some instance (*Mediola*) the cells at different regions (root and shoot tip) show different sizes of chromosomes. In general, the chromosomes of plants are larger than those of animals. Monocotyledons have larger chromosomes than dicotylendons.

Structure of the chromosome: The following are the parts usually associated with the chromosome.

Pellicle and matrix: Light microscopic studies have reveled that each chromosome is surrounded by a membranous structure called pellicle. This membrane encloses a jelly like substance called matrix. The presence of matrix has been shown in *Luzula campestris* (Juncaceae). Electron microscopic studies however have revealed, that there is nothing like pellicle enveloping the chromosome.

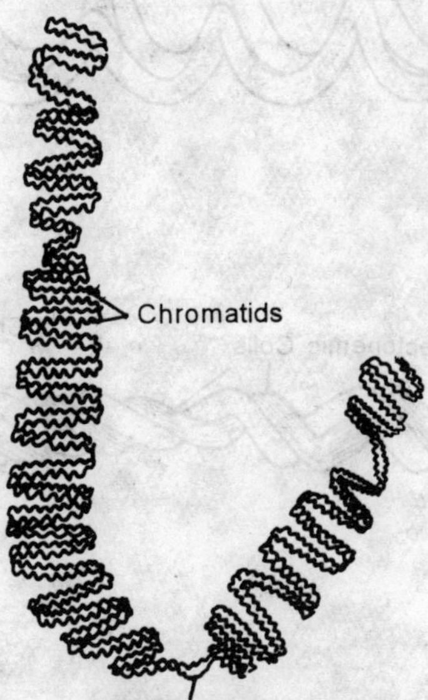

Fig. 1.2 Internal structure of a Metacentric chromosome

Chromonema: When the nucleus is resting, the chromosomal material will be in the form of a network called chromonema. At the time of cell division these break up and condense to form chromosomes. The metaphase chromosomes have two thread like structures called chromatids. The two chromatids are held together at the region of primary constriction (for details see later in the same chapter). Treatment of chromosomes with a protein hydrolysing enzyme (trypsin) will revel each chromatid to have two subchromatids; but there is no. Unanimity on this. Various chromosome models have proposed with chromonema coils ranging from 2 to 32. The chromonema coils (subchromatids) were first observed by Baranetzky (1880) in the pollen mother cells of *Trades-*

cantia. It was Vejdovsky (1912) who first gave the name chromonema to these coils. The number of chromonema fibrils not only varies with different species, but at different growth stages in the same cell.

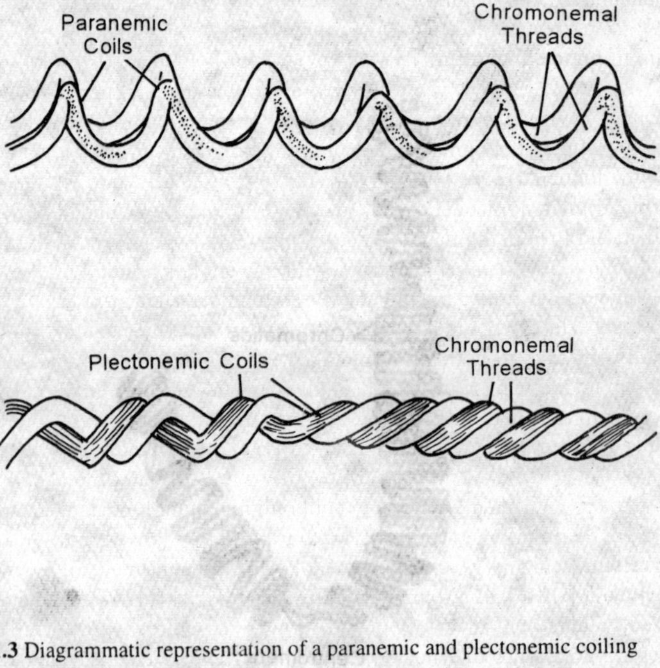

Fig. 1.3 Diagrammatic representation of a paranemic and plectonemic coiling

The chromonemal fibrils are coiled around each other in two different ways. These are- *paranemic* coils and *plectonemic* coils.

The paranemic coils are loosely coiled and the threads are easily seperable from one another. The plectonemic coils however are intertwined very closely so that they are not easily seperable.

The extent of coiling in the chromonema fibrils during cell division depends on the length of the chromosomes. These coiling are of three categories-

Major Coils: Each coil consists 10-30 gyres

Minor Coils: These are at right angles to the major coils and have many more gyres than major coils

Standard or somatic coils : These are coils found in the chromonema of cells during mitosis. They are similar to major coils..

Chromomeres: The chromonema in the chromosomes particularly during mitotic prophase appears to contain alternating thick and thin regions giving a bead like appearance in a chain. The thick bead like regions are called chromomeres and the region in between the two chromomeres is called the interchromomere. The position of chromomeres in the chromonema appears to be specific to a given chromosome in an individual.

What are chromomeres? Cytologists have given various interpretations; while some regard it as the regions of condensed nucleoplasmic material, others interpret it as regions of superimposed chromonemal coils. This view is supported by electron microscopic observations. Previously the chromomeres were mistaken for genes.

The chromomeres were first described by Balbiani in 1876 and later by Pfitzner in 1881. Belling (1928) had equated chromomeres with the genes in liliaceous plants. But McClintock (1944) has shown that genes are located in the inter chromomere region of the chromonema. However in case of polytene chromosomes the chromeric bands are associated with certain regions.

Centromere (primary constriction):

The stained chromosomes reveal a certain region or constriction which stains lightly. This is called centromere or kinetochore or primary constriction. The centromeres occupy a constant position relative to the ends of chromosomes and are responsible for the various shapes of chromosomes.

The centromere has a diameter of 0.2m and may occupy various positions on the chromosome. It (centromere) consists of spherules or granules (Kinosome) and interchromomeral fibrillae. The number of fibrillae per centromere varies. The chromonema are less coiled in the centromeric region and hence stain lightly. (Lime-de-Faria 1955).

Most of the organisms have their chromosomes with a single centromere (monocentric). Certain other chromosomes have two centromeres (dicentric). In hemipteran insects however the chromosomes have many centromeres (polycentric). Such a centromere is said to be a diffused centromere. The proof for centromere being diffuse comes from the fact that when such a chromosome is fragmented by X-Rays, each bit functions like a typical chromosome, replicates itself and participates in the division cycle. Diffused centromere has also been reported in *Luzula purpurea* of the family Juncaceae (Brown 1954). According to some cytologists, chromosomes with a diffused centromere are more primitive than the ones with a localized centromere. The polycentric chromosomes found in the nematode *Ascaris megalocephala* are unique in the sense they are restricted to only those cells which constitute the germ line (Swanson 1957).

The position of the centromere in the chromosome (which is constant to a given chromosome) varies ie., it may occupy different positions. Based on this, four morphological shapes have been identified in chromosomes. These are-

1. Metacentric:

The Centromere occupies a middle position with reference to the length of the chromosome. The two arms thus resulted are almost equal in length. During anaphasic movement in cell division, metacentric chromosomes appear 'V' shaped. eg: *Trillium,. Tradescantia* etc.

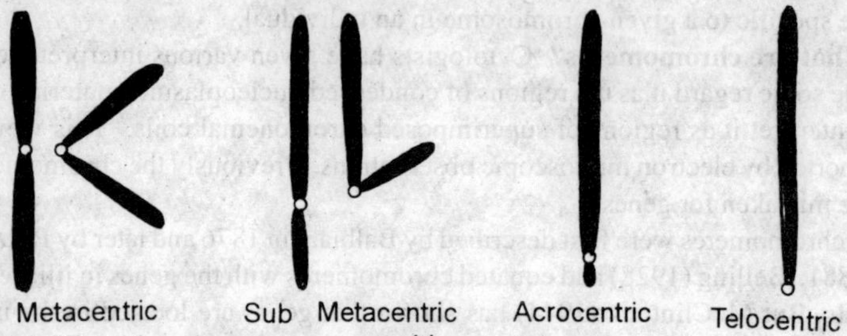

Fig. 1.4 Types based on centromere position

2. Sub metacentric:

When the centromere is located some distance away from the middle region of the chromosome, the position is said to be median and the chromosome sub metacentric. As a result, one arm of the chromosome will be shorter than the other. During anaphasic movement, sub metacentric chromosomes appear 'L' shaped eg: Human beings.

3. Acrocentric:

In this case, the centromere is situated almost near one end of the chromosome. As a result, one arm of the chromosome will be extremely short and the other one very long. The centromere is said to occupy a subterminal position. eg: Grass hoppers

4. Telocentric:

When the centromere is situated exactly at one end, the chromosome will be having only one long arm. Telocentric chromosomes are very rare. Truly telocentric chromosomes have been identified by Marks (1957) in certain plants, protozoa and certain mammals. Cleveland (1949) has also reported in certain protozoa inhabiting the digestive tract of wood termites, the occurrence of telocentric chromosomes. The earlier report regarding chromosome IV of *Drosophila melanogaster*, being telocentric has been disproved as a second arm has been seen.

Acentric chromosome:

In certain instances, chromosomes may undergo fragmentation resulting in one fragment without a centromere. Such a chromosome fragment is said to be Acentric.

Arm ratio:

The ratio of length between the long arm and short arm is said to be the arm ratio, which is constant for a given chromosome. The arm ratio is less in metacentric chromosomes and high in acrocentric chromosomes.

Centromeric chromomeres:

The centromere should not be mistaken or equated with a chromomere as the former stains lightly and is a constriction while the latter is a swelling on the chromosome found throughout its length. Metaphase chromosomes which have two chromatids exhibit in their centromere four granular structures. These granules called centromeric chromomeres are about 0.5m in size and are arranged in a square.

Functions of centromere:

1. Centromere provides mobility to the chromosome. Chromosomes lacking in centromere fail to orient themselves properly on anaphase plate, lag behind and finally get eliminated.

2. Centromeres decide the shape of the chromosome.

3. Centromeres attach themselves to the spindle fibres during anaphasic movement.

4. The centromere divides along the longitudinal axis of the chromosome at the beginning of anaphase. Very rarely the centromere may divide at right angles to the chromosome giving rise to the pieces of centromere, each bit holding the arms parallel to each other and genetically alike (because of duplication)as they represent the replicated parts of a single chromatid. Such chromosomes with genetically identical arms (having same genes on both arms) are called *isochromosomes* (Darlington 1939).

Secondary constrictions:

These are lightly staining areas on the chromosome other than centromere. One or more secondary constrictions may be found in the chromosomes. The part of the chromosome (arm) present beyond the secondary constrictions is called a *satellite*. There may be one or two satellites depending on the number of secondary constrictions. Chromosomes with satellites are called SAT- Chromosomes (Sine Acid Thymonucleinico - without DNA). Generally in a genome at least one chromosome will have a satellite.

Secondary constrictions associated with the nucleolus are called nucleolar organisers. Usually in a diploid organism two homologous chromosomes have additional constrictions where nucleolus will be located during the interphase stage of the nucleus. During prophase, the nucleolus disappears to appear again in the identical area.

Telomeres: The terminal part of chromosomes are called telomeres which have polarity, preventing the tips of other chromosomes from fusing with it. Telomeres however attach themselves to the nuclear envelope during interphase.

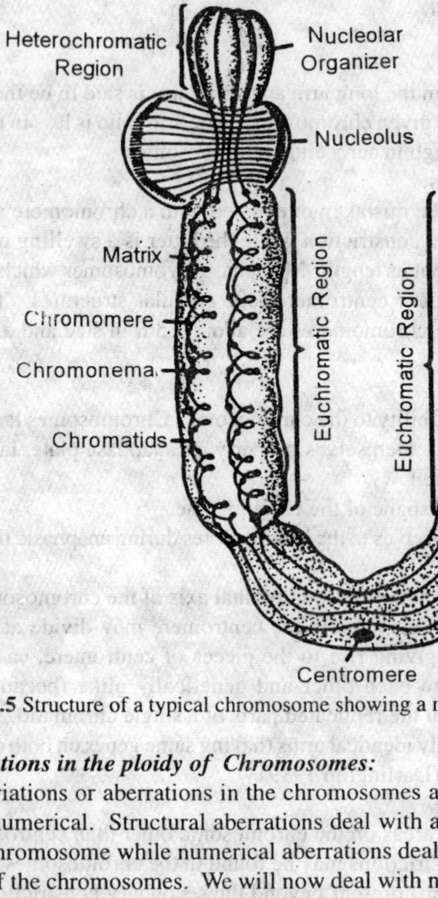

Fig. 1.5 Structure of a typical chromosome showing a nucleolar organiser

Variations in the ploidy of Chromosomes:

Variations or aberrations in the chromosomes are of two kinds - structural and numerical. Structural aberrations deal with alteration in the structure of the chromosome while numerical aberrations deal with alteration in the number of the chromosomes. We will now deal with numerical aberrations.

NUMERICAL ABERRATIONS

In the biological hierarchy, species form the basic genetic unit capable of surviving and maintaining its structure. This constancy (subject of course to mutational influences) of character in a species has as its basis, the chromosomal background. The number and kinds of chromosomes in a species are sufficiently constant to delimit it from others. But as the genes mutate or change, so also chromosomes are subject to variations. The process of alteration in chromosome numbers seen very often in plants and to a less extent in animals is called numerical aberrations.

Numerical aberrations in chromosomes produce two types of cells or individuals. These are 1) those whose somatic genome (chromosome complement) is the exact multiple of the basic number çharacteristic of the species and 2) those in which the somatic number is an irregular multiplication of the basic number; occasionally there could be reduction in the basic number also.

The first category is said to be *Euploid* and the second one is called *Aneuploid.* Euploid may be Haploid (monoploid), diploid, triploid, tetraploid and polyploid. Haploid is the basic number of the species with one set of chromosomes in the body cells, diploids have 2 sets, triploids have 3 sets, tetraploids have 4 sets and further multiplication series are refered to as polyploid.

Individuals whose genome complement is aneuploid have one or two extra chromosomes added or one or two chromosomes deleted from the normal number of diploid in the body cells. For instance a single chromosome may be missing from the diploid complement when it is called 2n-1: The following is the classification of numerical aberrations.

ANEUPLOIDY: In Aneuploidy, there is either reduction or partial increase of the chromosome number. Reduction is said to be hypoploidy, while increase is said to be hyperploidy. Hypoploidy is due to loss of one or two chromosomes. If a single chromosome is lost from a diploid complement it is said to be Monosomy (individuals are called monosomics) and is symbolically designated as 2n-1. If two chromosomes are lost (ie. both the homologues in a pair) it is said to be Nullisomy (2n-2). Similarly the addition of a third chromosome to one of the bivalents is called Trisomy (2n+1) and addition of 2 chromosomes to a bivalent is said to be tetrasomy (2nucleic acids+2).

Monosomy: In monosomic diploid individuals, (in the genome) one of the pairs of homologues would undergo a loss of one chromosome and it becomes a univalent while all others are bivalents. The diploids cannot generally withstand such a loss, while polyploids can withstand it. This is due to the fact that polypliods (see later in the same chapter) have more than two homologues for every pair. Theoretically the number of monosomics possible in an individual equals its haploid number. In common wheat where there are 21 pairs of chromosomes; there are 21 possibilities for monosomics. E.R. Sears has artifically induced these 21 monosomics in a wheat variety called Chinese spring. Monosomics have been induced in cotton (Endrizzi et al) and tobacco (Clausen and Cameron).

While diploids with monosomics do not survive, there are instances as in tomato, where monosomics could be produced. Here the chromosome number is 2n=24. Double monosomics (2n-1-1) have two homologues lost from two different bivalents. During meiotic prophase together with the normal bivalents there will be two separate univalents indicating that both these have lost their partners. Similarly in triple monosomy one homologue each is lost from three different bivalents and three univalents appear at the meiotic prophase.
Aneuploidy has been noticed in human beings also. A classical instance is the Turner's syndrome named after the discoverer H.H. Turner (1938). It occurs in .23 infants for every 1000 births.

Nullisomy: In this, an entire pair of homologous chromosomes will be missing (2n-2). This should not be mistaken with double monosomy where 2 chromosomes are lost from two different pairs. In nullisomics, the metaphase plate at meiosis I will apparently show no abnormality and only a comparison with the normal can detect the loss of one whole pair of chromosomes. Nullisomics have been identified in wheat and other plants.

Trisomy: In trisomic individuals one of the pairs of homologues will have an additional member. Trisomic individuals exhibit a trivalent together with the normal bivalents during meiotic metaphase. These are designated as 2n+1. There could be a double trisomy in which case two of the bivalents will have one additional member. At the metaphase of meiosis I, there will be two trivalents. The number of possible trisomy in individuals equals the number of haploid set. For eg. in barely where the haploid number is 7, seven trisomics are possible.

In plants, trisomy has been studied extensively in *Datura*, maize, tomato and *Nicotiana.* In animals, trisomy has been reported in *Drosophila* and others including human beings.

Blackslee and his associates have described a series of trisomy in the Jimson weed (*Datura stramonium*) belonging to the family Solanaceae. The base chromosome number of this species is 12(n=12) and 12 primary trisomics are possible. In this case instead of 12 bivalents at the metaphase of meiosis I, there will be 12 trivalents. Each one of this is associated with a distinct fruit character and can be separated from the others. Blackslee has named these trisomics (as represented by the fruit character) as follows - *Rolled, Glossy, Buckling, Elongate, Echinus, Cocklebur, Microcarpic, Reduced, Poinsettia, Spinach, Globe* and *Ilex.* Secondary and teritary trosomics have also been reported from *Datura.*

In *Drosophila,* trisomy has been reported for the 4th chromosome; the fly can survive this and breed also.

Trisomy has also been reported in human beings leading to many abnormalities, both in the physical features as well as in mental faculties. Down's syndrome (mongoloid idiocy) was first described by Langdon Down in 1866. Affected individuals are short in stature, have slanting eyes, broad skulls, and stubby hands. They have low mental faculties. It was Lejuene (1959) who first associated this disorder with a chromosomal abnormality. Cytological investigations have revealed that in Down's syndrome, chromosome 21 is produced due to primary non disjunction occuring at meiosis in the mother. The two chromatids do not separate and as a result the egg has both the chromatids of 21, while the first polar body receives none. The incidence of Down's syndrome occurs quite frequently (1:700) among European people. (see later for details)

Other trisomic abnormalities in human beings include the Patau syndrome (Patau 1960), Edward's syndrome (Edward J.H, 1960) and Kleinfeter's syndrome (H.F. Kleinfelter 1942). The first two are autosomal trisomics while the third one is a trisomy for the sex chromosomes.

Tetrasomy: In tetrasomy, a particular bivalent has two extra members; in other words there will be four homologous chromosomes in one pair, Symbolically tetrasomy is represented as 2n+2. tetrasomics have been indentified in wheat (E.R. Sears).

EUPLOIDY: Euploid individuals could be haploids, diploids or polyploids and represent the multiplication of the entire genome.

Haploids: Haploid or *n* number of chromosomes in the body cells may be normal or abnormal. In lower plants and animals, haploid condition is normal in the sense, here the individuals have only one set of chromosomes in the body cells, while diploid condition is represented only in the zygote. This is of normal occurrence. Haploidy is of interest when it occurs in diploid individuals. Here normally the body cells have 2 sets of chromosomes. Reduction division takes place and haploid gametes are formed. These gametes however have no morphological expression ie. they cannot directly develop into new individuals. They fuse with another gamete resulting in the formation of diploid individual. Rarely however haploid individuals are produced as during parthenogenesis when the egg directly develops into a new individual without fertilization. Haploids are of two kinds *euhaploids* and *aneuhaploids.* While the former are derived from euploids the latter are derived from aneuploids.

Origin and production of Haploids: In some insects haploid individuals are produced naturally due to parthenogenesis. In plants also parthenogenetic development of egg results in haploid individuals. Such haploids have been seen in cotton, tomato etc.

In rare instances, pollen tube, or microspores may develop into new individuals. Such individuals are called androgenic haploids as they are produced from the male tissue. Occasionally synergids, antipodals etc may also develop into haploid embryos.

Haploids can artificially be obtained by tissue culture of the anther, pollen, X ray treatment, colchicine treatment, distance hybridization etc.

Anther culture has yielded excellent haploids in *Nicotiana* and Potato. Similarly interspecific corsses may also result in haploid individuals. According to Kasha (1970), in a cross between *Hordeum vulgare* X *H.bulbosum;* chromosomes of *H.bublosum* are eliminated in early zygotic division. Embryosacs of these can be cultured to obtain haploid individuals.

Diploidy: The presence of 2 sets of chromosomes in the body cells is called diploidy. This is a normal condition in most of the higher plants and animals. In lower organisms however the body cells have only one set of chromosomes-haploid.

POLYPLOIDS: In polyploid individuals, the body cells have more than two sets of chromosomes .

Characters of polyploids: Polypoid individuals are generally larger in size and stronger in stature than their diploid counterparts. They are distributed widely in the plant kingdom and it has been estimated that quite a large number of flowering plants are polyploids.

Classification of polyploids: There are mainly three categories of polyploids-

1. Autopolyploids
2. Allopolyloids and
3. Autoallopolyploids

Autopolyploids: Here the same chromosomes set is multiplied. For instance if the diploid set in AA, auto polypoid would be AAAA. Autopolyploids are produced when there is somatic doubling of chromosomes.
If for example during mitotic cell division, after the duplication of chromosomes if the two groups fail to migrate to opposite poles, a polypoid cell would result.

Autopolyploids are widely distributed in the flowering plants. The common Indian grass *Cynodon dactylon* is an autotriploid and survives mainly by vegetative propagation.

Allopolyploids: These are produced by crossing two different species with no chromosome homology

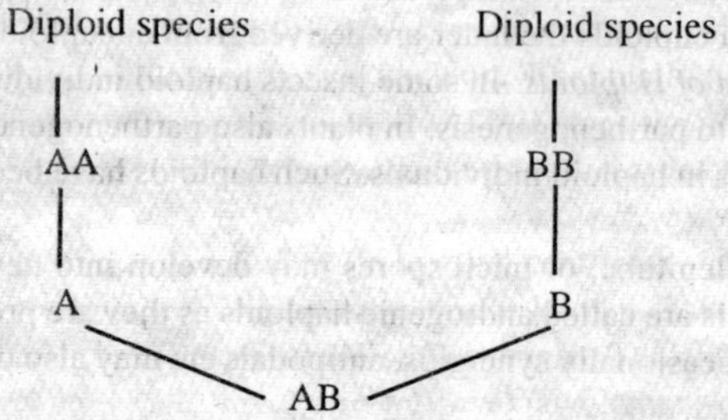

The F_1 hybrid AB would be sterile as the chromosomes of A and B are completely non homologus and can not pair. If the chromosomes of the F_1 hybrid are doubled they become AABB and the individual is called an allopolyploid. This individual is fertile as A can pair with A and so also B. As distinct from autopolyploids, allopolyploids have multiplication of different genomes (chromosomes).

A classical example of allopolyploid is provided by *Raphanobrassica* reported in 1927 by the Russian scientist Karpachenko. He crossed *Raphanus sativus* and *Brassica oleracea* and obtained an F_1 hybrid which was completely sterile. The sterility was due to complete non homology between the genomes of *Raphanus* and *Brassica*. But some plants of *Raphanobrassica* were fertile obviously due to the doubling of the chromosomes.

Autoallopolypoid: These, represent a higher series of polypoidy (octoploidy) in which there are 8 sets of homologues. It can be genetically represented as AAAA BBBB. This is produced by crossing two different species as follows-

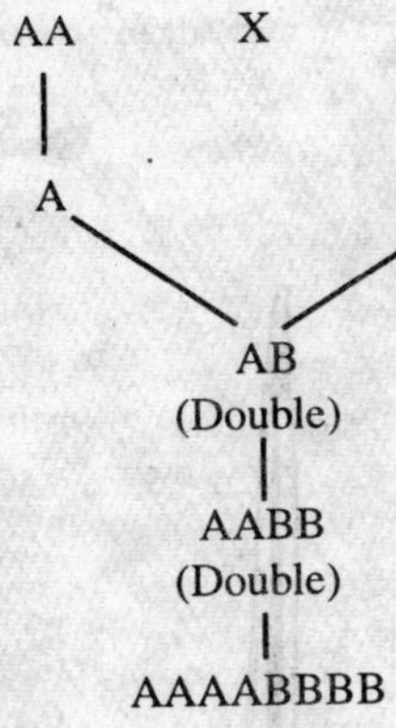

Down's syndrome: This abnormality in human beings was first reported by Down and later studied by Lejuene, Gautier and Turpin in 1959. Also known as Mongoloid idiocy, this syndrome is characterized by mental retardation and other clinical symptoms.

Some of the other morphological manifestations of this abnormality are- mongoloid facial appearance, webbed neck, a flat nasal bridge, presence of epicanthic fold (small fold of the skin over the inner corner of the eye as seen in Mongoloid race), brush field spots around the Iris, short fingers and a gap between the first and second toe. The most striking feature of this syndrome however is mental retardation. Individuals suffering from Down's syndrome will have an IQ between 25-50 as against the normal IQ of 80. Dermatoglyphic pattern (lines on the palm) shows a simian (as in monkeys) crease. Frequently all 10 fingers have ulnar loops.

Down's syndrome is a result of abnormality in chromosome no. 21. In this pair, instead of there being two homologus chromosomes, there will be three chromosomes. This condition is known as *trisomy.* Individuals with Down's syndrome will have 47 chromosomes instead of the normal 46. This abnormality results due to non separation of the duplicated chromosome during meiosis. As a result, some eggs will have 24 chromosomes instead of 23. If these are fertilized by sperms (which have normal 23 chromosomes) the resultant-zygote will have 47 and the individual will show Down's syndrome.

A detailed clinical study of Down's syndrome has been made in Western countries. It has been found out that trisomy of chromosome 21 occurs in one in 700 births and it is the most common of autosomal trisomics.

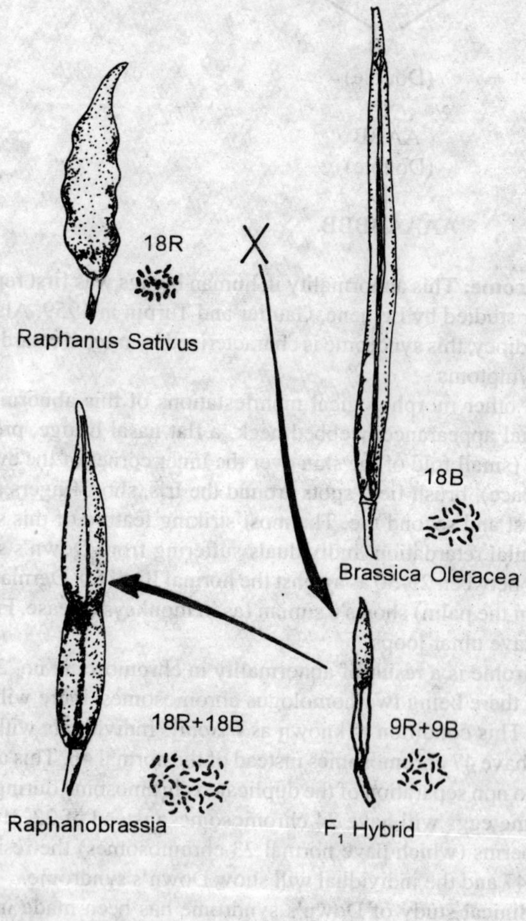

Fig. 1.6 Origin of Raphanobrassica

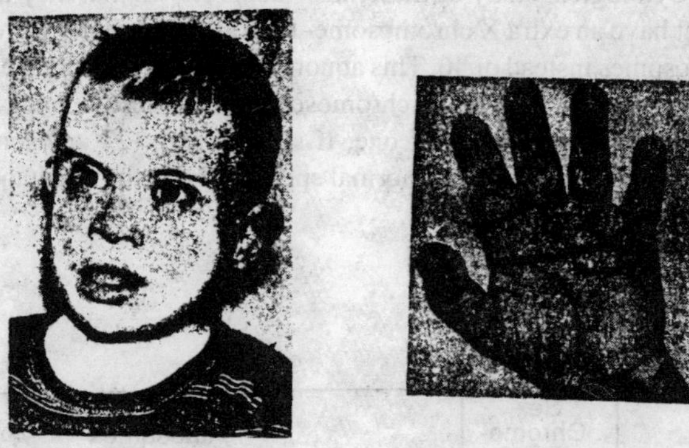

Fig. 1.7 Facial feature of a Down syndrome (left), Hand of a child with Down syndrome showing simian crease (right)

There is a relationship between the increased age of the mother and the incidence of Down's syndrome among the children. Down's syndrome individuals live for varying number of years. A Down's male is usually sterile, while females rarely produce children.

There is no cure for Down's syndrome as it is a genetic disorder.

Kleinfelter's Syndrome: This is another chromosome abnormality. While Down's syndrome is due to autosomal abnormality, Kleinfelter's syndrome is due to the abnormality of sex chromosomes.

Individuals exhibit abnormality only during adulthood, childhood being normal. Affected individuals are males. The symptoms are-

(i) Absence of spermatogenesis

(ii) Excessive secretion of gonodotropins in the urine

(iii) Genitalia underdeveloped
(iv) Sparse body hair
(v) Some degree of breast development
(vi) Mental retardation
(vii) Effeminate behaviour

Cytological study of these individuals shows that they are basically males, but have an extra X chromosome- they are XXY. They have a total of 47 chromosomes instead of 46. This abnormality occurs during meiosis due to the non separation of the two X chromosomes. As a result, some eggs will have 2X chromosomes instead of one. If such an egg (22 autosomes + 2X chromosomes) is fertilized by a normal sperm (22+Y) the resultant individual will be a Kleinfelter's Syndrome.

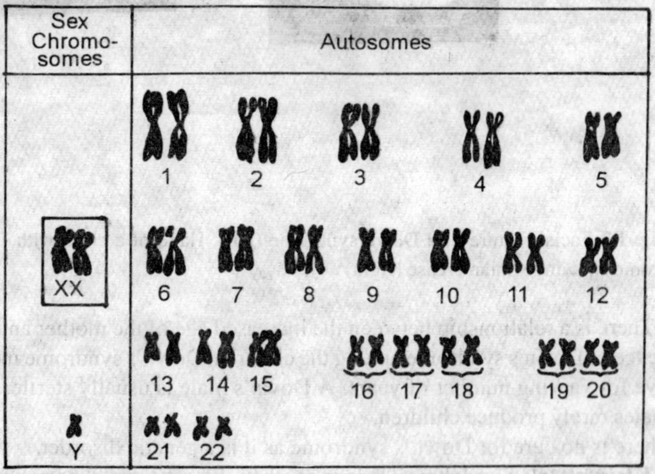

Fig. 1.8 Chromosome compliment of a person suffering from Klinefelter's syndrome

Turner's Syndrome: This is seen in females and is characterized by-
(i) Dysgenesis of gonads
(ii) Sexual infantilism

(iii) Webbed neck
(iv) Prominent ears
(v) Increased carrying angle of arms (cubitus valgus)
(vi) Dystrophy of the nails and hypoplastic nipples

Individuals with Turner's syndrome have only 45 chromosomes ie., 44 (22 pairs) autosomes and only one X chromosome instead of the normal 2 X chromosomes (XO). This arises due to meiotic abnormality when an egg does not receive any X chromosome. If such an egg is fertilized by a sperm carrying an X chromosome, the result will be 44 autosomes and only one X chromosome. One in 5000 births (in west) is estimated to be affected by this syndrome.

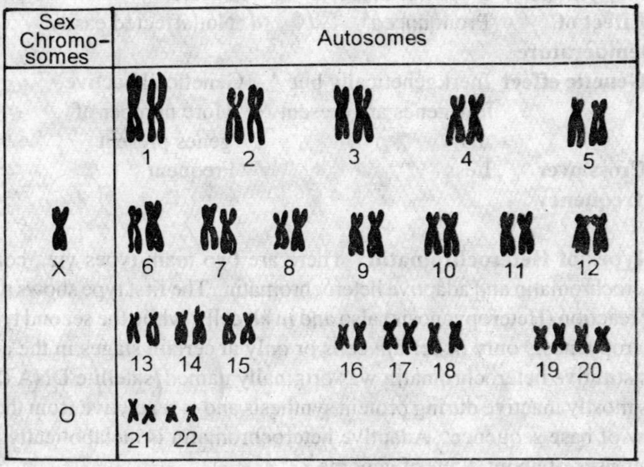

Fig. 1.9 Chromosome compliment of a person suffering from Turner's syndrome

Euchromatin and heterochromatin: Based on the staining reaction two regions may be identified in the chromatin material. These are Euchromatin and Heterochromatin.

The euchromatin stains positively with the DNA specific stains (basic fuchsin) indicating a concentration of DNA. This region is genetically active and stains lightly. The euchromatic regions are supposed to represent areas of less condensation..

Heterochromatin stains more deeply than euchromatin and represent highly condensed regions on the chromosome. In the interphase nucleus, the heterochromatic regions form condensed structures called chromocentres or false nucleoli. The following table summarises the differences between euchromatin and heterochromatin

	Heterochromatin	Euchromatin
1.Staining	Stains deeply	Stains less deeply
2.Coiling	Found in the region of more coiling and condensation	Found in the region of less coiling and condensation
3.Replication	Late replication replication in late 'S' phase	Early replication replication in early 'S' phase
4. Acetylation	Does not occur	Occurs
5. Effect of temperature	Pronounced	Not affected easily
6. Genetic effect	Inert genetically but few genes are present	Genetically active. More number of genes present
7. Cross over frequency	Less	Frequent

Types of Heterochromatin: There are two main types viz., constitutive heterochromatin and adaptive heterochromatin. The first type shows dark staining reaction (Heteropycnosis) also and in all cells, while the second type shows heteropycnosis only in certain cells or only at certain stages in the cell cycle. Constitutive heterochromatin was originally named satellite DNA (S-DNA). It is mostly inactive during protein synthesis and is repetetive from the point of view of base sequence. Adaptive heterochromatin is metabolically inert and comprises of about 2.5% of genome.

Heteropycnosis: This is the staining phenomenon seen in the chromosomes with reference to heterochromatic and euchromatic regions. The heterochromatic regions which stain deeply are said to exhibit heteropycnosis as against euchromatin which stain lightly. However it should be noted that the heterochromatin always does not stain deeply with reference to euchromatin. While the euchromatic region is always constant in its behaviour (non heteropycnotic) towards staining (has constant coils), the hetrochromatic region may stain deeply or lightly and this is known as differential heteropycnosis. Differential heteropycnosis is of two types-positive and negative.

The areas of heterochromatin showing positive heteropycnosis are more condensed and deeper staining than euchromatic region. In areas of positive heteropycnosis there is more of nucleic acids and proteins. Chromosomes with positive heteropycnosis in prophase appear like metaphase chromosomes. The areas of heterochromatin showing negative heteropycnosis are possibly less condensed and stain lighter than euchromatic region. Chromosomes with negative heteropycnosis have a hazy outline and metaphase chromosomes appear like prophase chromosomes.

CHROMOSOME MODELS: The chromosome consists of DNA and proteins, but the way in which the two are arranged in the chromosome is a matter of debate. The main aspects of the chromosome structure are a) the number of strand or strands of DNA b) whether the DNA molecule is continuous through out the length of the chromosome or whether it is interrupted and c) what is the mode of association of protein and DNA.

Various chromosome models have been proposed and these fall into two basic categories viz, Multiple strand models and single strand models. The following model is discussed in this book.

Nucleosome model: This model proposed by Woodcock (1973) suggests that the chromatid is a string of beads made up of repeating units. These beads called nucleosomes are oblate particles with a diameter of 110A°. They have a core of histones 40A° high and 80A° wide, around which is found DNA in a super helical structure. Finch and Clug (1976) have proposed an arrangement of chromatin in the chromosome model of nucleosomes. The coiled DNA according to them is in the string of nucleosomes. The nucleosome string is then coiled into 300A wide solenoid, which in turn coils to form a supersolenoid with a diameter of 4000A°. This super solenoid structure is the unit fibre (chromonema seen in light microscope) whose coils constitute the chromosomes.

The exact nature of the chromosome arrangement of proteins and nucleic acids may be said to be still an open question.

UNUSUAL CHROMOSOMES: The above discussion on chromosomes pertains to normal ones. There are however abnormal or gaint chromosomes encountered in many instances. These are - The polytene chromosomes in the salivary glands of fruit flies, the lamp brush chromosomes in the oocytes of amphibians, and the supernumerary or B chromosomes.

Polytene Chromosomes: These multithreaded chromosomes were first reported by Balbiani in 1881 while observing the squash of salivary cells of *Chironomous.* Subsequently they have been seen in the larvae of dipterans (*Drosophila* etc). The name polytene was suggested by Kollar due to the appearance of many threads. Besides salivary glands the polytene chromosomes also occur in rectal epithelium and malphigian tubules.

The nuclei in the salivary gland cells are much larger than the normal ones and are about 25 m in diametre with the giant chromosomes larger by 50-200 times than the normal chromosomes. The polytene chromosome in the salivary cells of *Drosophila melanogaster* (Painter 1933) are about 200m in length. The enormous size is due to the duplication of chromonema which do not separate. According to an estimate, the polytene chromosomes have 1000 times more DNA than the normal somatic chromosomes.

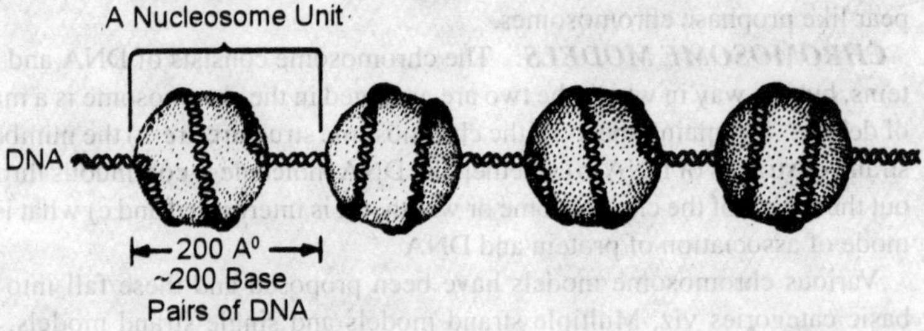

Fig. 1.10 Nucleosome model

Structure: Each polytene chromosome is transversely straited by alternating bands and interband regions. In *Drosophila,* over 5000 bands have been found in the four chromosomes of the salivary gland cells. The bands take an intense stain (Feulgen) while the interband regions stain less intensely.

The bands have chromomeres arranged at right angles to the long axis of the chromosome. The band actually represents regions of tight coils of the chromonema. Van Herwerden (1911) interpreted these bands as gyres of a continuous spiral. This view was later supported by Kostoff (1930). The bands have a high DNA content and absorb U.V. light at 2600 A°.

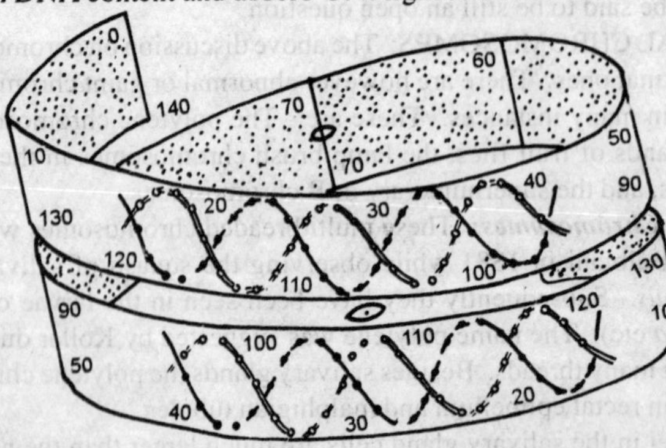

Fig. 1.11 Schematic representation of DNA super helix in the nucleosome

The interbands stain lightly and they are negative to feulgen stain indicating a low concentration of DNA. The chromonema in the interband region are less coiled and absorb very little U.V. light.

Painter (1933) and Bridges (1936) have shown that in *Drosophila* the bands are clearly associated with the genes. Association of certain defects due to mutated genes have also been associated with the bands. Certain recent studies however have shown that together with the bands, the interband regions also have genes.

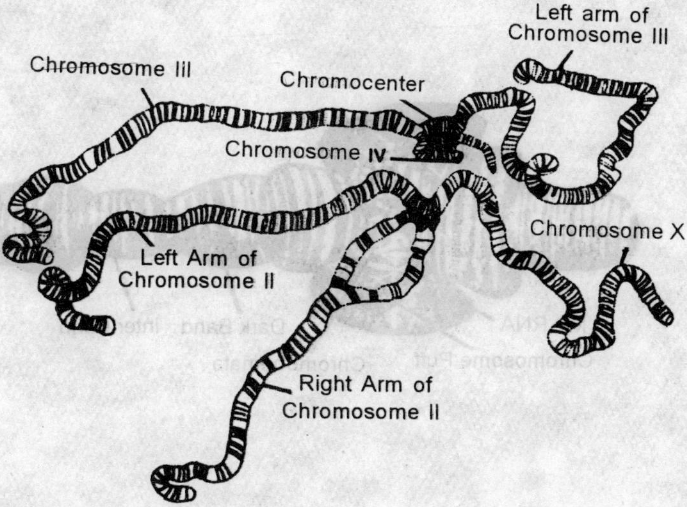

Fig. 1.12 Polytene chromosomes of Drosophilla

Chromosome puffs: There are certain interesting structures associated with the bands in the giant chromosomes called chromosome puffs or Balbiani rings. The swelling in bands at certain stages are given the name chromosome puffs. It is quite possible that these puffs are associated with the metabolic activities and represent areas of active RNA synthesis. According to Beerman and Bahr (1954), the coils in the puff region open out to form many loops. The larger chromosome puffs are called Balbiani rings and they show a high DNA and mRNA content.

Lamp brush chromosomes: The giant chromosomes found in the oocytes of fish, amphibians, reptiles and aves look like the brush used to clean the chimney of a kerosene lamp-hence the name lamp brush chromosomes. These are very large and are characterised by loops arising out of the main axis laterally. Discovered first by Flemming is 1882, in amphibian oocytes, they were studied in great detail by J.Rucket (1892)in the oocytes of shark. Rarely lamp brush chromosomes have also been found in some invertebrates (*Sepia*,

Echinaster, insects etc) but they are not as large as those of vertebrates. It seems they are present in certain plants also (Grun 1958).

In some organisms, lamp brush chromosomes as long as 1000m and 20m wide have been reported during early prophase I. Towards the end of the prophase, they get reduced in size. There are reports of lamp brush chromosomes as long as 5900 microns in some salamander oocytes. These chromosomes are elastic and can be stretched.

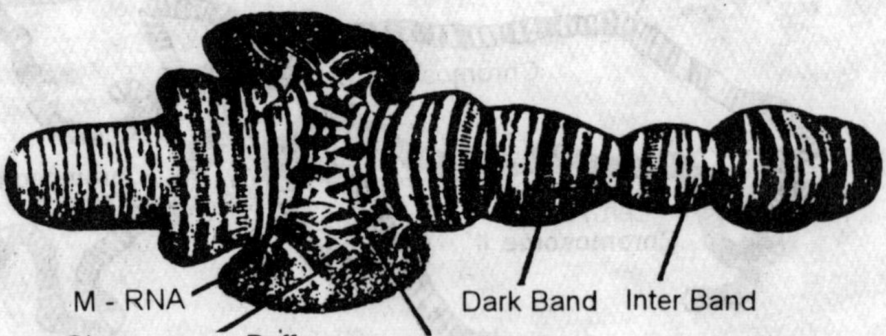

Fig. 1.13 Polytene chromosomes of Dipteran Insect

Structure: There are two main parts viz., the main axis and lateral loops. The main axis has two chromatids (the homologues have four) and is rich in DNA. The axis is in continuity with the loop axis. The tips of the axis are without loops and are the regions of telomeres. The centromeric region is also without loops. At certain regions along their length the chromonema of the axis gets coiled and forms the chromomeres. The chromomeres are found in pairs. Electron microscopic studies of the lamp brush chromosomes of the salamander, Triturus (Miller and Beatty 1969) have shown the axial fibre containing dense granuels which are probably the molecules of the enzyme RNA polymerase.

The loops are of two main types viz. *typical* and *special.* Most of the loops which are typical consist of a bent central axis from which are given out RNA fibrils. These RNA fibrils are of progressively increasing length as a result one side of the loop is pronouncedly thick. The special loops are symmetrical and have granules at the end of RNA fibrils. Each loop axis varies from 30-50A. Axial fibre is made up of DNA, while the matrix has RNA and proteins.

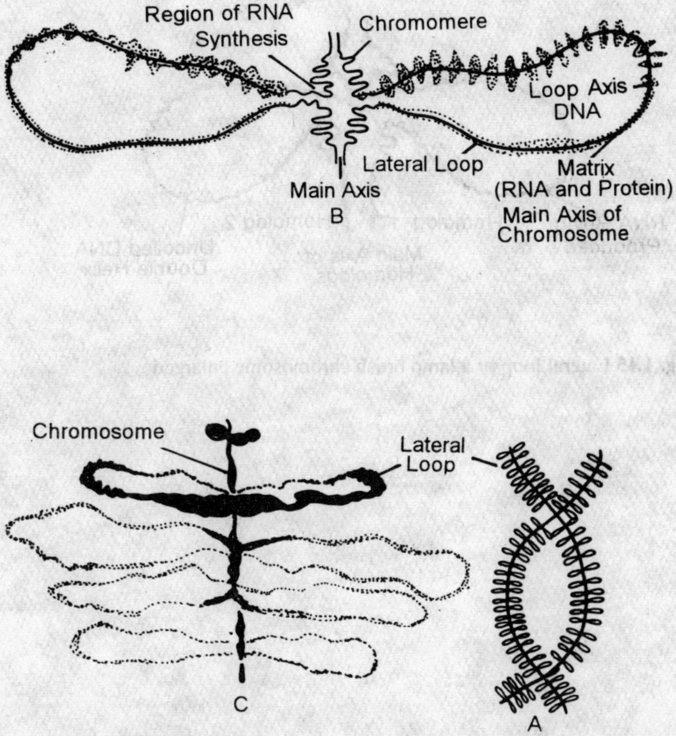

Fig. 1.14 Lampbruxh chromosomes. A-Paired homologues with two chiasmata, B-Variation in loops, C-A chromosome with an attached loop.

Lamp brush chromosomes are involved in the synthesis of RNA and proteins. Each loop is believed to represent one long operon consisting of repetetive cistrons (structural genes). Each gene locus codes for RNA. The loop is supposed to synthesise at a high rate because of repetetive gène sequence. There are also reports that the lamp brush chromosomes help in the formation of yolk material in the egg.

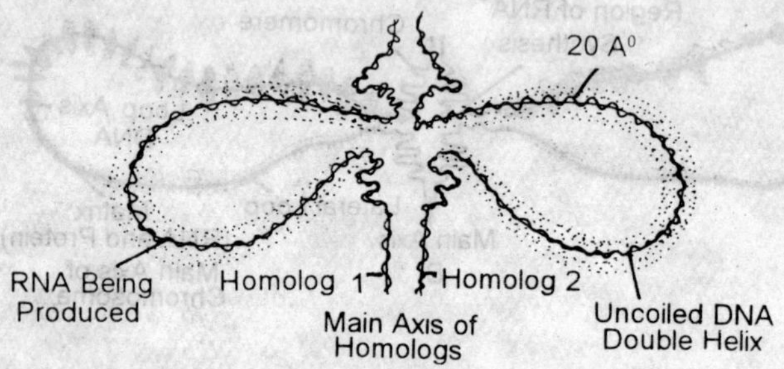

Fig. 1.15 Lateral loop of a lamp brush chromosome enlarged

Two

Nucleic Acids

These are the most important constituents of the nucleus playing a vital role in the inheritence of characters from one generation to the other. There are two types of nucleic acids - Deoxyribose nucleic acid (DNA) and Ribose nucleic acid (RNA). While both constitute the genetic material in different organisms, it is predominantly the DNA which is the genetic material in most of the oragnisms. The following are some of the important milestones in the discovery of Nucleic acids.

1. Meishcer (1869) isolated nucleic acids and called them nuclein.
2. Hertwig (1884) suggested a heriditary role to the nuclein.
3. Kossel (1884) recognised purines and pyrimidines in the nucleic acids.
4. Feulgen (1912) devised basic fushsin dye positive only to DNA.
5. Astbury and Bell (1938) showed that the bases in DNA molecule are arranged one above the other.
6. Avery, Macleod and Macarty (1944) presented evidence for DNA to be the genetic material.
7. Chargaff (1950) showed that DNA contains equal amount of purines and pyrimidines.
8. Wilkins et al (1950) identified the molecular distance between the bases.
9. Furberg (1952): DNA is a coil of single chain.
10. Pauling and Corey (1952): DNA is a helix of three chains.
11. Watson and Crick (1953) demonstrated the double helix model for DNA molecule.
12. Meselson and Stahl (1958) demonstrated the semi conservative replication in DNA.
13. Jacob and Monad (1961) postulated the presence of mRNA in protein synthesis.

Deoxy Ribose nucleic acid (DNA): DNA is the main genetic material in almost all organisms except for certain viruses. In prokaryotes like *Escherichia coli*, the genetic material is a single giant molecule of *DNA*. While *DNA* is mainly concentrated in the nucleus, it has also been reported in cytoplasmic organelles like chloroplast, mitochondria etc.

Chemical composition of DNA: DNA extract from many living organisms has revealed it to be a complex macromolecule. It is a long chain polymer composed of monomers (units) called nucleotides. Each nucleotide inturn is composed of three sub units. These are - a) nitrogen base, b) sugar and c) phosphate.

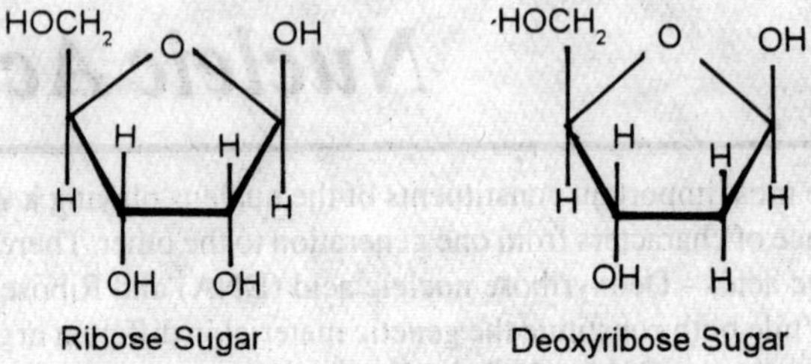

Fig. 2.1 Pentose sugars of nucleic acids (left ribose; right deoxyribose)

Nitrogen bases: The nitrogen bases are of two kinds viz., pyrimidines and purines. Both pyrimidines and purines are heterocyclic compounds, with a basic six membered ring in which one or the other carbon position is replaced by nitrogen.

The pyrimidines have a basic six membered ring of benzene in which, position 1' and 3' are replaced by nitrogen. There are two kinds of pyrimidines-*cytosine* and *thymine.* Structurally cytosine has a single six membered ring with a hydroxyl group at the 2' position and an amino group at the 6' position (*2 hydorxy, 6 amino pyrimidine*). Thymine also has a single six membered ring with hydorxyl groups at the 2' and 6' position and a methyl group attached to the 5' position (*2,6 hydroxy,5 methyl pyrimidine*).

The purines have a double ringed structure. Attached to the six membered benzene ring is a 5 membered imidazole ring. The imidazole ring is joined to the hexagonal ring at 4' and 5' positions. The benzene ring has nitrogen in the 1' and 3' position (replacing Carbon).

There are two types of purines viz. *Adenine* and *guanine.* Adenine has an amino group at the 6' position (*6 aminopurine*), while guanine has an aminogroup at the 2' position and a hydroxyl group at the 6' position (*2 amino, 6 hydroxy purine*).

Sugar: The sugar molecule in DNA is basically a ribose sugar. The ribose sugar is a *pentose sugar.* It is a 5 membered ring. The ring consists of four carbon atoms while the 5' C atom is outside the ring. Positions 1', 3' and 5' in the ring have OH groups while the 2' carbon position has only H. There is no O; hence the sugar is said to be Deoxyribose (no oxygen in the second carbon position).

Phosphate: It has one central phosphorus with four oxygen atoms.

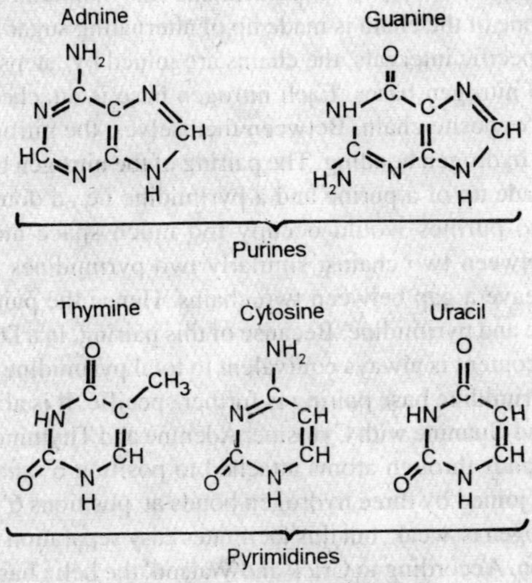

Fig. 2.2 Nitrogen bases of nucleic acids

Nucleosides: A nitrogen base coupled with sugars forms a nucleoside. DNA has four nuleosides - a)Deoxycytidine b) Deoxythymidine, c) Deoxyadenosine and d) Dexoyguanosine. The sugar molecule is attached to the nitrogen base to the 3' position in the benzene ring in the case of pyrimidines and to the 9' position of the imidazole ring in the case of purines.

Nucleotides: A nucleotide is derived from the nucleoside by the addition of phosphoric acid molecule to the latter. The phosphate group is usually linked to the two sugar moleucles above and below. The linkage is with the 3' C atom of one deoxyribose sugar and with the 5' C atom of another. DNA has four nucleotides- *Deoxycytidilic acid, Deoxy thymidilic acid, deoxy adenylic acid* and *Deoxyguanilic acid.*

Polynucleotide: DNA is said to be a polynucleotide chain with a number of nucleotides joining together with the help of DNA polymerase.

Molecular model of DNA structure : A molecular model for DNA structure first proposed by J.D. Watson and F.H.C Crick (1953) has been universally accepted. They were awarded Nobel prize for their discovery. The work of Crick and Watson has been supported by X-ray diffraction picture obtained by Wilkins and his associates.

According to this model, a DNA moleucle is composed of two chains that are spirally coiled and form a double helix, much like a spiral stairway. The distance between the two chains is uniform and it is maintained by base pairing. The back bone of the chain is made up of alternating sugar and phosphate molecules. At specific intervals, the chains are joined by 'steps'. Each step is made up of two nitrogen bases. Each nitrogen base is attached to the sugar molecule of the opposite chain. Between themselves, the nitrogen bases of a set are joined by hydrogen bonding. The pairing of the nitrogen bases is always specific. It is made up of a purine and a pyrimidine i.e., a double ring and a single ring. Two purines would occupy too much space and can not be accomodated between two chains; similarly two pyrimidines would form a short step and leave a gap between two chains. Hence the pairing is always between a purine and pyrimidine. Because of this pairing, in a DNA molecule, the total purine content is always equivalent to total pyrimidine content.

The purine, pyrimidine base pairing is further specific. It is always Adenine with Thymine and Guanine with Cytosine. Adenine and Thymine are joined by two hydrogen bonds through atoms attached to position 6' and 1'. Cytosine and Guanine are joined by three hydrogen bonds at positions 6' 1' and 2'. the bonding of hydrogen is weak; but this facilitates easy separation of two strands during replication. According to Crick and Watson, the helix has a diametre of 20 A° and there is one complete turn of the helix at every 34 A° and has a stack of 10 nucleotides at every turn.

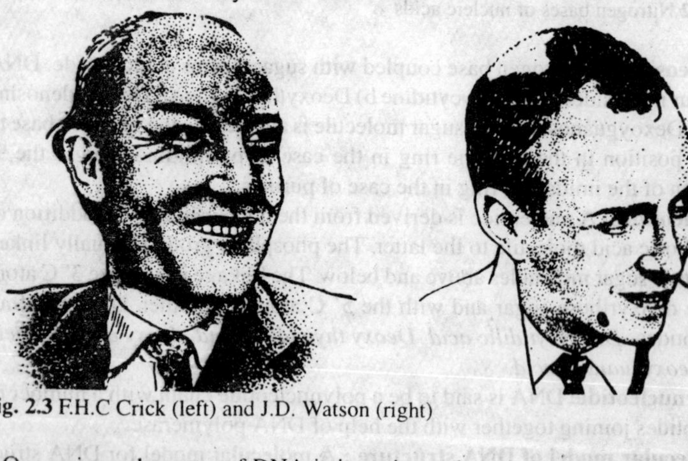

Fig. 2.3 F.H.C Crick (left) and J.D. Watson (right)

One unique character of DNA is its unity and yet infinitismal variety. The DNA structure mentioned above is uniform in all organisms, whether it is man or virus. At the same time it is also true that DNA is the genetic material that decides the characters of individual organisms. Then how does DNA decide the individuality? The answer lies in the base pair sequence. The combination of A/T, G/C is uniform, but their sequence is not; for instance in one individual it may be AT, GC, AT, AT, AT, in another- GC, GC, AT, GC, AT etc. There is no limit to the variety in which base pair sequences are arranged. This is the secret of individual differences. In no two individuals, the base pair sequnece is same.

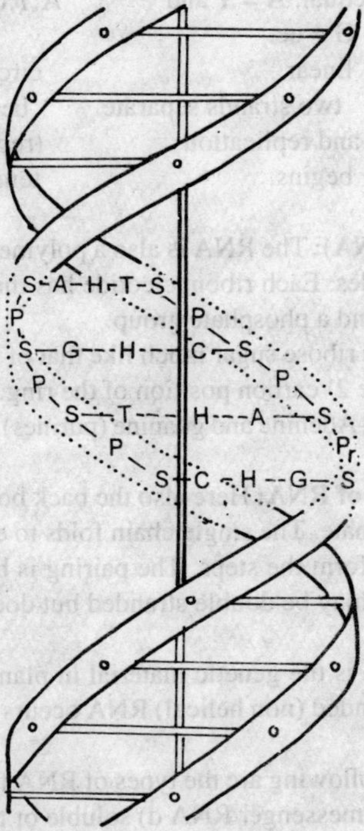

Fig. 2.4 The double helix model of DNA

Unusual bases in DNA: The normal bases in DNA are A, T, G and C. In some viruses live PBS1 and PBS2 uracil replaces thymine. In some bacteriophages C is replaced by 5 hydroxy methyl cytosine.

Single strand DNA: Although DNA in most organisms is double stranded, in some phage viruses (ϕ x 174), single stranded DNA is found. The following table summarises the differences between doulbe stranded DNA and single stranded DNA.

Character	Double strand DNA	Single strand DNA
1. U.V. absorption	Constant from 10-80°c and then rises rapidly	Steady increase from 20° to 90°c.
2. Action of formaldehyde	Resistant	Not resistant
3. Ratio of base pair	equal. A = T and G = uc.	A:T:G:C is 1.33:0.98:0.75
4. Shape	linear	circular
5. Replication	two strands separate and replication begins.	becomes double stranded (replicative form) temporarily.

Ribonucleic acid (RNA): The RNA is also a polymer of nucleotides. These are called ribonucleotides. Each ribonucleotide has three parts viz. a pentose sugar, a nitrogen base and a phosphate group.

The pentose sugar is a ribose sugar much like that of DNA, but with oxygen (present) attached to the 2' carbon position of the ring.

The nitrogen bases are-Adenine and guanine (purines) and cytosine and uracil (pyrimidines).

Molecular strucutre of RNA: Here also the back bone is made up of alternating sugar and phosphate. The single chain folds to enable the base pairs to meet here and there to form the steps. The pairing is between A-U and C-G. Occasionally the RNA may be double stranded but does not form a helix like DNA.

Single stranded RNA is the genetic material in plant viruses, poliomyletis viruses etc. Double stranded (non helical) RNA occurs as the genetic material in reoviruses.

Types of RNA: The following are the types of RNA found -a) Genetic RNA b) Ribnosomal RNA c) messenger RNA d) soluble or transfer RNA

Genetic RNA: In certain viruses, RNA is the only genetic material, DNA being absent. This RNA is self replicating; it is called RNA dependent RNA synthesis. The viral RNA directly functions as a messenger and produces RNA polymerase enzyme as well as proteins.

Ribosomal RNA (rRNA): This is found in the ribosomes. It constitutes about 80% of total RNA of the cell. The base pairs of rRNA are complementary to that part of the nuclear DNA from where they are produced. rRNA consists of a single strand folding upon itself at some regions. The folded regions have paired bases, while the other regions have unpaired bases. Due to this in rRNA, there is no purine pyrimidine equality.

There are three types of rRNA based on sedimentation and molecular weight.

These are a) rRNA with molecular weight over a million, e.g. 21s-29s rRNA, b) rRNA with molecular weight less than a million. e.g. 12s-18s rRNA, c) rRNA with very low molecular weight (40,000) e.g. 5s rRNA.

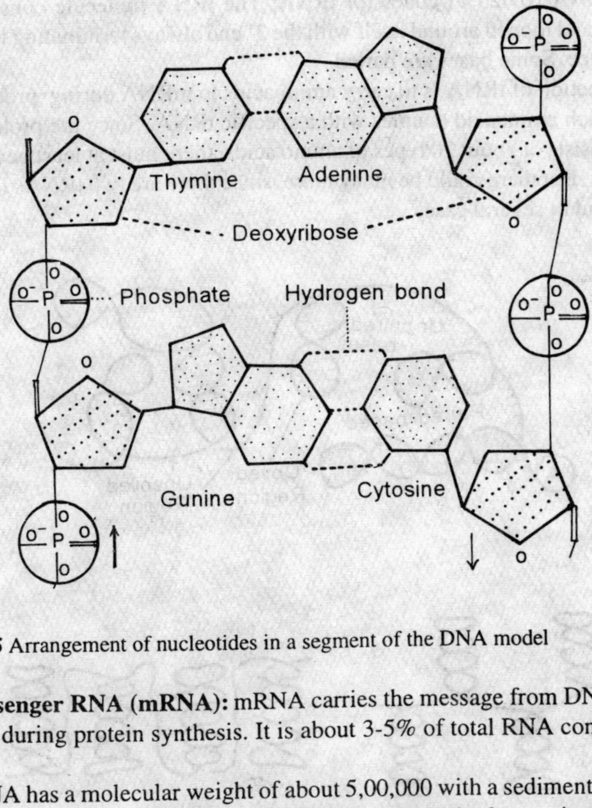

Fig. 2.5 Arrangement of nucleotides in a segment of the DNA model

Messenger RNA (mRNA): mRNA carries the message from DNA to ribosomes during protein synthesis. It is about 3-5% of total RNA content of the cell.

mRNA has a molecular weight of about 5,00,000 with a sedimentation coefficient of 8s. This however is highly variable. mRNA is always single stranded and there is no base pairing, even though random coiling may be observed. The mRNA molecule is produced from the coding strand and its base sequence is complementary to the coding strand. The following are some of the general features of mRNA.

1. *Cap:* This is a blocked methylated structure found at the 5' end of RNA.

2. *Non coding region:* This follows the cap and consists of 10-100 nucleotides. There are no codes for proteins in this region.

3. *Initiating codon:* This region begins the coding.

4. *Coding region:* Codes for proteins and has about 1500 nucleotides.

5. *Termination codon:* The coding region is followed by a terminating codon, which stops translation.

Transfer RNA (tRNA or sRNA): The transfer RNA also called soluble RNA, is next in quantity to rRNA. It has a low molecular weight of about 25,000-30,000 and the sedimentation co efficient is 3.8s. It is made up of about 70-90 nucleotides.

tRNA is synthesised in the nucleus on a DNA template. Only a minute fraction of DNA (0.025%) codes for tRNA. The tRNA molecule consists of a single strand looped around itself with the 3' end always terminating in a C-C-A sequence. Some bases are paried.

The function of tRNA is to carry aminoacids to mRNA during protein synthesis. Each aminoacid couples with a specific tRNA. Since the protein molecule consists of about 20 types of amino acids, there must at least be 20 types of tRNAs. But there could be many more, since there are two tRNAs for every amino acid in several cases.

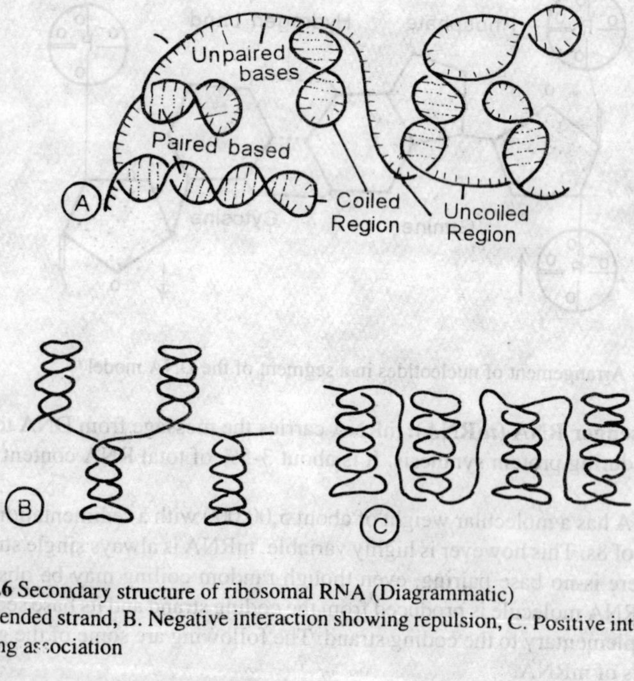

Fig. 2.6 Secondary structure of ribosomal RNA (Diagrammatic)
A. Extended strand, B. Negative interaction showing repulsion, C. Positive interaction showing association

Secondary structure of tRNA

This refers to the arrangement of the chain of the molecule to enable it to function in a proper way. Many models have been proposed for the secondary structure of tRNA. Of these, clover leaf model proposed by Holley (1966) is most widely accepted.

According to the clover leaf model, the single chain of tRNA is folded on itself to form five arms. As a result of this folding, the 3' end and 5' end lie adjacent to each other. Each arm consists of a basal stalk, and a loop. The doulbe stranded stalks show the typical base pairing. There could be an exceptional mismatch.. The loops do not have base pairing. One of the arms which does not have a loop is called the acceptor as this is the binding site for amino acids. The seoncd arm is called the DNA arm. This has a synthetase site which recognises the amino acid activating enzyme. The fourth is a variable arm; it has a stem, but the loop may or may not be present. The T ψ arm has a ribosome recognition site.

There is also a tertiay structure to the rRNA. This is brought about by H bonding between bases, and between bases and sugar phosphate backbone.

REPLICATION OF DNA:

DNA has two important properties - Heterocatalytic function i.e. to order the synthesis of other molecules like proteins and Autocatalytic function i.e. to reproduce or to make copies of itself. The second function of DNA is unique and remarkable. The following is a brief account of replication of DNA.

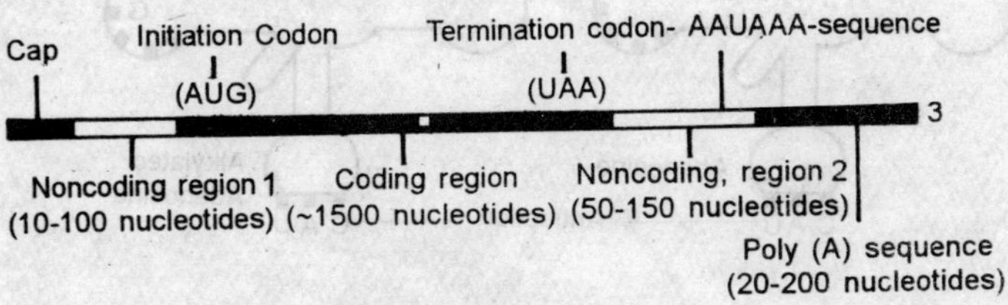

Fig. 2.7 General features of eukaryotic mRNA

DNA replication usually takes place during interphase between two mitotic cycles. According to Watson and Crick whose model has a beautiful inbuilt mechanism of replication, the two coils of the molecule get separated due to the dirsuption of hydrogen bonds between base pairs. The two strands separate and unwind. The replication usually starts at a specific point called origin. According to some, the replication begins with an incision made by an enzyme. The unwinding of the coils separates the base pairs. The nucleus consists of many free nucleotides. These are attracted to the complementary nucleotides on the strands and polymerization begins. The original helix acts as a template. The nucleotides freshly added to the template form their own helical back bone with the help of sugar and phosphate.

The initiation of DNA synthesis requires an RNA primer. This is synthesised by the DNA close to the replication site. This is catalysed by RNA polymerase.

The nucleotides are added to the template with the help of DNA polymerase. The RNA primer is degraded while another DNA polymerase enzyme releases the DNA strand. The synthesis orginally of RNA primer is necessary since DNA cannot initiate by itself the formation of a new chain. Replication is discontinuous and produces short fragments of DNA called *Okazaki fragments.* In some instances a continuous replication may also take place. The okazaki fragments later join to form long strands with the help of the enzyme *polynucleotide ligase.* Replicaion may be in one direction or in both directions.

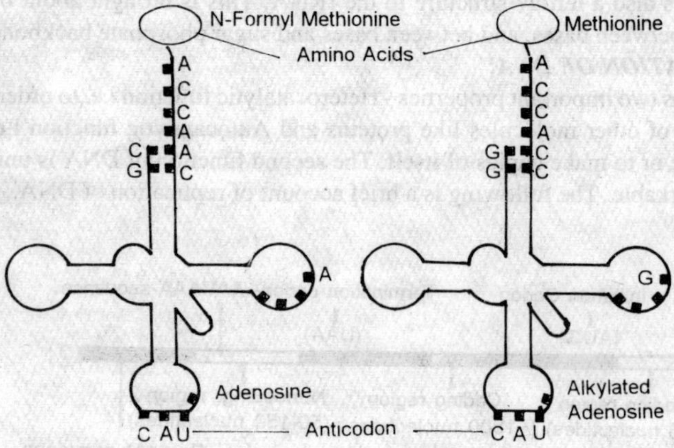

Fig. 2.8 Secondary structure of tRNA (Clover leaf model)

Models of replication in eukaryotes

The linear DNA molecule in eukaryotic organisms can theoretically replicate according to any one of the following types suggested by Delbruck. These are *Conservative, Semi conservative* and *Dispersive.* According to the conservative model, of the two molecules of double helices of DNA formed after replication, one would be completely new and the other completely old. The parent helix remains unchanged.

In the semi conservative model supported by Crick and Watson, each strand of the two molecules of DNA formed after replication would have one old strand and one new strand. This is quite possible because the original DNA strand acts as a template and the newly formed chain binds to the old strand.

The dispersive model of replication suggests that the DNA helix breaks at several points forming many fragments. Each fragment would replicate and the fragments rejoin randomly to form two double helices. Each double helix has a mosaic of old and new fragments of DNA.

The available evidences seem to suggest that the semi conservative hypothesis is true for DNA replication. When the two helices of them unwind, each helix can act as a template and hence the new molecules formed must be hybrids ie. each having one old and one new strand. Taylor et al (1951) demonstrated by autoradiographic methods that chromatids during prophase have half old and half new material.

The experimental evidence of Meselson and Stahl (1958) have given conclusive support for the semiconservative hypothesis. They cultured *E.coli* in a medium containing N^{15}.

After several generations, they obtained a culture with a completely N^{15} DNA. These were then transfered to an ordinary nitrogen (N^{14}) medium. After one stage of replication, the DNA in the daughter cells were analysed. In each cell the DNA had both N^{14} and N^{15} in each molecule indicating their hybrid nature. This is exactly what is to be expected in a semiconservative mode. The method is called semiconservative because, the whole of the original strands are not retained after replication, but only half of it (one strand) is conserved.

Meselson and Stahl's experiment disproves both conservative and dispersive modes of DNA replication.

GENES

Gregor John Mendel was the first person to have thought about the physical basis of Heredity. He postulated that characters from parents to offspring are passed on in the form of disease particles or units of heredity, which he called 'factors' Mendel's factors or particles of inheritance are now called 'genes'. The term gene was first coined by Johanson in 1909.

After Mendel's laws were rediscovered in 1900, it was clearly establised that there is a definite correlation between the hypothetical hereidatry factors and chromosomes and it was also found out that chromosomes were the only link between one generation and the other. In general, the following are the correlations between genes and chromosomes.

a. Genes are carried singly in gametes like the chromosomes.

b. Like the chromosomes genes are brought together in the zygoe during fertilization

c. Genes seggregate like chromosomes during Metosis

The number of genes in any organism far exceeds the number of chromosomes. In 1902 Sutton and Bovert explained that many genes are present in a single chromosome. For example in *Drosophilla* (a fruit fly), the chromosomes are only four paris. But genes are in thousands. Morghan proposed that each chromosome carries hundreds or thousands of genes and the genes are arranged on the chromosomes on specific *loci* in a linear fashion.

Genes are composed mainly of DNA and with this discovery it is no longer possible to think of a gene as a physical particle, hence these days the gene is understood as an operational unit. There are two types of definitions for genes the classifical and the modern.

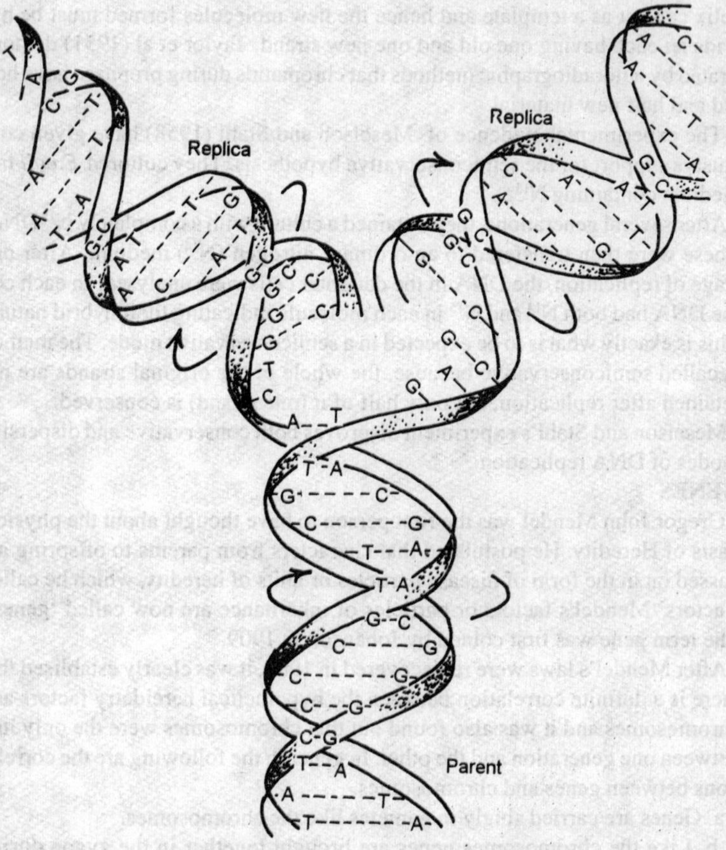

Fig. 2.9 Replication of DNA molecule as proposed by Crick and Watson

Classical definitions : The gene was thought to be hereditary unit by the following three criteria.

a. A gene is a unit of function occupying a definite locus on the chromosome and governing the expression of a particular character

b. A gene is a unit of transmission and seggregation of a characters from one generation to the next

c. A gene is also a unit of mutation as it can give rise to changed characters by means of mutation.

Modern definitions. After the discovery of DNA and elucidation of its molecular structure, a gene has been defined and understood in the following sense.

A. Cistron. A portion of DNA which codes for a single polypeditide chain is caled a cistron. This is comparable to the operational or functional unit of the gene.

B. Recon. The characters of recombination and seggregation are exhibited by the gene. A recon is a unit of recombination.

C. Muton. This is the smallest segment or unit of DNA that undergoes alteration in the base sequence and thus brings about a phenotypic change.

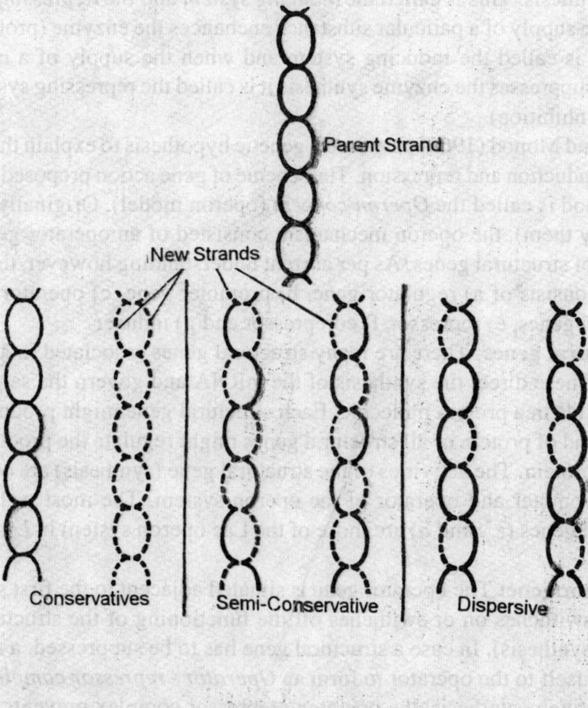

Fig. 2.10 Models of DNA replication- Conservative, semiconservative and dispersive

Genetic regulation of protein synthesis in prokaryotes: (Lac operon concept)

All the activities of an organism, be it unicellular or multicellular, are ultimately controlled by the genes. In multicellular organisms, where cells differ widely in structure and function all of them carry the same genes, since all of them are traceable to a single cell viz., zygote. This poses an interesting question as to the function of genes. How do similar genes function differetnly? The answer to the question lies in differential gene action. In other words, not all genes are active at all times. The genes are switched on and switched off at different times. This differential gene action is also responsible for the synthesis of different types of proteins (enzymes) at different times. Not all enzymes are required at all times. Enzymes required during early stages of development (seed germination) may not be required at later stages (flowering process) and *vice versa.* The metabolic products produced may induce or repress the enzyme synthesis. This is called the *Inducing system* and the *Repressing system.* When the supply of a paticular substance enchances the enzyme (protein) synthesis, it is called the inducing system and when the supply of a metabolic product suppresses the enzyme synthesis it is called the repressing system (end product inhibition).

Jacob and Monod (1962) proposed a genetic hypothesis to explain the mechanism of induction and repression. The scheme of gene action proposed by Jacob and Monod is called the *Operon concept* (operon model). Originally (as suggested by them), the operon mechanism consisted of an operator gene and a number of structural genes. As per current understanding however, the operon system consists of a) regulator gene, b) promoter gene, c) operator gene, d) structural genes, e) repressor, f) corepressor and g) inducer.

Structural genes: There are many structural genes associated in an operon system. These direct the synthesis of the mRNA and govern the sequence of amino acids in a protein molecule. Each structural gene might produce a particular kind of protein or all structural genes might regulate the production of a single protein. The activities of the structural gene (synthesis) are controlled by the promoter and operator of the operon system. The most well studied structural genes (*z,y* and *a*) are those of the Lac operon system in *Escherichia coli.*

Operator gene: The operator gene is situated adjacent to the first structural gene. It swithches on or swithches off the functioning of the structural gene (protein synthesis). In case a structural gene has to be suppressed, a repressor attaches itself to the operator to form an *Operator - repressor complex.* In the case of protein synthesis, the operator - repressor complex prevents the transcription by blocking the movement of RNA polymerase.

Promoter gene: The promoter gene is continuous with the operator gene and is believed to lie left to it. It is suggested that RNA polymerase binds to the promoter site during transcription. Three regions have been recognised in the promoter site. These are a) recognition site, initial binding site and the mRNA inititaion site (operator site).

(i) *Recognition site:* Also called the *cga* site (catabolic gene activator site), it consists of certain palindromic sequnces of DNA. These symmetrical sections of DNA are recognised by proteins having symmetrically placed sub units. This site, also called the *CRP* site (cyclic AMP receptor protein site) binds a CR protein to the promoter gene and thus facilitates the binding of the enzyme RNA polymerase. It has been found that in *E.coli* CRP combines with cAMP (cyclic adenosine monophosphate) forming a CRP - cAMP complex which binds to the promoter enchancing the binding of RNA polymerase and activates transcription. This regulation is called *positive control.*

(ii) *Initial binding site:* This consists of seven bases (DNA) to which the RNA polymerase binds.

(iii) RNA initiation site: The site where transcription begins is called initiation site. This is the region overlapping with the operator region.

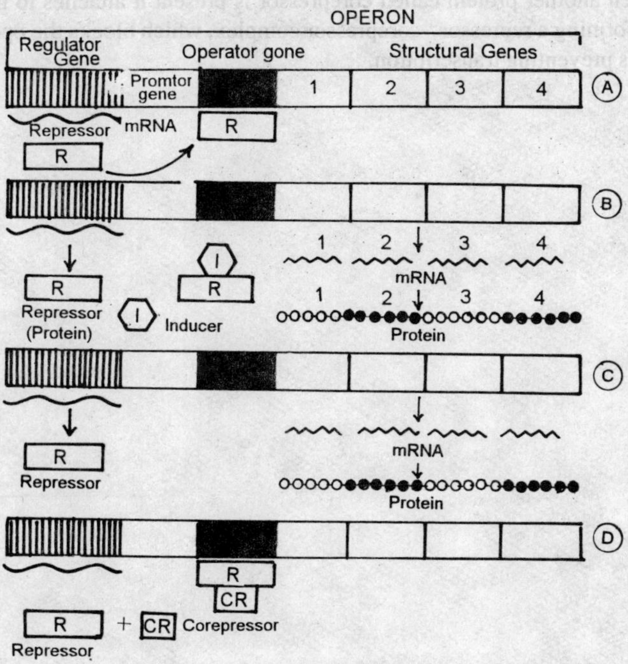

Fig. 2.11 Genetic regulation of protein synthesis in prokaryotes (the Operon system of Jacob and Monod, 1961)

Regulator gene: The regulator gene directs the activity of the operator gene by producing inhibitor proteins called repressors. This repressor protein binds to the operator gene and blocks the path of RNA polymerase, thus preventing transcription. If an inducer is present in the system, it binds to the repressor which undergoes conformational change and becomes inactive. As the inactive repressor cannot bind to the promoter, the strucutral genes get activated and protein synthesis continues.

Repressor, inducer and *corepressor*: This is a protein whose function is to attach to the operator gene to prevent the process of transcription. In an inducible system (an inducer is a metabolic product which promotes protein synthesis), if an inducer is absent, the repressor blocks protein synthesis by attaching itself to the operator gene. When an inducer is present, it (inducer) binds to the repressor rendering it inactive and transcription continues.

In a repressible system, if only a repressor is presnt, it is inactive by itself and cannot attach to the operator gene and hence transcription continues. However when another protein called corepressor is present it attaches to the repressor forming a repressor - corepressor complex, which blocks the operator gene thus preventing transcription.

Three

Genetic code and biosynthesis of proteins

Proteins are the fundamental biological constituents of the protoplasm. They form 15% of the body living organisms and are important building material of the cells. They play a significiant role in the physical and chemical activities of the cells. They are involved in the structural and functional organisation of the cell. The structural proteins form the various cellular organelles. The functional proteins like enzymes and hormones control the metabolic, biosynthetic, biogenetic, bio-energetic, growth regulating, sensory and reproductive functions of the cell. The 'gene actions' take place by means of enzymes. The enzymes are the important catalysing agents in most biochemical reactions. Thus proteins through their particular structure, composition and enzyme activity form the most important biochemical compounds through which 'hereditary traits' are expressed in the organism.

Proteins are giant molecules formed of polypeptide chains of several hundred amino acids. About 20 and odd number of amino acids are known to occur in the cells. Amino acids are formed of a basic amino (NH_2) and an acidic carboxyl (COOH) groups. The variety of cells in an organims and multiple variety of organisms represent myriads of portiens. Different types of proteins are synthesised by different cells of the body. Different combinations of amino acids produce different proteins.

Synthesis of Protein: The process of protein synthesis is the most complicated biological activity. Basically it is governed by the *genetic information* carried in the genes (DNA) of the chromosomes. Specific amino acids are linked in particular sequence or order to form a polypetide chain of protein. The various cell organelles and chemical compounds involved in the synthetic process are

1. Deoxyribonucleic acid......DNA
2. Three type of ribonucleic acids.......RNA
3. Amino acids
4. Enzymes and
5. Ribosomes

DNA and genetic code: DNA is the master molecule of the cell and architect of the living organisms. It directs the synthesis of proteins with 4 types of nucleotides arranged in specific sequences. The *sequence of bases* form *ge-*

netic message; because it is the sequence of bases in the DNA molecule that determines the sequence of amino acids in the peptide chain of protein. The 4 bases (thymine cytocine, guanine and adenine) of the 4 nucleotides form the 4 alphabets of the *genetic language.* It is a *coded language* or cryptogram.

A set of bases that determines the position of amino acid in the protein molecule is called a *codon* or *code.* In other words, the group of nucleotides that specifies one amino acid is a codon. The 4 alphabets of the genetic code are *A* (adenine) T/U (thymine/uracil) C (cytosine) and G (guanine).

How many bases form a code ? A single code of one base will form only four codes which will be inadequate to specify the sequence of 20 amino acids. Similarly a doublet code is insufficient as it can specify only (4x4)=16 amino acids. But a *triplet code* is most suited as it has 64 (4x4x4=64) possiblities and so can be determine more than 20 amino acids. That is to say, the 4 letter genetic alphabets in different combinations of 'triplets' can form 64 possibililties. For example the 4 alphabets of the English word *SEAT* can be rearranged in different ways to form different 3 letter words like SET, SAT, SEE, EAT, SEA, ASS, ATE, TEA, each having a different meaning. Similarly ATCG can form different combinations. The codon AAG determines the amino acid leucine, CUA-valine, UGC-cystine, GGA-glycine etc.

Francis, H Crick gave the first experimental proof for the 'triplet codes' in 1961. In 1968, Drs. H.G. Khorana, Nirenberg and Holley worked out all the 64 possible triplet codes and were awarded the Nobel Prize.

Modern concept of protein synthesis. The biogenetic process which assembles and links the specific amino acids in specific sequence to form protein molecules is depicted as the *central dogma* of protein synthesis.

Replication Transcription Translation

DNA —— DNA —————— RNA—————— Proteins.

Transcription : At the time of protein synthesis under the activity of certain enzymes the DNA helix unwinds and separates. One of the DNA strand acts as the template and synthesises a particular RNA strand as its complimentary strand by attracting the 4 types of ribonucleotides found in the nuclear medium. The base sequence on the RNA strand are complimentary and determined by those on the DNA strand. In the process, the genetic message of the DNA (base sequence) is *transcribed* on the *m* RNA strand. This is called *transcription* and is catalysed by an enzyme, *DNA dependent RNA Polymerase.*

The mRNA with transcribed genetic code diffuses out of the nucleus and arrives at the site of protein synthesis - the *ribosomes.* The separated DNA strands recoil to form the double helix.

First Letter	Second Letter: U	C	A	G	Second Letter
U	UUU-Phen.	UCU-Ser	UAU-Tyr.	UGU-Cys	U
	UUC-Phen.	UCC-Ser	UAC-Tyr.	UGC-Cys	C
	UUA-Leu	UCA-Ser	UUA-Amber	UGA-Nonsense	A
	UUG-Leu	UCG-Ser	UAG-Ochre	UGG-Trip	G
C	CUU-Leu	CCU-Pro	CAU-H	CGU-Arg.	U
	CUC-Leu	CCC-Pro	CAC-H	CGC-Arg	C
	CUA-Leu	CCA-Pro	CAA-Glu	CGA-Arg	A
	CUG-Leu	CCG-Pro	CAG-Glu	CGG-Arg	G
A	AUU-Ile	ACU-Thr	AAU-Aspn	AGU-Ser	U
	AUC-Ile	ACC-Thr	AAC-AspN	AGC-Ser	C
	AUA-Ile	ACA-Thr	AAA-Lys.	AGA-Arg	A
	AUG-Met	ACG-Thr	AAG-Lys.	AGG-Arg	G
G	GUU-Val	GCU-Ala	GAU-Asp	GGU-Gly	U
	GUC-Val	GCC-Ala	GAC-Asp	GGC-Gly	C
	GUA-Val	GCA-Ala	GAA-Glu	GGA-Gly	A
	GUG-Val	GCG-Ala	GAG-Glu	GGG-Gly	G

Fig. 3.1 The codon dictionary

Translation of the genetic code takes place by the coordination of ribosomes, mRNA, rRNA, tRNA amino acids and enzymes. The mRNA strand runs through a cluster of 7 or more activated ribosomes, forming *polyribosomes.*

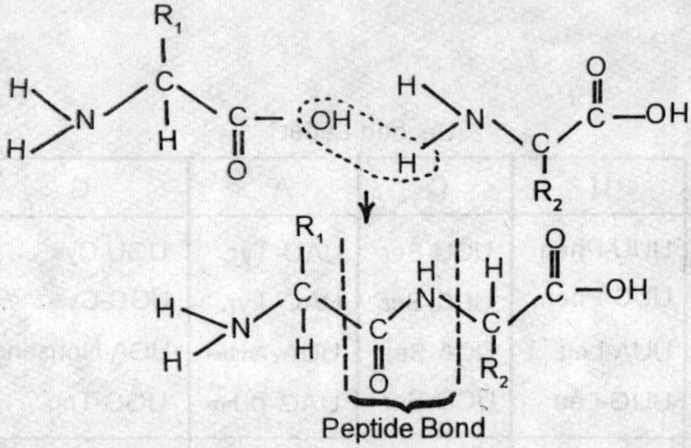

Fig. 3.2 Formation of peptide linkages

Meanwhile the amino acids in the cytoplasm get activated and charged with energy. This is promoted by ATP. The *activated amino acids* become chemically bound with specific tRNA at its end under the action of an enzyme. This is called *amino acid* tRNA complex. Many such 'complexes' are transferred to the polyribosomes.

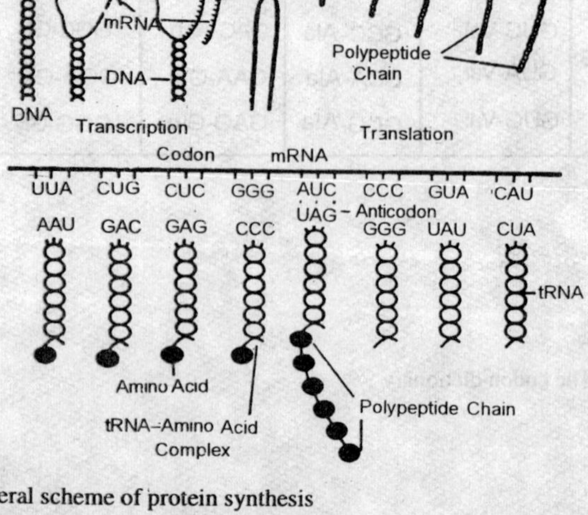

Fig. 3.3 General scheme of protein synthesis

Each tRNA has a 'binding site' bearing triplet bases called the anticodons, which are complimentary to specific triplet codon, on the RNA. Chemical binding occurs between the complimentary codon of mRNA and anticodon of different tRNA complexes. As the ' b i n d i n g' goes on, the amino acids carried by each tRNA link with one another forming a peptide chain of protein molecule. By the time the mRNA passes off the last ribosome of the polyribosome, the mRNA and peptide chain of protein are released under the action of R1 and R2 *factors.* The ribosome and tRNAs are also released.

Role of RNA: Three types of RNA namely, mRNA, tRNA and rRNA take active part in the protein synthesis. mRNA carries coded information form the DNA to the ribosomes. tRNAs transport amino acids from amino acid pool to the ribosomes. Amino acids are attached to the sites of codon. rRNA has only general functions to perform.

Mechanism of Genetic code and assembly of Amino acids

DNA——— ——— ——— ———

Transcription {C G A A C A G T A A A A T C G

{G C U U G U C A U U U U A G C

codons

m RNA-Genetic Code ————————————

{ tRNA COA ACA GUA AAA UCG

Anti codons

Translation

Aminoacids Alanine Cystine Hystidine Phenyl alanine Serine Peptide chain

The Wobble hypothesis: The triplet code of codon is degenerate i.e., there are many more codons than there are amino acids. Only 20 amino acids are involved in protein synthesis, while there are 64 possibilities or types of codons (4^3). Hence more than one codon can code for an amino acid. It is interesting to find out as to why the codes are degenerate? There could have been only so many codons as there are amino acids. But there is a problem here. If there were to be only one code for an amino acid, what happens if there is a mutation at that site in DNA ? Obviously the code changes and the amino acid cannot be incorporated in the protein leading to abnormalities. On the other hand, if there are alternate codes, they act as a protection against mutuation in a specific base. Hence degeneracy of codes is a must.Regarding tRNA present in cytoplasm, since they have to bind themselves to a specific codon, there should have been as may tRNA's as

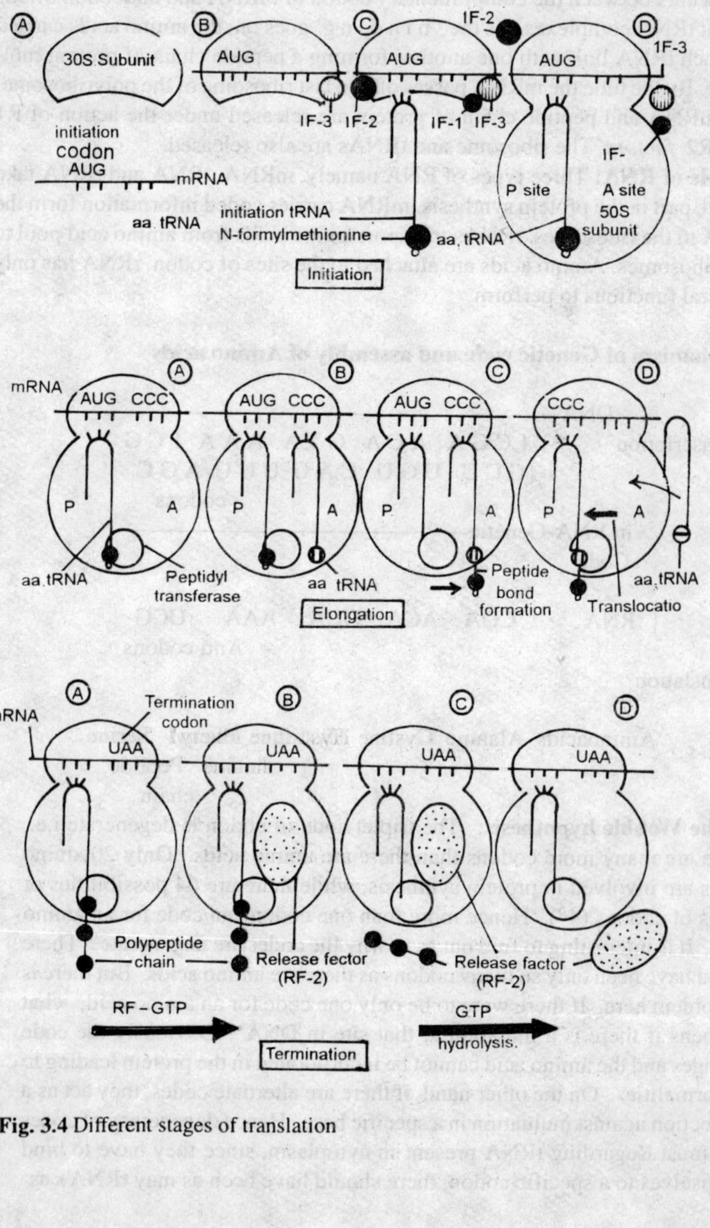

Fig. 3.4 Different stages of translation

there are codons: But the number of tRNA is only as many as there are amino acids. This means, the anticodons of tRNA's must be able to 'read' more than one codon of mRNA. How is this reading possible when bonding between base pairs is highly specific ?

Crick (1966) proposed ' the Wobble hypothesis' in order to solve this apparent dilemma. According to this hypothesis, only the first two bases of the codon have a precise pairing with the bases of the anticodon of mRNA, while the pairing between the third bases of codon and anticodon may Wobble (non specific). The pairing in the third base is ambiguous. Thus a single tRNA can pair (bind) with more than one mRNA codon differing in only the third base. For example the anticodon ICG of tRNA can read codons AGC and AGU of mRNA. The pairing between the first two bases is normal, while between G and U is against the normal pairing pattern. This unusual bonding between G and U is called Wobble pairing.

The degeneracy of the code is not totally random. Multiple codons for an amino acid always carry the same bases in the first two positions in a triplet with only the third 'wobbling'. For example all the four codons for Valine have the first two bases-GU and for

alanine it is Gloeocapsa

Valine - GUC, GUC, GUA and GUG

Alanine- GCU, GCC, GCA and GCG

Four

Heredity and mendelism

Genetics is a branch of biology that deals with heredity and variation. The above definition would appear to limit the area of study to the transmission of characters from parents to offspring and to the study of the variability in traits that may occur from one generation to the next. The life blood of the science of genetics is the variation that occurs within and among plant and animal species. Thus the word heredity is simply the name given to the means by which living organisms reproduce their kind. An understanding of heredity, therefore requires a knowledge of the various processes involved in reproduction. Today with the descriptive elements of reproduction largely clarified and behind us, we are going after the very atoms of heredity. We want to know about their arrangements, their architecture and their activities, the chemical transformations and the physical forces that provide for the continuation of life.

Heridity can be defined as the study of transmission of development potentialities, (genes) and their expression from generation to generation. In general, heredity explains the similarity between the parents and offspring. The similarity is due to transmission or inheritance of factors (genes) from the parents to offspring.

Heritable characters are transmitted to the offspring through a very narrow bridge of gametes. In the act of fertilisation the gametes from two parents come together. Thus the characters expressed by the offspring will be the sum total of the characters of both the parents. Heredity is determined by the genetical content of the individual. But the environment also plays an important role. The genetical composition of an individual is called *genotype* and its external appearance is called *phenotype*. The external appearance is the result of interaction between the genetical constituent and the environment.

Variations are the character differences between individuals. These differences or variations may be due to heredity or environment or even due to both.

JOHN GREGOR MENDEL

The scientific study of the 'Material basis' of heredity began in the later part of the 19th century with the work of an Austrian monk John Gregor Mendel. He had studied science at the university of Vienna and was working as a teacher of natural science at Brunn. In his monastry gardens, he carried out his famous experiments on garden peas from 1857 to 1865. The results of these classic experiments and Mendle's conclusions which are basic to the study of hered ity, were published in the annual proceedings of the Natural History Society of Brunn, in 1865. His work was not given much importance at that time as the scientific world was very busy with the controversy raised by the publication of Darwin's book, "Origin of Species". Mendel during his later years conducted hybridization experiments on many plants and bees. Gradually he became more and more absorbed in the administration of the monastry and he died in the year 1884.

Then in 1900 three men, viz., Hugo de Vries of Holland, Karl Correns of Germany and Eric Von Tschermak of Austria arrived at the same result as Mendel had arrived earlier. Thus heredity', which in turn, provided the inspiration for the research that continues at an accelerated pace today. Since Mendel's work forms the basis of genetical study he is referred to as the 'founder of genetics'.

Fig. 4.1 John Gregor Mendel
The father of Genetics

Mendel's work: Mendel grew in his garden 34 different varieties of peas. He first made sure that everyone of these varieties bred true to its type. For his experiments, he selected those varieties which contrasted sharply in certain definite traits. For example, he selected a tall variety and a short one, or a red flowering variety and a white flowering variety. He crossed these plants differing sharply in their characters and obtained the seeds which were later planted to produce offspring. These plants, thus obtained from the crossing of two original pure strains, were called hybrids. In genetical language a hybrid is defined as an individual which results from the crossing of two individuals differing in at least one set of characters. (Crossing of 2 plants or animals differing in at least one set of characters is called hybridization experiment.) The hybrids, on attaining maturity, were allowed to self-fertilize to produce the second generation. The plants of unlike characters, with which the hybridisation is first made, constitute the parental or P generation and the progeny of hybrids obtained from this cross is called the first filial or F_1 generation. When the F_1 hybrids are self-fertilized or inbred, the progeny resulting therefrom, is called second filial or F_2 generation. When an F_1 hybrid is crossed with one of the parents, the operation is called Back Crossing.

As an illustration, we shall describe one of Mendel's own experiments. He crossed a tall variety of Pea with a dwarf variety. The resulting seeds when planted, always produced tall plants(F_1). It did not matter as to whether the tall parent provided the pollen and dwarf variety acted as the seed bearing plant or viceversa. The character of the dwarf did not appear in the F_1 plants which were allowed to self-fertilize. He found that the F_2 generation has a mixed progeny of tall and dwarf plants. The talls always accounted for approximately 3/4 of the total F_2 generation. Thus there was a ratio of 3:1 (3 tall and 1 dwarf). This showed that dwarfness disappeared only in the F_1 generation but was not completely destroyed since it reappeared among F_2 population. Mendel symbolised the characters by letters to explain clearly the process. For example tall character as t and dwarf character as t.

Mendel repeated his experiments with the other characters of pea plants and on other plants also. The results of his experiments led him to formulate the laws of heredity.

Red flowered x White flowered
Pea plant Pea plant
RR rr

gametes R r

F_1 generation Rr
Red coloured Rr X Rr self pollination
pea plant

	R	r
R	RR	Rr
r	Rr	rr

F_2 generation

1 RR—|
2 Rr—| RED COLOUR
1 rr = White coloured

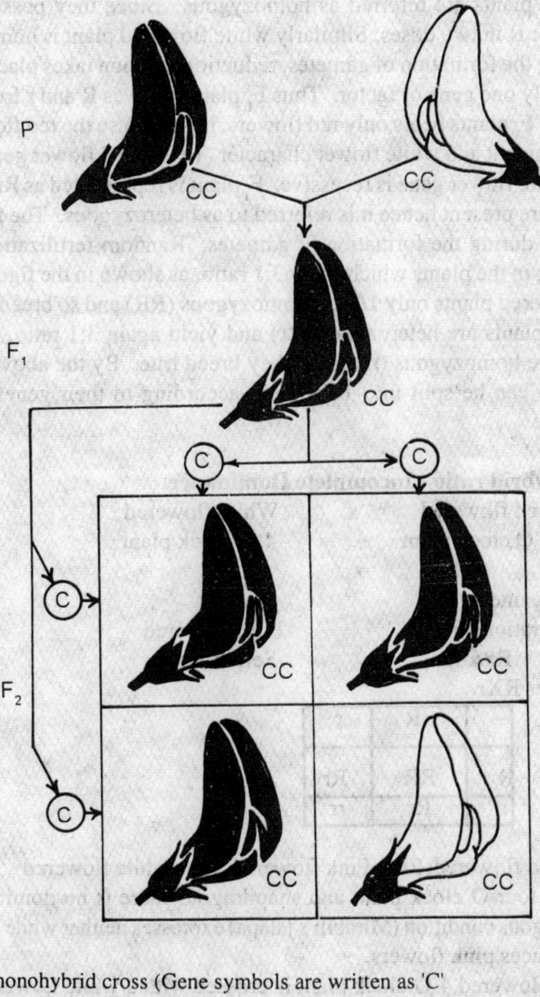

Fig. 4.2 The monohybrid cross (Gene symbols are written as 'C' and C instead of R and r)

Monohybrid ratio: It is the cross between two plants or animals differeing in only one trait. In this corss, only a pair of contrasting characters are observed, for example red flower and white flower character; and smooth with wooly character of hair etc. These types of contrasting traits or characters are said to be alleles. Alleles are the two alternative genes at the same locus on homologous chromosomes.

Red flowered pea plants are crossed with white flowered pea plants. The parental plants are referred as homozygous. Since they possess red flower character R in two doses: Similarly white flowered plant is homozygous.

During the formation of gametes, reduction division takes place and gametes carry only one gene or factor. Thus F_1 plant receives R and r from each of the parents. F_1 plants bears only red flowers, it is because the red flower character R is dominant and white flower character , i.e. the red flower gene is dominant over white flower gene is recessive. F_1 plant is represented as Rr-two different factors are present hence it is referred to as heterozygous. The factors R and r separate during the formation of gametes. Random fertilization takes place resulting in the plants which show 3:1 ratio, as shown in the figure. Out of 3/4 red flowered plants only 1/3 are homozygous (RR) and so breed true; remaining 2/3 plants are heterozyous (Rr) and yield again 3:1 ratio, remaining 2/3 plants are homozygous (rr) since they breed true. By the above observation. 3:1 ratio can be split into 1:2:1 ratio according to their genetical nature or genotype.

Monohybrid ratio - Incomplete Dominance.

Red flowered 4'O clock plant x White flowered 4'O clock plant

RR x rr

gametes R r

F_1 generation Rr Pink flowered

Rr x Rr Self cross

Gametes RXr

	R	r
R	RR	Rr
r	Rr	rr

1RR=Red flowered, 2Rr=Pink flowered, 3rr= white flowered

In the four O'clock plant and snapdragons there is no dominance that its, heterozygous condition (Mirabili's jalapa) expresses neither white nor red flower but produces pink flowers.

A red flowered 4'O clock plant is crossed with a white flowered 4'O clock Plant. The F_1 generation plant which is heterozygous yield's pink flowers. It appears in this case, as if the characters have been blended. But when F_2 generation is raised both red and white flowered plants together with pink flowered plants appear in the ratio of 1 Red; 2 Pink :1white.

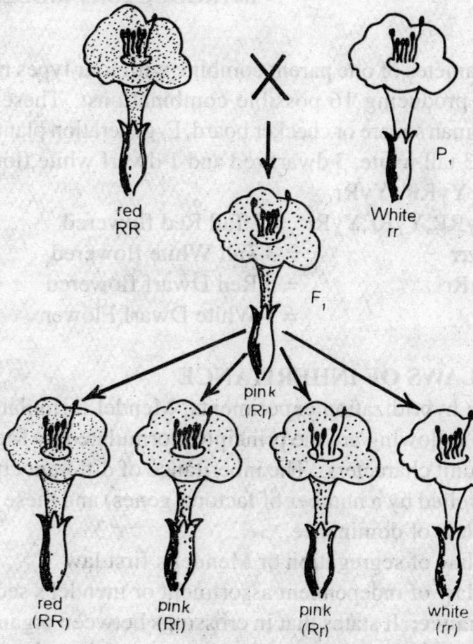

Fig. 4.3 Incomplete dominance in Mirabilis jalapa

In F_1 generation, in plant Rr, neither of the genes are dominant, both express their characters. For the expression of red colour, R is to be present in two sets similarly for white colour, r is to be present in two doses. Since both R and r genes are present pink colour is expressed in F_1 generation.

In the case of Andalusain fowls also incomplete dominance can be observed. A cross between black and white feather coloured fowls produce blue gray coloured fowls in the F_1 generation. When the F_1 generation gray fowls are inter-crossed among themselves, the F_1 generation shows 1:2:1 ratio.

Dihybrid Cross: It is the cross between two plants or animals differing in two pairs of contrasting characters. This cross involves two characters seperable in inheritance like flower, colour and height of the plant, colour of the hair and colour of the skin etc.,

A homozygous pea plant with yellow round seeds is crossed with a green wrinkled seeded pea plant (homozygous). In peas, yellow seed character Y is dominant over green y and round seed character R is dominant over wrinkled r (each allele when studied seperately shows 3:1 ratio). All the F_1 generation plants are round and yellow seeded. Yellow round seeded parent produces one type of gametes YR and green wrinkle seeded parent also produces one type of gametes yr and gametes from both the parents undergo fertilisation and result in F_1 plant Yy Rr. Because of the dominance, Round and Yellow colour are expressed in the F_1 generation. F_1 plants are self-pollinated or inter-crossed. They produce four types of gametes YR, Yr, yR and yr. Since all the four genes separate, they combine in all possible combinations.

Four types of gametes of one parent combine with four types mof gametes of the other parent producing 16 possible combinations. These can be easily represented in Roman square or checker board, F_1 generation plants show 9:3:3:1 ratio. 9 tall red, 3 tall white, 3 dwarf red and 1 dwarf white flowered plants.

YYRR. YYRr, YyRR, YyRr
YYRr,YyRr,YyRR,YyRr,YyRr = 9 Tall Red flowered
Yyrr, Yyrr, Yyrr = 3 Tall White flowered
yyRr, yyRr, yyRr = 3 Red Dwarf flowered
yyrr = 1 White Dwarf Flower

MENDELIAN LAWS OF INHERITANCE

Based on these hybridization experiments, Mendel formulated the laws of inheritance. The following are the principles formulated by Mendel.

1. Principle of unit characters. The inheritance of different characters by an organism is controlled by a number of factors (genes) and these occur in pairs.
2. Principle of law of dominance.
3. Principle of law of segregation or Mendel's first law
4. Principle or law of independent assortment or mendel's second law.

Law of Dominance: It states that in crossings between organisms (pure for contrasting characters of a pair) only one character of the pair appears in the F_1 generation. In Mendelian termology a character that shows up itself in F_1 is the dominant and the white is recessive.

Mendel studied seven differentiating traits in peas. These traits and the dominant trait in each case are given below:

S. NO.	TRAIT STUDIED	DOMINANT	RECESSIVE
1.	Seed form	Round (R) Smooth	Wrinkled (R)
2.	Cotyledon colour	Yellow (Y)	Green (y)
3.	Flower colour	Red (R)	White(r)
4.	Pod shape	Inflated (I)	Constricted(i)
5.	Pod colour	Green (G)	Yellow (g)
6.	Flower position	Axial (A)	Terminal(a)
7.	Stem length	Tall(T)	Dwarf(t)

Such pairs of constrasting characters are called allelomorphic pairs or allelic pairs and each member of a pair is said to be the allele of the other. It may be mentioned that the gametes do not carry characters as such but they carry some hereditary units, which are responsible for the appearance of characters in the offspring. These hereditary units have been called factors or determiners. They are now called genes. In general, the ability to synthesize an active enzyme qualifies a gene as a dominant. The presence of dominant gene in the diploid cell also usually permits the formation of a specific enzyme at a rate sufficient to allow growth.

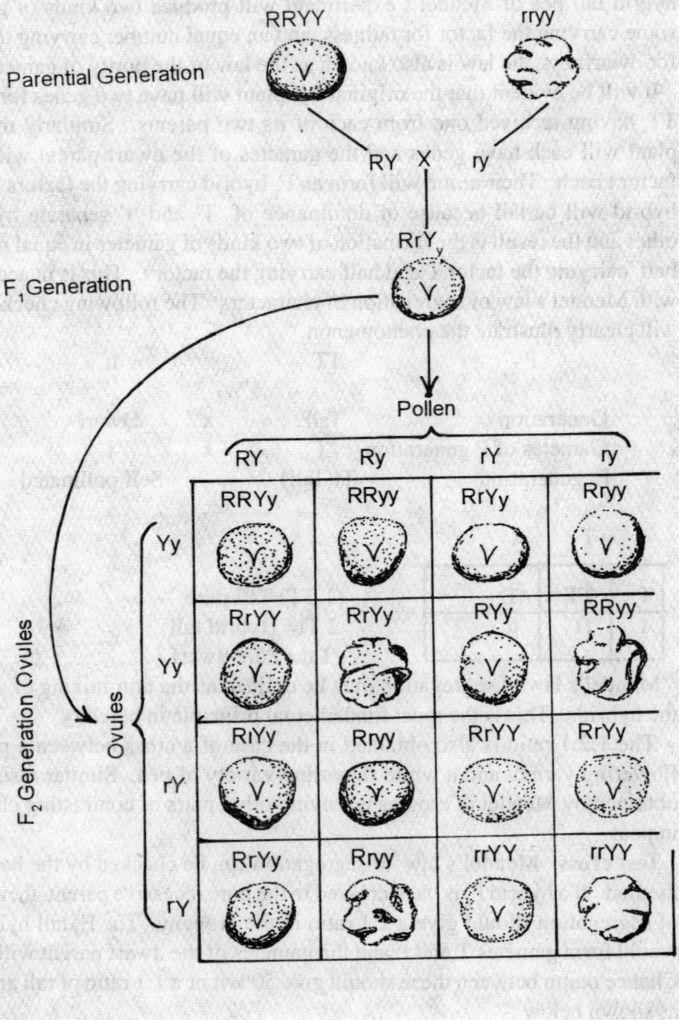

Fig. 4.4 Mendel's dihybrid cross

Law of Segregation: The law of segregation of characters states that when a pair of allels are brought together they segregate without mixing and separate from each other when the hybrid forms gametes. According to this law, F_1 hybrid tall pea of Mendel's experiment will produce two kinds of gameters some carrying the factor for tallness, and an equal number carrying the factor for dwarfness, the law is also known as the law of the purity of gametes.

It will be evident that the original tall plant will have two genes for tallness TT, having received one from each of its two parents. Similarly the dwarf plant will each have genes and the gametes of the dwarf parent will have a factor t each. Their union will form an F_1 hybrid carrying the factors Tt. This hybrid will be tall because of dominance of 'T' and 't' separate from each other and the result is the formation of two kinds of gametes in equal numbers, half carrying the factor T and half carrying the factor t. This is in accordance with Mendel's law of segregation of characters. The following checker board will clearly illustrate the phenomenon.

P	TT		tt
Generation	Tall	x	Dwarf
Gametes of P generation	T	x	t
F_1 generation	Tt(Tall)	-	Self pollinated

	T	t
T	TT	Tt
t	Tt	tt

1 TT=Tall pure
2 Tt= Hybrid tall
1 tt= Pure dwarf

Mendel's law of segregation may be defined as the non-mixing of alleles in the hybrids. This is the most fundamental principle in heredity.

The 1:2:1 ratio is also obtained in the case of a cross between a pure red-flowering variety and a white flowering variety of pea. Similar results were obtained by Mendel in crosses involving other pairs of contrasting characters in peas.

Test cross: Mendel's law of segregation can be checked by the back cross method. If a hybrid F_1 is back crossed to the pure recessive parent, then the law of segregation should give a 1:1 ratio in the progeny. The F_1 tall hybrid (Tt) would form gametes T and t: and the gametes of the dwarf parent will be all t. Chance union between these should give 50% it or a 1:1 ratio of tall and dwarf as shown below :-

F_1 hybrid	Tt	x	tt,
	tall		double recessive
gametes	T	Tt	— Tall
	t	tt	—Dwarf
		1:1	

The Law of Independent Assortment: This law states that "the genes for 2 or more different traits are distributed independently and the distribution is independent of the other traits. In other words each pair of different characters in hybrid union is inherited independently of the other differences in the original parental stocks. When 2 or more pairs of alleles are studied, each allele acts independent of the other crossing. Factors or genes expressing different characters exhibit independent assortment and shuffling without being influenced by other factors.

Because of the independent assortment two new types are produced in the F_2 generation in the dihybrid cross. The determining genes are present in separate chromosomes. Law of independent assortment can be studied by back cross of F_1 with double recessive parent.

In pea plants, smooth seed character is dominant over wrinkled yellow cotyledon character Y is dominant over green y.

F_1 Hybrid pea plant with Smooth seed and yellow cotylendon (ss yy)	x	Double recessive parent with Wrinkled seed and green cotylendon (SsYy)

F_1 hybrid produces four types of gametes SY, Sy, sY and sy. The double recessive plant produces only one type of gametes sy.

	SY	Sy	sY	sy
sy	SsYy	Ssyy	SsYy	ssyy

1 Ss Yy = Round Yellow
1 Ss yy = Round Green
1 ss Yy = Wrinkled Yellow
1 ss yy = Wrinkled Green

It is clear from the back cross that four types are produced in 1: 1:1:1 ratio. The two new types are produced due to the independent assortment of genes (alleles) during gamete formation in F_1 hybrid and subsequent random fertilisation.

Multiple alleles

All Mendelian characters have only two alternate expressions or have only two alleles. But this may not be true (having only two alleles) in all cases, for a gene, once having undergone mutation, may mutate again to have one more altered from or allele and in their case, there will be three alleles. Theoritically a gene 'a' may mutate any number of times leading to the formation of many variants or multiple alleles. A multiple allelic series of a gene represents a series of gene mutations in the same locus leading to a variety of phenotypic expressions seggregated in different organisms. Multiple alleles may be defined as a series of gene mutations from wild to a number of mutated forms all of them affecting the same trait. Dawson and Whitehouse have used the term panalleles to denote multiple alleles. Multiple alleles should not be mistaken with multiple factors where non allelic genes present in different loci affect a trait; further many pairs of genes are involved in each individual to delimit a character. In multiple alleles, the genes occupy the same locus on the chromosome in different organisms producing variants of a trait due to sequential mutations. In multiple alleles only one pair of genes is involved in each individual to delimit a character. Multiple alleles are different from the diallelic series of Mendel in that the former produce many variants, while the latter produce only two variants.

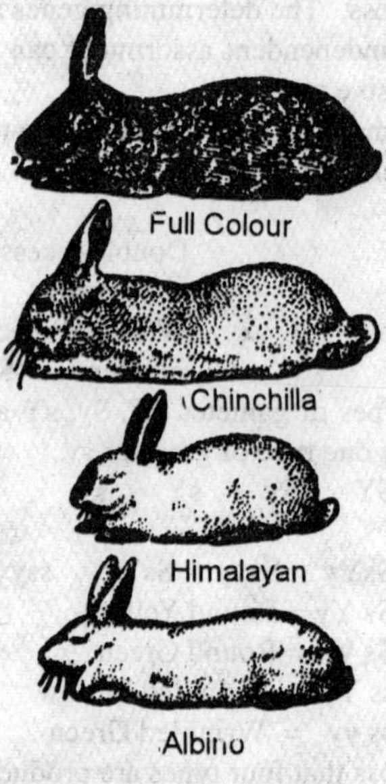

Fig. 4.5 Coat colour in rabbits

Salient features of Multiple alleles.

1. They occupy the same gene locus on the chromosome

2. In diploid organisms, as there are only two homologus chromosomes, there will be only two allelic states of the multiple sėries.

3. In haploid gamėtes multiple allelic series is represented by a single allelic state.

4. No crossing over in known to occur between multiple alleles as they occupy the same level of the homologus chromosomes. In a cross between two alleles, the F_2 generation reveles the e combination of alleles.

5. A single multiple allelelic series affects only one trait-eye colour, shape, skin colour etc.,

6. Suppose there is a series of five genes in the multiple allelic series starting with the wild A, Ab, Ac, Ad, Ae etc., the last one representisng the last mutation (Ae) on the gene series, the wild gene (A) is dominant over all others and the last mutation (Ae) is recessive to all others. In the series starting with A, all genes are recessive to their previous allele while dominant over the subsequent allele (A, Ab, Ac, Ad, Ae).

The fact that A is wild can be proved because it is dominant over all others and fact that Ae is the last mutant can be proved by its being recessive to all others.

In a cross between two representatives of the allelic series, that which is dominant is old and that which is recessive is recent (from the point of view of mutation)

Gene Symbols for Multiple Alleles

The wild gene is described first and is given the symbol of a capital letter (A,B,C, etc.), the next in the sèries has the same capital letter but is superscribed on top with the suitable small letter. This small letter is the identifying mark for the allele. The series of alleles have the same capital letter but with different small letters on top. We shall study some examples of multiple allelic series both in plants and animals. The following traits have been proved to be multiple alleles in different organisms-

1. Coat colour in Rabbits
2. Eye colour In Drosophilia
3. Sexual incompatibility in Nicotiana
4. Blood groups in Man

Blood Groups in Man:

Blood can be transfused from one individual to the other in case of emergency. Bl‸od transfusions were tried as early as 18th century in England and France. But some cases were successful and in some instances the receipient died soon after transfusion. On testing it was noticed that the blood had coagulated or clumped. However this did not happen in all instances. In some instances the blood of two individuals mixed smoothly. It was Karl Landsteiner who in 1900 found the reason behind clumping or smooth mixing. He noticed that the plasma and red blood cells (RBC) of a person when recombined after seperation mixed smoothly so also blood from some persons. But in certain other instances the plasma and the RBC did not mix, the RBC clumped. Landsteiner worked in detail on this and came to the conclusion that the mixing or clumping depends on the antigen and antibody reaction present in the blood. He discovered that there are two antigens in the blood. A and B and two serum antibodies that clump or agglutinate them. Correspondingly the blood of all human beings could be classified into four types based on the antigen present in the RBC. These are-

1. Antigen A- Blood group A
2. Antigen B- Blood group B
3. Both Antigen A and B- Blood group AB
4. No Antigen- Blood group O

Before we proceed on further discussion we wil! try to understand what is meant by antigen - antibody reaction.

[Blood is a connective tissue in fluid form. It has two main parts - cells and plasma. The cells are RBC, WBC and platelets. The plasma consists of protein, fibrinogen and the serum.

When the serum is injected into the blood stream of another organism, a substance is produced in the blood which reacts with the serum protein and neutralizes it. The substance produced by the blood cell is known as antibody and the protein in the serum which causes the production of antibody is called antigen. Antibody is also called Agglutinin (because it agglutinates) and antigen is very specific and is often compared to the lock and key mechanism.]

Landsteiner discovered A,B and O blood groups. His students Von Decastello and Sturli discovered the AB groups. Besides these, scientific investigations have also proved the existence of two other blood groups viz MN blood groups and Rh factors. MN blood groups are typical Mendelian characters. They are controlled by a single gene.

A, B, AB, and O groups: In these blood groups, some of them contian natural antibodies against certain others. It has been pointed out already, there are two antigens (agglutinogen) A and B and there are two antibodies (agglitinin) A and B. Persons of blood group A contain within the serum of their blood an antibody which agglutinates RBC of blood group B. Group B serum has antibody against A. The AB group serum however has no antibodies while O group has both antibodies for A and B. Hence O group cannot receive blood from any group except itself. But it has no antigens and can be given to any other group (it does not cause the production of antibodies in blood types, the following table gives the antibody and antigens of various blood groups.

Blood group	**Antigen (RBC)**	**Antibody (Serum)**
A	A	B
B	B	A
AB	A,B	None
O	NONE	A,B

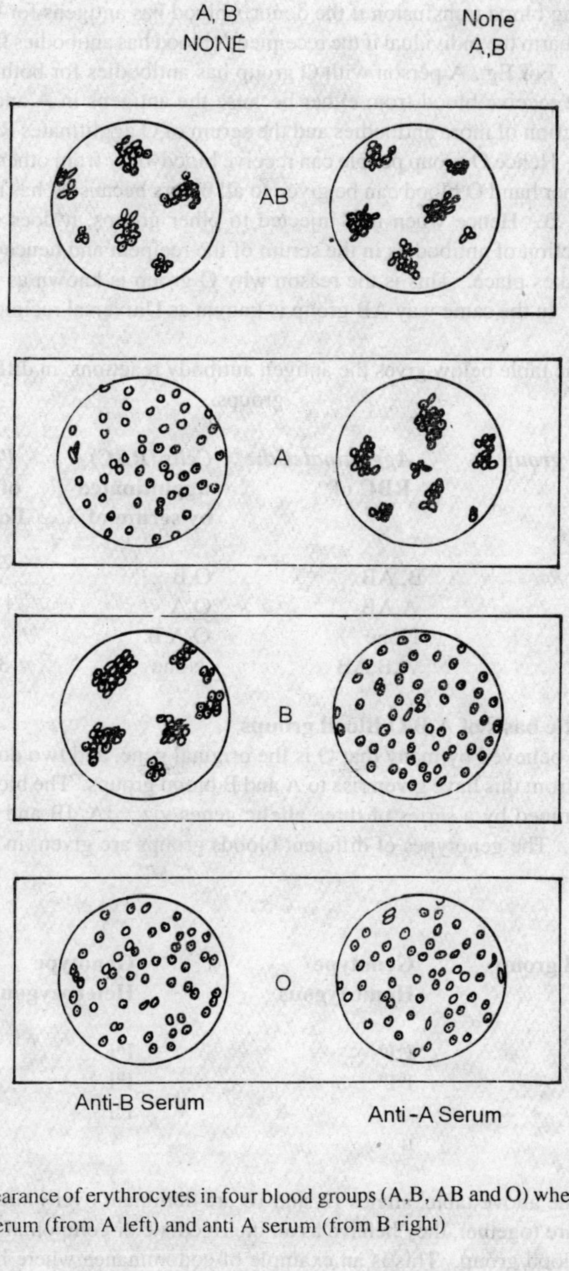

Fig. 4.6 Appearance of erythrocytes in four blood groups (A,B, AB and O) when tested with anti B serum (from A left) and anti A serum (from B right)

During blood transfusion if the donor's blood has antigens for the receipent, it will harm the individual if the receipient's blood has antibodies for the donor's blood. For Eg., A person with O group has antibodies for both A and B and cannot receive blood from either because the antigens in A and B cause the production of more antibodies and the serum of O agglitinates RBC of both A and B. Hence O group people can receive blood only from other O group. On the other hand O blood can be given to all others because it has no antigen for A and B. Hence when it is injected to other groups, it does not cause the production of antibodies in the serum of the rceipent and hence no agglutination takes place. This is the reason why O group is known as the Universal donor. In the same way AB group is known as Universal recipient.

The table below gives the antigen antibody reactions in different blood groups.

Blood group	***Agglutinates the* RBC of**	***Cells (RBC)* agglutinated by serum of**	***Percentage* of White Population**
A	B, AB	O,B	3.7
B	A,AB	O,A	13.4
AB	None	O,A,B	5.7
O	A,B,AB	None	37.3

Genetic basis of A,B,O Blood groups

It is believed by many that O is the original gene, and two dominant mutations from this have given rise to A and B blood groups. The blood groups are determined by a series of three allelic genes viz., IA, IB and i (Bernstein 1925). The genotypes of different bloods groups are given in the following table.

Blood group	**Genotype Homozygous**	**Genotype Heterozygous**
A	I^AI^A	I^Ai
B	I^BI^B	I^Bi
AB		I^AI^B
O	ii	

In the above table, alleles IA and IB are dominant over i allele. But when they are together, they behave as an intermediate or compound producing the AB blood group. This is an example of codominance where both the genes have equal expression. It will be interesting to observe the possible genotype of offspring and their phenotype (blood group), when marriage occurs between people belonging to different blood groups. The following table gives the genotypes and phenotypes of parents and off springs in different combinations.

Genotypes of parents	Blood groups of parents	Blood groups of offspring	Genotype of offspring
1) $I^AI^AxI^AI^A$	A x A	A	$I^A I^A$
2) I^AI^A x I^Ai	A x A	A	$I^A I^A$, $I^A i$
3) $I^A i$ x $I^A i$	A x A	A,O	$I^A i$, ii
4) $I^B I^B$ x $I^B I^B$	B x B	B	$I^B I^B$
5) $I^B I^B$ x $I^B i$	B x B	B	$I^B I^B$, $I^B i$
6) $I^B i$ x $I^B i$	B x B	B,O	$I^B I^B$,ii
7) $I^AI^BxI^AI^B$	AB x AB	A,AB,B	I^AI^A,$I^B I^B$,I^AI^B
8) $I^AI^BxI^AI^A$	AB x A	A,B,AB	I^AI^A,I^AI^B,I^Ai,I^Bi
9) $I^AI^BxI^Ai$	AB x A	A,B,AB	I^AI^A,I^AI^B,I^Ai,I^Bi
10) I^AI^Bxii	AB x O	A,B	I^Ai,I^Bi
11) $I^AI^BxI^BI^B$	AB x B	AB,B	I^AI^B,I^BI^B
12) $I^AI^BxI^Bi$	AB x B	AB,A,B	I^AI^B,I^AI,I^BI^B,I^BI
13) $I^AI^AxI^BI^B$	A x B	AB	I^AI^B
14) $I^AI^AxI^BI$	A x B	AB,A	I^AI^B,I^AI
15) $I^AIxI^BI^B$	A x B	AB,B	I^AI^B,I^AI
16) I^AIxI^BI	A x B	AB,A,O	I^AI^B,I^AI,II
17) I^AI^AxII	A x O	A	I^AI
18) I^AIxII	A x O	A,O	I^AI,II
19) $IIxI^BI^B$	O x B	B	I^BI
20) $IIxI^BI$	O x B	B,O	I^BI,II

Sub divisions of a A,AB and b:

The A group RBCs are divided into two sub groups namely A_1 and A_2. The latter is not very common and is found in about 20% of the group. There is no agglutination reaction between A_1 RBC and A_2 serum and vice versa. But both A_1 and A_2 are agglutinated by B and O serum. The B group has two types of antibodies. The A_1 antibody agglutinates both A_1 and A_2 corpuscles and the other A_2 strongly agglutinates only A_1, but agglitination agsinst A_2 is very weak. Similarly the AB group is also known to have two to four sub groups, A B, A_2 B, A_3B, A_4B etc.,

Blood group inheritance and parentage: In case of disputed parentage when blood groups of parents and children are known it can be easily decided (see table) whether a given child is born to particular parents or not. When the parent's blood groups are known, their genotype can be worked out and the child's genotype and blood group can be matched. The following table gives (in cases of disputed parentage) the blood group of mother, child and the unlikely blood group of father ie., if the blood group of mother and child is of a particular type, the disputed father cannot have the genotype given in the table in that he is not the father. It should be noted however that in instances when both the babies and families are of the same group it will not be possible to decide (if the disputed father also has the matching blood group) parentage on the basis of blood group.

Blood group of child	Blood group of mother	Blood group of unlikely father
O	O	AB
O	A	AB
O	B	AB
A	O	O,B
A	B	O,B
B	O	O,A
B	A	O,A
AB	A	O,A
AB	B	O,B
AB	AB	O

Rh factor:

Landsteiner and Wiener (1940) discovered one more type of antigen (agglutinogen) in the RBC's of human beings. It is named after the Rhesus Moneky (*Macaca rhesus*) and called the Rh factor or Rh antigen. It is present in about 95% of Indians (about 85% of whites also have it). However the serum does not have the corresponding antibody in the human beings. When a blood of rehsus monkey is injected into the rabbit, the rabbit produces antibodies against it. When this antiserum (serum with antibody) is tested against human blood agglutination takes place in about 85% cases. In about 15% no agglutination takes place. Persons whose blood agglutinates with Rh serum of monkey are called Rh positive and where no agglutination takes place they are called Rh negative. Six groups of Rh factors have been identified in human beings C, c, D, d, E and e. Out of these D and d are commonest. D is a Mendelian gene hence is dominant over d. Almost all human beings (85%) who are Rh positive belong to D group while Rh negative people belong to d group.

The importance of Rh factors can be ascertained from the following examples.

a) Assume that a Rh negative individual gets a blood transfusion from a Rh positive individual. At the first instance nothing will happen as there are no natural antibodies in the serum of Rh negative individual. But the presence of Rh blood will induce antibody production. Nothing will happen however as there are no sufficient antibodies. Should the same Rh negative individual require another transfusion from Rh positive blood, the donor blood will show agglutination due to the antibodies already present. This may be harmful to the recipient.

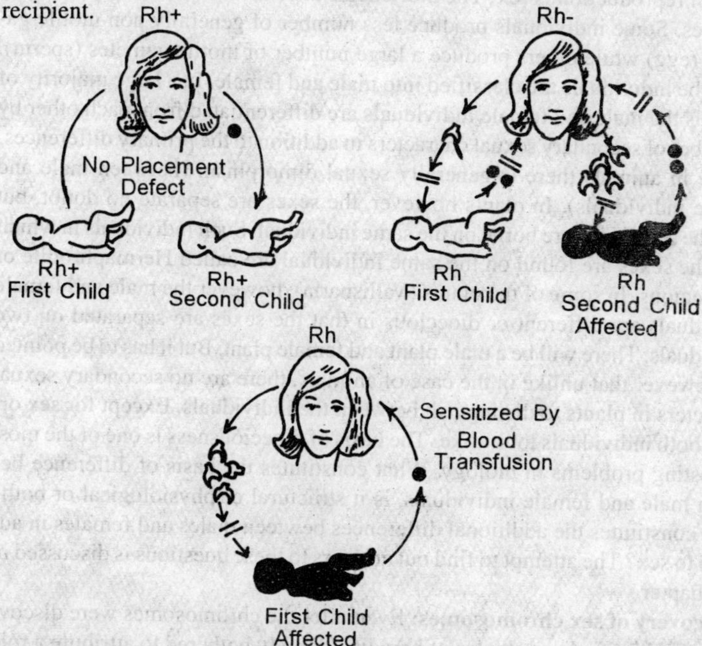

Fig. 4.7 Effect of Rh factor on the child birth

b) A Rh negative woman marries a Rh positive man. The child of this couple obviously belongs to Rh positive as the gene D is dominant over d. In the first pregnancy however the child is not harmed eventhough the maternal blood and the child's blood mix. Once again the mother's Rh negative blood will not produce enough antibodies to haemolyse the child's blood. But with each succeding conception, the antiobodies in the mother's blood accumulate, sometimes the thired or even the second child itself may have its blood (Rh positive) haemolysed leading to erythroblasts from the bone marrow enter the blood (to compensate for the haemolysed RBC) This codition is called *Erythroblastosis foetalis*. Such a baby is generally still born or even when born highly anaemic. It can be saved with total transfusion of Rh negative blood.

Five

Sex determination

In a large majority of higher organisms (specially among animals), the most striking difference between the individuals of same species from the point of view of reproduction is sex. The individuals differ in their mode of producing gametes. Some individuals produce less number of generally non motile gametes (egg) while others produce a large number of motile gametes (sperm). Thus the individuals are classified into male and female. In a large majority of animals the male and female individuals are differentiated from each other by a number of secondary sexual characters in addition to the primary differences. Hence in animals there is generally sexual dimorphism (between male and female individuals). In plants however, the sexes are separate no doubt, but both the sex organs are borne on the same individual. Such individuals in which both the sexes are found on the same individual are called Hermaphrodite or monoecious. In some of the plants (Vallisnaria), however the male and female individuals are different or dioecious in that the sexes are separated on two individuals. There will be a male plant and female plant. But it has to be pointed out howevei that unlike in the case of animals, there are no secondary sexual characters in plants to distinguish between the individuals. Except for sex organs, both individuals look alike. The basis of dioeciousness is one of the most interesting problems in biology. What constitutes the basis of difference between male and female individuals, is it structural or physiological or both? What constitutes the additional differences bewteen males and females in addition to sex? The attempt to find out answers to these questions is discussed in this chapter.

Discovery of sex chromosomes: Even after the chromosomes were discovered and identified as vehicles of heredity, nobody bothered to attribute a role to them in the determination of sex. It was beleived that physiological and environmental parameters are decisive in the sex determination of the offspring. The X chromosome was first observed by the German biologist Henking (1891) during spermatogenesis in male bugs. He called it X body. But he did not speculate on the funciton of the X body. Mc Clung (1902) described the chromosome numbers in the male and female grasshoppers; the male had 23 chromosomes, while the female had 24 chromosomes. Wilson and Stevenes (1905) followed the path of the X bodies during ooogenesis and spermatogenesis in *Protenor* (a genus of insects). They came to the reaization that the so called X body is the X chromosome which influences the sex. Wilson and Stevens further pointed out, that the females had two identical X chromosomes while the males had one X and another smaller chromosome Y. The X and Y chromosomes may be identified both cytologically as well as genetically. The discov ery of sex chromosomes made many geneticists to think in terms of the chromosomal basis of sex determination. However our present knowledge of sex determination clearly reveals that besides the chromosomes, several other factors also govern the functioning of the sex. In the paragraphs below we shall discuss, some of the mechanisms of sex determination.

Genetic determination of sex: The male and female individuals possess different structural attributes in terms of chromosomes to decide the sex of the new born. Genetic determination of sex is based on the chromosomes deciding the sex of the offspring. In these individuals the chromosomes are of two categories - *Autosomes* and *sex chromosomes*. Autosomes are responsible for all the somatic characters, while the sex chromosomes decide the sex besides carrying genes for some somatic characters also. In sexually dimorphic individuals (including man) there is a pair of chromosomes meant to govern the sex, these are called X and Y or Z and W chromosomes. Whatever may be their nomenclature their role is same. The chromosomal mechanism of sex determination varies in diffenent individuals, although essentially in all these cases, sex chromosomes decide the sex of the offspring. The following table summarises the variations in chromosomal determination of sex.

Adopted from Principles of Genetics - Sinnott, Dunn and Dobzhansky

Variations in chromosomal sex determination:

Organisms	**Heterogametic sex**	**Gametes**		**Zygotes**	
		sperms	**eggs**	**Females**	**Males**
Drosophila, man etc.,	Male	x,y	All x	xx	xy
Protenor, grasshopper etc.	Male	x-o (nil)	All x	xx (nil)	xo
Birds moths etc.	Female	All X	x,y	xy	xx
Fumea (a moth)	Female	All X	x,o (nil)	xo (nil)	xx

[Note: In the above table 'O' indicates the absence of the sex chromosome. Xo means only one sex chromosome]

Structure of the sex chromosome : The X and Y sex chromosomes exhibit certain structural differences as revealed by cytological studies. In most of the organisms the X chromosome is straight, rod like and larger than the Y chromosome. The Y chromosome is smaller in size with one end slightly hooked or curved as in *Drosophila.* In man however no such curve is found in the Y chromosome . X chromosomes have a large euchromatic area and hence carry many sex linked somatic characters. The Y chromosome has less of euchromatic region and carries but few sex linked somatic characters.

In dioecious organisms chromosomal sex determination is of basically two catagories - Heterogametic males and Heterogametic females.

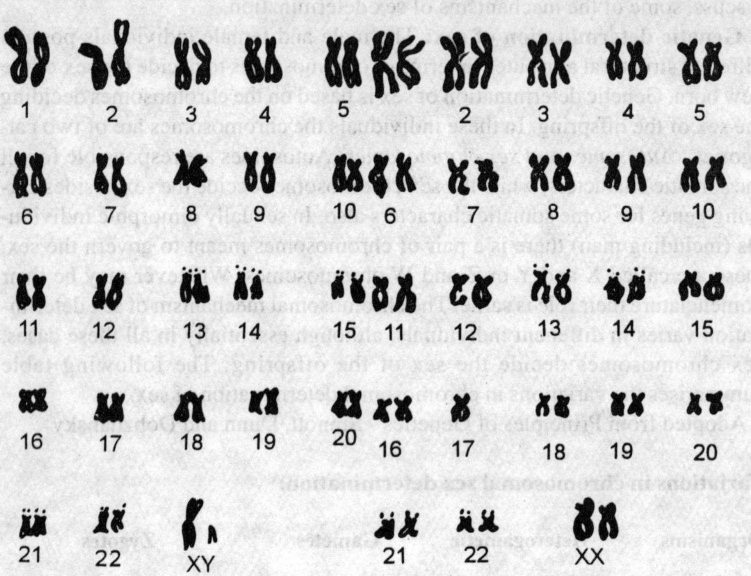

Fig. 5.1 Human chromosomes at metaphase Male (left) Female (right)

Heterogametic males: The male individual produces two types of sperms. 50% of the sperms carry one type of sex chromosome (X) the rest carry another type (Y) of sex chromosome. All the females carry only one type of sex chromosome (X). Hence the females are said to be homogametic. The sex of the new born depends upon the type of sperm fertilized the egg. The female gamete has no role in deciding the sex in the sense it always carries only one type of a chromosome. Sex determination in heterogametic males belongs to any one of the following types.

1. XX - XY Type: In man, *Drosophila* etc., where male is the heterogametic sex, the male and female have the following types of sex chromosome (this type was first studied in milkweed bug Lygaeus turcicus, hence it is also called as the Lygeus type.

Organism	Sex	Autosomes	Sex chromosome
1. Drosophila	Male	3 pairs	XY
2. Drosophila	Female	3 pairs	XX
3. Human being	Male	22 pairs	XY
4. Human being	Female	22 pairs	XX

In the XX - XY type, the male produces two type of sperms. 50% sperms would carry X chromosome and the rest carry Y chromosome. All the eggs are of only type in that all of them carry a single X chromosome . The sex of the new born depends on the type of sperm (carrying X or Y) fertilizing the eggs. It is as follows.

Sperm	Egg	Sex of the offspring
X	X	XX- FEMALE
Y	Y	XY- MALE

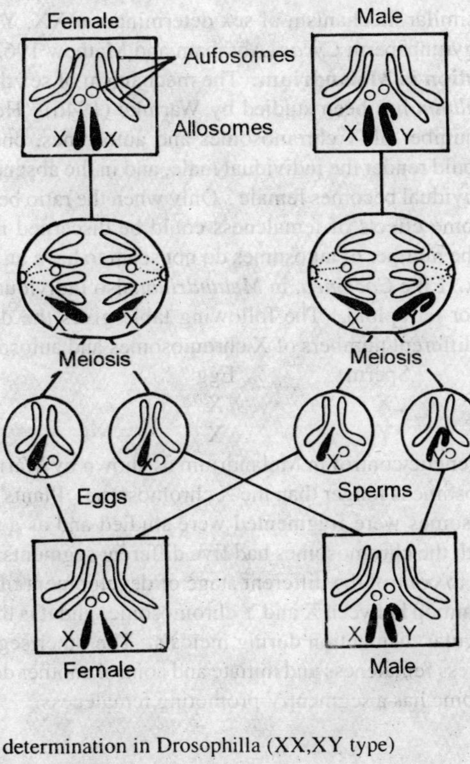

Fig. 5.2 Sex determination in Drosophilla (XX,XY type)

Mechanism of Sex Determination in Plants

In many of the lower plants the mechanism of sex determination is restricted to single gene and it does not bring about any morphological variation between the mating types. In individuals like *Mucor,* Yeast *Aspergillus* etc., sex is governed by the two alleles of a single gene A and a. The Zygote will have both these genes. After meiosis, when segregation takes place two races are produced A and a. Sexual reproduction is possible only between the races carrying the opposite genes.

In most of the flowering plants, which are monoecious, it is more a question of sex differentiation than of sex determination. The cells possess both the potentialities (male and female) and possible hormonal influence will differentiate the male and female sex organs . (A detailed account of sex determination in plants is available in Discussions in cytogenetics by C.R.Burnham.)

In most of the dioecious plants, sex determination is of thė XX, XY type. In some Bryophytes like Sphaerocarpus, Marchantia etc., the diploid spore mother cells have X and Y chromosomes. During reduction division and spore formation these two seggregate and consequently two types of spores are formed. 50% of the spores will have X chromosome and they develop into female gametophytes. A similar mechanism of sex determination(XX, YY) has been reported for the gymnosperm *Cycas* (Abraham and Mathew 1962).

Sex determination in Melandrium: The mechanism of sex determination in *Melandrium album* has been studied by Warmke (1946). Here also irrespective of the number of X chromosomes and autosomes, one or more Y chromosomes would render the individual male, and in the absence of Y chromosome , the individual becomes female . Only when the ratio between X and Y reaches 4:7, some effects of femaleness could be discerned inspite of the presence of Y. The number of autosomes do not seem to have an effect on the expression of sex. Like *Coccinea,* in *Melandrium* also individuals are either diploid, triploid or tetraploid. The following table gives the details of sex expression with different numbers of X chromosomes and autosome sets.

Sperm	Egg
X	X
Y	X

The details of genetic control in Melandrium as shown by Warmke indicate that the Y chromosome is longer than the X chromosome . Plants in which the X and Y chromosomes were fragmented were studied and as a result it was found out that both the chromosomes had five different segments. These segments are known to control the different stage of development of sex organs. Segment IV is common between X and Y chromosomes and it is this that helps in pairing and regular disjunction during meiosis. The other segments on Y chromosome supress femaleness and initiate and control another development. The sex chromosome has a segment V promoting femaleness.

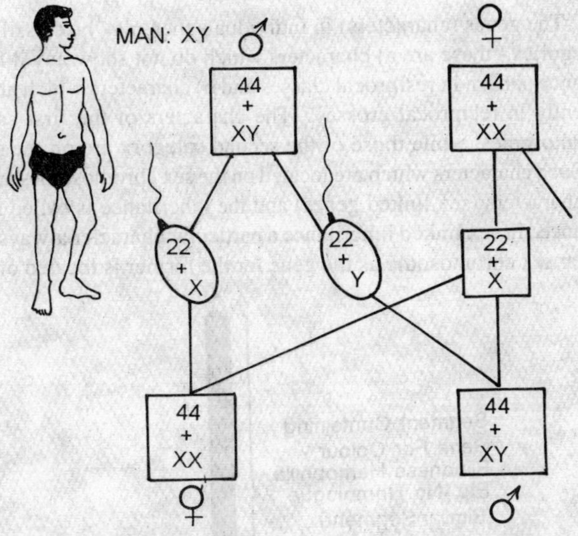

Fig. 5.3 Sex determination in human beings

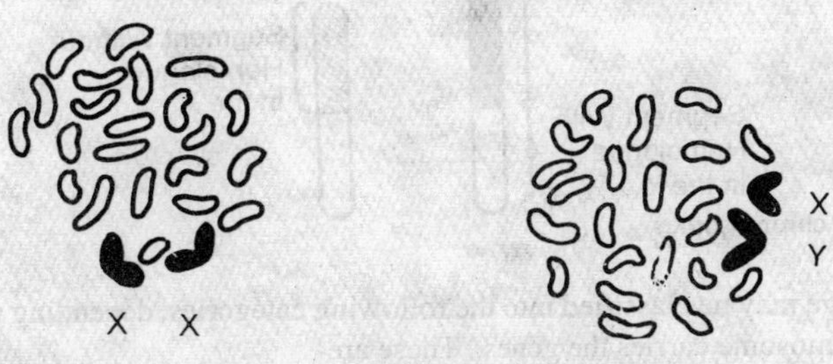

Fig. 5.4 The chromosomal sex determination in Melandrium

SEX LINKED INHERITANCE

The laws of inheritance as propounded by Mendel indicate that reciprocal crosses always yield the same result in hermaphrodite individuals. In other words in a cross between tall and dwarf plants the resuitant offspring would be tall irrespective of the fact whether tall parent is male or female. This means that tallness or dwarfness are inherited irrespective of the sex of the individual and obviously the genes responsible for these characters must be situated on the autosomes. On the other hand there are certain characters which seem to be associated with a particular sex or sex chromosome and such characters are not inherited in the same pattern in reciprocal classes.

The genes (characters) in individuals then may be classified under two categories - these are a) characters which do not show any difference in inheritance pattern in resiprocal classes and b) characters which are inherited differently in reciprocal crosses. The characters of the first category are on the autosomes, while those of the second category are on the sex chromosomes. Such characters which are located on the sex chromosomes are called sex linked characters (sex linked genes) and the inheritance is called sex linked inheritance. In sex linked inheritance a particular character always follows a particular sex chromosome as the gene for the former is located on the latter.

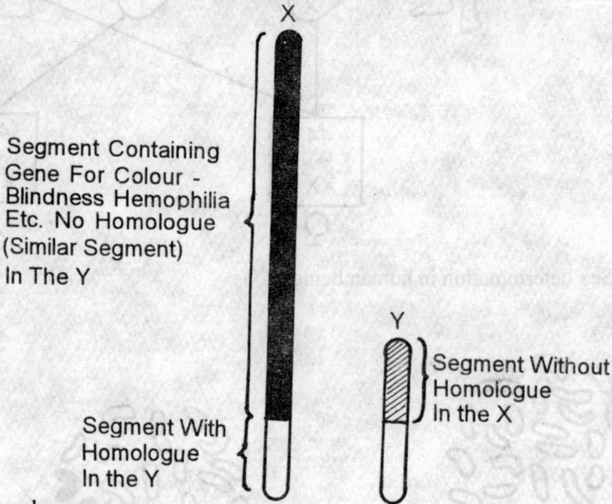

Fig. 5.5 Sex chromosomes

Sex linkage may be classified into the following categories, depending upon which chromosome carries the genes. These are-

1. XY linked characters (genes)
2. Y linked characters (genes)
3. X linked characters (genes)

Inheritance of Y linked Characters: As has already been pointed out, some genes are exclusively linked to Y chromosome . A few instances of these in man are - a) Porcupine man in whom the entire surface of the skin will be warty and coarse (*Ichtyosis hystrix*) b) hypertrichosis hair on the lobes of the ear c) Kerotoma dissipatum-skin disorder affecting the hand and feet d) Occurrence of Webbed toes in some individuals etc.

In plants, the Y chromosome contains genes for maleness and its absence induces sterility. Unfortunately not much detail is available on the Y linked somatic characters in plants.

Inheritance of sex linked or X linked genes: The genes or characters linked to the X chromosome are generally termed sex linked as they are the most prevalent sex linked characters. In order to understand the pattern of inheritance of X linked characters one should understand the inheritance pattern of the X chromosome.

In female individuals (which are chromosomes, while the males (which are hetrogametric) have only one X chromosome in the body cells. During reproduction the eggs will have only one X chromosome where as the sperms will be 50% X and 50% Y. When an egg is fertilized by an X sperm the resultant female (XX) individual will have one X from the mother and the other X from the father. When a Y sperm fertilizes the egg the resultant male offspring will get its X chromosome from the male individual again goes to the female (daughter). In this way an X chromosome and the genes linked with it will alternatively go to male and to female. The following chart summarises, the above explanation.

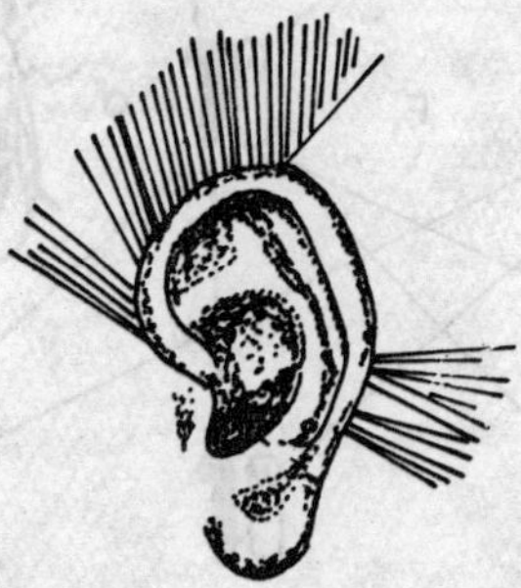

Fig. 5.6 Hyperitrichosis of human car

This type of inheritance of X chromosome which alternatively finds place in the male and female generations is called the criss cross inheritance . We shall study a few examples of this type of inheritance in Drosophila and man.

Sex linked inheritance in Drosophila: In *Drosophila* the normal eye colour is red. Morgan in one of his cultures found a white eyed individual. On breeding, this proved to be a pure line. When the white eyed flies were crossed with the wild (red) type, the results of reciprocal crosses were different i.e. white male x red female : white female x red male. The result seemed to be dependent on the sex of the parent through which the trait was introduced into the cross rather than following the Mendelian pattern. Morgan assumed this to be due to the sex linked character of the gene for eye colour. The gene for normal eye colour (red) W is present on the X chromosome . White eye colour is a recessive gene occupying the same locus (w). Hence for white eye colour to

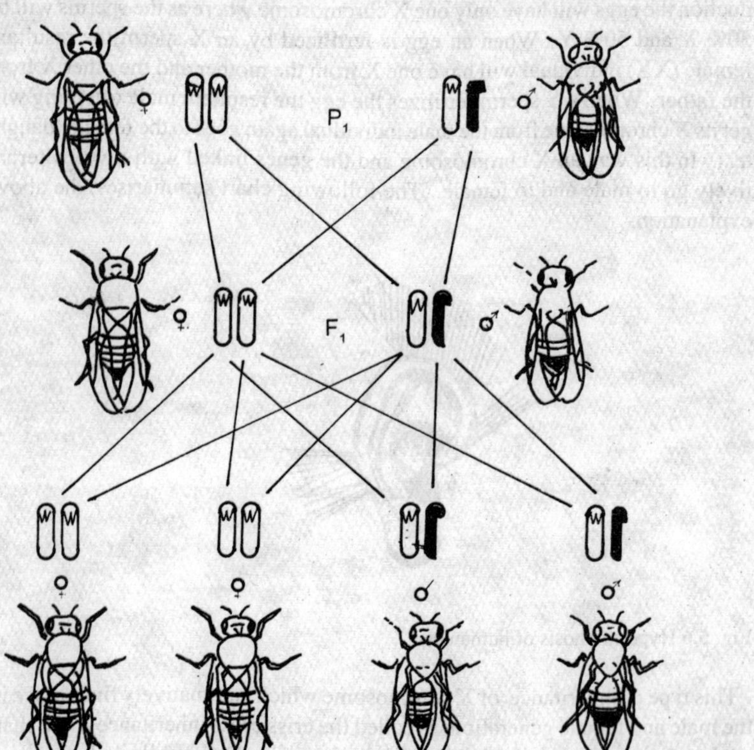

Fig. 5.7 Mechanism of Sex linked inheritance in Drosophilla (Red eyed female x White eyed male)

appear in the females the genes should be double homozygous(should be present in both the X chromosome). In the heterozygous condition in female (Ww) one X carries the W (red) gene, while the other X carries the w (white) gene. Phenotypically the heterozygous females are red eyed, but they are called carriers as they carry the recessive gene. The males however can not be carriers, for the gene for red or white being present on the only one X chromosome they have, their genotype and phenotype are one and the same. Even a single dose of the recessive gene (w) is enough to produce the white eye colour. This can be understood by the reciprocal crosses as follows.

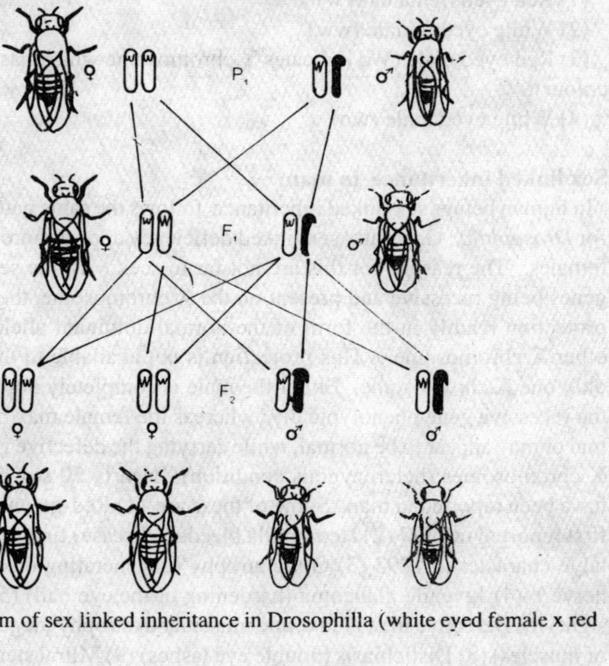

Fig. 5.8 Mechanism of sex linked inheritance in Drosophilla (white eyed female x red eyed male)

1. Red eyed male X White eyed female : When a red eyed male (Wo) is crossed to a white eyed female (ww) in the F_1 generation all the females would be red eyed and all the male individuals are white eyed. This is because the white eyed recessive gene is present on both the X chromosomes in the female parent. The F_1 male offspring gets its X (w) from the mother and Y from the father. Since Y chromosome does not carry the gene for eye colour the individual would be white eyed. The female offspring (F_1) having got one X (W) from the father and one X (w) from the mother would be heterozygous (Ww) genetically, but phenotypically show red eye colour. When the F_1 heterozygous female is crossed with the white eyed F_1 male, in the F_2 generation both red and white eyed (1:1) individuals are found in the female individuals, while in the male progency also red and white eyed individuals appear in equal proportion (See fig)

Red eyed female x white eyed male : When a pure red eyed female (WW) is crossed with a white eyed male (wo), the F_1 offspring will have both females and males red eyed. The females however are carriers (heterozygous) as they have both Ww genes on their two X chromosome . When these two F_1 individuals are crossed (heterozyous red eyed female X white eyed male), among the female (F_2) progeny (1) 50% will be red eyed (heterozyous) (2) 50% will be white eyed (homozygous double recessive). Among the male progeny (3) 50% will be red eyed and (4) 50% white eyed (see fig). The genotype of these flies will be as follows.

(1) Red eyed female (Ww)

(2) White eyed female (ww)

(3) Red eyed male (Wo indicates Y chromosome which has no gene for eye colour)

(4) White eyed male (wo)

Sex linked inheritance in man:

In human beings sex linked inheritance follows the same pattern as explained for *Drosophila.* Generally sex linked deficiency appear more in males than in females. The reasons for this are not far to seek. All the sex linked mutant genes being recessive and present on the X chromosome, the female gets the protection readily in the form of the normal dominant allele present on the other X chromosome . This protection is not available to the male as it has only one X chromosome. Either the male is completely normal or expresses the recessive gene phenotypically, whereas the female may be complete normal or may appear to be normal, while carrying the defective gene in one of the X chromosomes (heterozygous condition). Nearly 20 sex linked characters have been reported in man. Some of these are - 1) Red-green colour blindness first reported in 1777 (2) Hemophila bleeders disease) first recorded as a heritable character in 1793 (3) Optic atrophy (degenerating disease of the optic nerve.) (4) Juvenile glaucoma (hardening of the eye ball) (5) Myopia (short sight) (6) Defective iris (7) Juvenile muscular dystrophy (degenerating disease of muscles) (8) Distichiasis (double eye lashes) (9) Mitral stenosis (Abnormal mitral valve of the heart (10) Mystagmus (involuntary oscillation of the eye ball (11) baldness and (12) Epidermal cysts. We shall study the inheritance pattern of a few of these traits.

Inheritance of Hemophila: Commonly called bleeder's disease, this refers to a condition, where blood on exposure to air will not clot, as a result hemopjhilic individuals may bleed to death even at the slightest injury. Hemophilia is a sex linked trait offen associated with males, females being very rare, (but genetically possible), The failure of the blood to clot is due to the absence of a protein antihemophilia globulin which is normally present in the blood of man. Hemophilic individuals lack this, as a result bleeding, when occurs may prolong for 1/2 hours to 20 hrs., leading to the death of the individual if proper measures are not taken. The production of the coagulating substance is governed by a gene present on the X chromosome . In hemophilics, this wild gene would have undergone mutation resulting in a hemophilic gene.

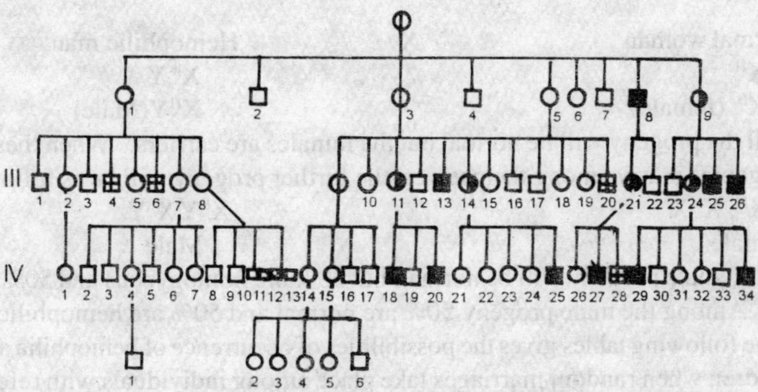

Fig. 5.9 Pedigree of Haemophilies in desendents of Queen Victoria

Hemophilia gained some notoriety in the past as the Royal disease, as some members of British royalty were hemophilics. One of the well known victims of the disease was the great grandson of queen Victoria who died due to an injury and subsequent non stop bleeding.

As has already been pointed out, hemophilia is found predominantly in males, but the inheritance is criss cross- From father to daughter to grandson. The interesting feature is the females generally do not suffer, but they will be carriers. Hemophilia can manifest in females only in double recessive conditions. Such an individual dies before adolscence. The chances of randum mating producing a hemophilic are extremely rare (in males it is about 1 in 10,000); it is estimated that if all the hemophilic males survive and undergo random mating, the mathematical probability of a homozygous female would be equal to the square of 10,000 or 1 in 1,00,000,000. The genetic manifestation of hemophilic gene is as follows.

The hemophilic gene is a recessive allele of the wild gene situated on the X chromosome only. There is no allele for this on the Y chromosome. Hence no cross over is possible between XY for hemophilia. If we denote the normal gene as H and hemophilic gene as h, the genetic constitution of different individuals will be as follows.

	Normal	**Hemophilic**	**Carries (Heterozygous)**
Male	X^HY	X^h,Y	
Female	X^H,X^H	X^h,xh	X^H,X^h

Let us study the inheritance pattern by studying a hypothetical example. When a hemophilic man marries a completely normal woman, the following are the progenies Produced.

Normal woman	X	Hemophilic man
X^HX^H		X^h,Y
X^HX^h, (female)		X^HY(Male)

All the progeny will be normal but the females are carriers. When these two progenies marry among themselves, the further progeny will be as follows.

X^HX^H, X^HX^h	X^HY,X^hY
Female	Male

All female progeny will be normal: but 50% are homogyzous and 50% carriers. Among the male progeny 50% are normal and 50% are hemophilics.

The following tables gives the possibilities of occurrence of hemophilia among children when random marriages take place among individuals with reference to the hemophilic gene.

	Parents genotype		Parents Phenotype		Offspring genotype		Offspring genotype	
	Male	Female	Male	Female	Male	Female	Male	Female
1	X^HY	X^HX^H	Normal	Normal	X^HY^H	X^HX^H	Normal	Normal
2	X^hY	X^HX^H	Hemophilic	Normal	X^HY	X^HX^h	Normal	Carrier
3	X^HY	X^HX^h	Normal	Carrier	X^HY	X^HX^H	Normal	Normal
4	X^hY	X^hX^h	Hemophilic	Carrier	X^HY	X^HXh	Normal	Carrier
5	X^HY	X^hX^h	Normal	Hemophilic	X^hY	X^HX^h	Hemophilic	
6	X^hY	X^hX^h	Hemophilic	Hemophilic	X^hY	X^hX^h	Hemophilic	Hemophilic

Note:

1. X denotes X chromosome; Y denotes Y chromosome
2. X^H denotes normal dominant gene on X chromosome
   X^H denotes heomphilic gene on X chromosome
3. Normal female means homozygous
   Carrier means heterozygous; normal phenotypically
4. Homozygous hemophilic female individuals do not survive beyond adolscence. Crosses involving them (5 and 6) are only hypothetical.

Colour blindness in Human beings: The red and green colour blindness is another sex linked recessive trait in man. Individuals suffering from this defect will not be able to distinguish between various shades of red and green and hence are colour blind.

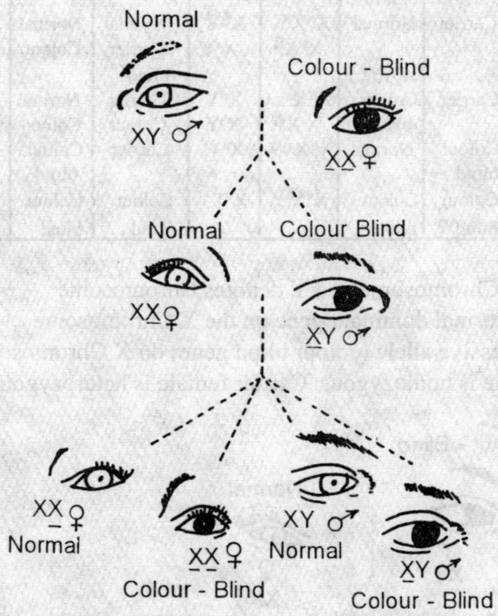

Fig. 5.10 Inheritnce of colour blindness in man (normal x colour blind female)

Individually red colour blindness is called *Protanopia* and green colour blindness is called *Deuteranopia*. Ishihara has designed a colour chart to identify these defects. It has been shown that the gene for colour blindness is a recessive allele of normal vision and is situated on the X chromosome. Like any sex linked defect, colour blindness also is more prevalent among males than among females. The Y chromosome does not carry any allele for vision.

A survey for colour blindness has not been conducted in our country, but the data available from USA give a figure of 8% for male and 0.5% for female in the total population. The following chart gives all the possible mating types and the possible progeny for colour blindness among human beings.

Parents Genotype		Parents Phenotype		Offspring Genotype		Offspring Phenotype		Total Offspring	
Female	Male	Female	Male	Female	Male	Female	Male	Female	Male
1 X^NX^N	X^NY	Normal	Normal	X^NX^N	X^NY	Normal	Normal	50	50
2 X^NX^N	X^nY	Normal	Colour blind	X^NX^n	X^nY	Carrier	Normal	50	50
3 X^NX^N	X^nY	Carrier	Normal	X^NX^N	X^NY	Normal	Normal	25	25
				X^NX^N	X^nY	Carrier blind	Colour	25	25
4 X^NX^n	X^nY	Carrier	Colour blind	X^NX^n	X^NY	Carrier	Normal	25	25
				X^nX^n	X^nY	Colour	Colour	25	25
5 X^nX^n	X^NY	Colour blind	Normal	X^NX^n	X^nY	Carrier	Colour blind	50 50	50 50
6 X^nX^N	X^nY	Colour blind	Colour blind	X^nX^n	X^nY	Colour blind	Colour blind	50	50

Note:

1. X denoted X Chromosome and Y denotes Chromosome
2. X n denotes normal dominant gene on the X Chromosome

X n denotes recessive allele (colour blind gene) on X Chromosome

3. Normal female is homozygous. Carrier female is heterozygous.

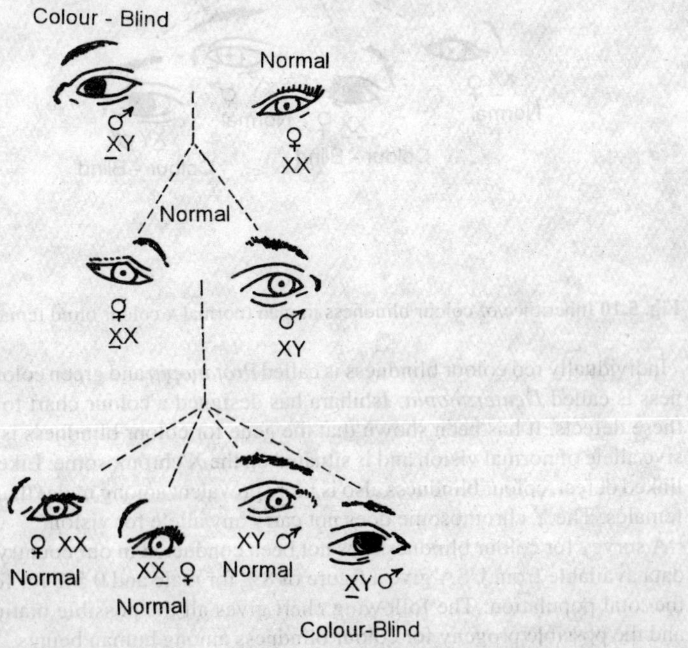

Fig. 5.11 Inheritance of colour blindness in man (normal female X colour blond male)

In all individuals where sex determining mechanism follows that of man or Drosophila (male being heterogametic and sex linked, recessive gene being located only on the X chromosome) the following are the salient features of its manifestation.

1. Sex linked recessive traits are phenotypically expressed more often in males than in females.

2. The trait appears among the female progeny only when it is found in the male parent.

3. The trait seldom is found both in father and son. It may be the case, when mother is homozygous.

4. The inheritance of the sex linked recessive gene is from father to daughter to grand son.

Six

Genetic diseases

Disease caused byv genetic disorders are called genetic diseases. Malfunctioning of a gene is called a genetic disorder. For instance a gene is supposed to produce a particular enzyme and that has to participate in a reaction producing a specific product. If for some reason the gene gets altered i.e., undergoes mutation, then the gene product - enzyme can not be produced and this may block a particular metabolic pathway resulting in the non formation of a product. If this product is an important one it leads to diseases known as genetic diseases.

The discovery of Beadle and Tatum (1940) in *Neurospora* that mutational changes in genes can be linked to losses of specific enzymes paved the way for further studies of many unexplained diseases in human beings.

A.E. Garrod (1908) was the first to study genetic diseases connected to distrubed metabolic pathways. He used the term *'inborn errors of metabolism'* to denote the genetic disease*Alkaptaneria* or dark urine disease. The phenomenon of inborn errors of metabolism may be understood clearly by understanding *metabolic blocks*.

Metabolic block: A biosynthetic pathway normally consists of several steps with each step and the formation of an intermediary product under the influence of an enzyme which in turn is controlled by a gene. If a mutation occurs in any gene, either the production of normal enzyme is stopped or a defective enzyme is produced. This defective enzyme instead of producing a normal character (phenotype) produces a drfective character. Such an alteration of normal phenotype due to alterarion of metabolic pathway is called *metabolic block.*

Metabolic blocks or metabolic errors produce a number of defective characters in human beings and provide an oppurtunity to understand the relationship between genes, enzyme and phenotypes. This helps us to unravel the genetic regulation of biochemical pathways. Such a field of study is known as biochemical genetics. Biochemical genetics may be defined as analysis of genetic control of biosynthetic pathways.

Defects in bio synthetic pathways are of various types. These are-

1. Defects in the metabolism of aminoacids.
2. Defects in the metabolism of lipids.
3. Defects in sugar metabolism.
4. Defects in the production of Haemoglobin etc.

In this book defects in aminoacid metabolism and haemoglobin are discussed.

Defects in the metabolism of aminoacids:

Metabolic pathways involving (a) phenylalanine-Tyrosine, (b) Tryptophan (c) Arginine serve as good examples to study the genetic defects and associated phenotypic alterations. Here we shall study the metabolic pathway involving the amino acids Tyrosine and phenyl alanine.

Metabolism of phenylalanine: Phenylalanine is an important constituent of proteins in our diet. During digestion of proteins, the amino acid (phenyl alanine) is released into the digestive tract and then gets absorbed by the body cells. Within the cell there are three pathways for the metabolism of phenyl alanine (fig)

(i) It may combine with other amino acids to form proetins.

(ii) can convert itself into phenyl pyruvic acid.

(iii) can convert itself into Tyrosine.

We shall now consider the possible abnormalities in the (ii) and (iii).

Phenyl ketonuria: The conversion of phenyl alanine to tyrosine is under the control of the enzyme phenyl alanine hydroxylase (present in liver) which is in turn regulated by the gene *P*. In homozygous recessive conditon genes *pp* will not produce the enzyme and hence there is accumulation of phenyl alanine as it cannot convert itself into tyrosine. As a result there is abnormal increase in the level of phenyl alanine and also its product (in another pathway) phenyl pyruvic acid. Both these- phenyl alanine and phenyl pyruvic acid acenmulate in the cells in large amounts and most of these is excreted through the urine or may get accumulated in the blood. This condition is known as phenylketonuria (Pku).

Normal persons will have a plalsma phenylalanine level of 1-2 mg/100ml while Pku persones will have 15-64mg/100ml.

Similarly the urine level phenyl alanine in normal humans is 30mg/100ml while it is as high as 300-1000mg/100ml in Pku's.

Complications of Pku include brain damage leading to mental retardation (with an IQ level below 20). High quantities of phenyl alanine seem to interfere with the development of brain cells. Persons suffering from Pku are called *Phenylpyruvic idiots* as they have very low IQ.

Effects of Pku will be noticed a few months after birth. Affected babies may show eczema of skin, defective enamel of teeth and bone deformations.

Treatment of Pku includes phenyl alanine free diet which can prevent brain damages. The diet however should include fixed amount of phenyl alanine also as it is an essential amino acid.

Tyrosine metabolism:

Tyrosine is another key amino acid involved in various metabolic pathways. It may be obtained by conversion of phenyl alanine or by the direct break down of proteins. In the normal human beings the metabolism of Tyrosine proceeds as follows -

a. It may combine with other amino acids and become part of cellular proteins.

b. It may combine with Iodine in the thyroid gland to produce thyroid hormones - Failure of this results in insufficient thyroid hormones producing a defect called *genetic goitrus cretenism*. This is characterised by physical and mental retardation.

c. Tyrosine gets converted into dihydroxy phenyl alaline (DOPA) which inturn leads to melanin formation. Failure of this leads to *Albinism*.

d. In the fourth pathway tyrosine gets degraded in a stepwise fashion to CO_2 and water.

(1 Tyrosine (2) Parahydroxy phenyl pyruvic acid (3) Dihydroxyphenyl pyruvic acid (4) Homogenetic acid (5) $CO_2 + H_2O$.

Failure of conversion of 2 to 3 (as shown above) leads to a disease called *tyrosinosis.* Faiiure of degradation of Homogentisic and leads to its excretion along with urine and the disease is known as *Alkaptoneuria.* Urine of Alkapataneurics turns black on exposure to air. Hence it is called black urine disease.

Albinism: As has already been pointed out Albinism is a result of failure of metabolism of Tyrosine. In normal human beings metabolism of Tyrosine leads to the formation of melanin pigmentation which gives the normal colouration to the skin. The normal metabolic pathway is as follows-

Tyrosine DOPA DOPAquinone melanin.

Albinism results due to the absence of the enzyme *Tyrosinase* (which normally acts on tyrosine and produces DOPA). There are two types of albinism.

(i) Ocular albinism - only eye pigments are absent.

(ii) genral albinism - no pigments in both eyes and skin.

In normal human beings the enzyme tyrosinase is present in the structures caled melanosomes present in melanocytes. In albinos the enzyme is not produced due to a defective gene.

Metabolic gefect in the production of haemoglobin

Haemoglobin in the blood is necessary for gaseous exchange particularly to absorb oxygen. A normal haemoglobin consists of four polypeptide chains each associated with a single haeme group. Out of the four chains two are called *alpha* chains and are common in children and adults. The remaining two chains are different at different ages in human beings. Each polypeptide chain is coded by a different gene.

Defects in the haemoglobin chain may be divided into the following groups-

(i) Thallassaemia caused by reduction in globin chain synthesis

(ii) Sickle cell anaemia.

(ii) Hereditary persistance of foetal haemoglobin.

We will study sickle cell anaemia in some detail.

Sickle cell anemia : The name is derived from the anaemic condition (destruction of haemoglobin) owing to abnormal condition of the haemoglobin. In patients suffering from sickle cell disease, most of the erythrocytes (RBC's) of the blood lose their normal dusc shape and become sickle shaped. These sickled cells increase the viscosity of the blood. Their presence in small vessels leads to blockage of capillaires. The sickle erythrocytes are fragile and easily get destroyed in the spleen. This leads to anaemic condition and a blood test will reveal the presence of sickle RBC's.

Sickle celled anaemia is caused due to abnormal haemoglobin called sickle haemoglobin. In sickle haemoglobin *alpha* and *beta* chains undergo structural variations due to replacement of amino acids. This in turn is due to mutations in the structural genes. Linus Pauling (1949) discovered the difference between normal haemoglobin and sickle haemoglobin. Tugram (1957) showed clearly that a single amino acid substitution is the difference between the normal and sickle haemoglobin.

Normal haemoglobin is controlled by the gene *ha* and the mutated form is *hs*. Disease manifests in the following combinations.

h*a* h*a* - Normal individuals

h*a* h*s* - Mild form of sickle cell anaemia (heterozygous)

h*s* h*s* - Severe form of sickle cell anaemia (homozygous)

Hameophilia : This is another irregularity of the blood and is a genetic disorder. Commonly called bleeder's disease, in this blood fails to clot on exposure to air. In haemophilics the gene to produce vitamin K necessary for clotting will be defective, hence even in a minor wound there is continuous bleeding which may be fatal. This is a sex linked character. (for details see chapter 1.5)

Seven
Biotechnology

Biotechnology is a hybrid science which has emerged in recent years as the most vital branch of biological science having the potential of changing the course of life itself.

Biotechnology may be defined as application of scientific and engineering principles to the biological systems to obtain a desired product or product of choice. In otherwords we can alter the process in a biological system and order it to produce a different product by introducing new materials.

Eventhough biotechnology as a scientific discipline is fairly recent, knowingly or unknowingly man has been practising biotechnology since time immemorial. It can be safely said that ever since the first toast (over a glass of wine) was proposed and the first loaf of bread was baked biotechnology is known to man.

Involvement of yeasts in alcohol production was known since the time of Louis Pasteur, though detailed studies were undertaken only in 1900s.

Presently biotechnologists are engaged in some of the most challenging works such as production of cheap drugs from microbes, production of growth hormones, gene therapy, transfer of N2 fixing genes to higher plants, transfer of disease resistant genes, monoclonal anitboides etc., all aiming at a better future for mankind.

The following table summarizes the main areas of biotechnological research.

Area of Research	Products
1.Recombinant DNA technique	enzyme, vaccines, growth hormones, antibiotics, interferon, insulin etc.
2. Ulitization of biomass	single cell protein (scp), alcohol, biofuels.
3. Tissue culture	fine chemicals, rapid multiplication of disease free plants, somatic embryos, monoclonal antibiotics.
4. N_2 fixation	production of biofertilizers.
5. Bioenergy	Hydrogen, alcohols, methane (biogas).
6. Fermentation technology	production of acids, alcohols, vitamins etc.
7. Enzymes	extraction of enzymes for use in industries, food processing, breweries.
8. Environmental engineering	effluent treatment, water recycling, biological control.

The various aspects of biotechnology may be studies under two main categories. These are-

1. Recombinant DNA Technology (Genetic engineering)
2. Tissue culture.

Genetic Engineering

Recombinant DNA technology or Genetic engineering as it is popularly known, is the newest tool that is available in the human arsenal of knowledge in man's quest for a better human life. What is genetic engineering or recombinant DNA technology?

Imagine a situation of chronic diabetic patients being able to get back the capacity of synthesis of insulin; patients suffering from incurable diseases getting an injection of gene therapy - lo and behold they are cured of their diseases; consider a situation which is at present in the realms of fantasy of making man producing his own food just like the green plants do from natural sources; or visualize a situation wherein revolution in agricultural technology would make the use of chemical fertilizers completely redundant or going to the extreme, culturing a human being of the desired trait on a medium of nutrients ie., man made to order. The scope is fantastic; possibilities are immense; it is as if the role of creator has been for once given to human beings. Well this is genetic engineering or recombinant DNA technology, the most powerful biological technique which the human beings can use either to make or mar the society. In its impact and potential it is equivalent to nuclear power which can be harnessed either for progress or for destruction. Using the techniques of genetic engineering you can produce varieties of agricultural crops which are disease free or you can introduce pathogens into the crops and totally destroy them. Genetic engineering used judiciously can be a harbinger of new revolution but used injudiciously may produce many prototypes of criminals, antisocial elements, boot leggers etc. In this chapter we shall briefly discuss some of the fundamental aspects of genetic engineering and how to use them for betterment of human life.

Genetic engineering may be defined as the method of artificial synthesis of new genes and their subsequent transplantation in the new genome of an organism, by molecular biological techniques. Recombinant DNA is achieved by constructing molecules outside the living cells by joining natural or synthetic DNA molecules that can replicate inside a living cell. Recombinant DNA technology involves several stages or steps in the cloning and multiplication of the desired genes.

Gene cloning : This involves the following steps -

a) Isolation of a specific gene or DNA sequence.
b) Insertion of this into a vector or vehicle.
c) Creation of a recombinant DNA molecule.
d) Introduction of the recombinant DNA into a suitable host.
e) Multiplication and identification of clones containing recombinant DNA molecule.

Isolation of specific gene: Identification and isolation of suitable gene or sequence of DNA is a prerequisite for the preparation of recombinant DNA and subsequent cloning. The DNA fragments to be cloned are isolated, chemically synthesized or transcribed from the isolated mRNA. Backwitch and his colleagues have perfected the technique of the isolation of a gene lactose operon from *Escherichia coli.*

Theoretically, identification of a gene involves, the reading of the message back in protein synthesis. During protein synthesis DNA produces mRNA and this inturn identifies and allows the alignment of amino acids to synthesize a protein. In the identification of a gene responsible for the production of a particular protein, the process has to be followed backwards. Let us take an instance of the human insulin which is a protein.

The steps involved in the identification of the gene or segment of DNA responsible for insulin are as follows.

a) Isolation and purification of the protein.

b) Sequential degradation of the protein: In sequential degradation, not only the qualitative composition of the amino acids are known, but even their sequence of arrangement in the polypeptide chain will be known. After ascertaining the sequence of amino acids the next step would be:

c) Identification of codes for each amino acid: This can be done easily as we know the triplet codes for each amino acid. This would give us a length of the mRNA segment equivalent to the length of the polypeptide chain. (For example if the protein consists of 25 amino acids, the length of the mRNA should be 75 nucleotides).

The nucleotide sequence in mRNA can be easily substituted for the complementary DNA segment. We now have a gene or a segment of DNA which governs the production of insulin.

The DNA segment can either be isolated from natural sources or synthesized artificially. Artificial synthesis of biologically active DNA was first conducted by Kornberg (1961) and his associates. They mixed together bionucleotides, enzymes, ions and a piece of DNA primer. After some time they noticed the synthesis of a new molecule of DNA. These new DNA molecules (the primer was originally from a virus) when inserted into a bacterial cell, produced new virus particles. Another Nobel Laureate H.G. Khorana synthesized a gene in a test tube with known sequence of nucleotides.

The newly created DNA segment can be tested for the presence of specific complementary nucleotides. Specific mRNA isolated from natural sources is used to hybridze with the DNA segment. If the hybridization is successful, it is an indication of the correct gene (segment of DNA). Genes for lac operator, somatostatin, insulin and growth hormone have been synthesized chemically.

Insertion of DNA segment into a vehicle:

The newly produced segment of DNA has to be transferred to the host cell, where it is supposed to function and produce its genetic effect. Theoretically it is possible for a new cell or organism to take up the naked pieces of DNA and incorporate it into its hereditary material. As a result, the recipient produces the genetic effect of the inserted gene. This however may not be successful. Very often an intermediary is used to accept the new piece of DNA and then it is finally transferred to the host cell. These intermediaries or vehicles are plasmids and viral DNAs.

Plasmids: These are covalently closed circular molecules of DNA occurring frequently in bacteria in addition to the main genome. These promote genetic transfer by conjugation, confer drug resistance and are capable of producing antibiotic toxins. Plasmids are of two types - Sex factor plasmids which are conjugative and non sex factor plasmids which are non conjugative. The plasmids replicate independently of the main genome. Since the plasmids are small they can easily enter and come out of the cell. Restriction endonuclease enzymes will form a nick in the plasmid and the foreign DNA molecule is inserted in between the broken ends. Ligase enzyme helps in the union of sticky ends, thus incorporating the new DNA segment. When the bacterium and the plasmids multiply, the new DNA segment also multiples and produces its own genetic effect.

Creation of a recombinant DNA molecule and introduction into the host

As has already been pointed out, the foreign DNA gets incorporated into the plasmid and becomes a part of it. But the segment of newly introduced DNA will carry out its own genetic functions and produces its characteristic expression. For instance if a gene for insulin is introduced into the plasmid, the bacterial cell starts producing insulin together with its own products. In this way a bacterial cell can be used to produce the drug or whatever product that we require.

Gene insertion in mammalian cells : Inserting a segment of new DNA has been attempted in a number of instances in mammalian cells. These are attempts to finally find an answer to genetic diseases in man.

Avonson and Todaro (1969) opined that the DNA isolated from simian virus 40 (Sv - 40), can be incorporated into the genome of human fibroblast cells. The incorporated viral genome can produce its own proteins. The viral genome is a circular double stranded DNA containing 5224 base pairs. The host cells incorporated with the viral genome exhibit characters like loss of sensitivity to contact, inhibition of cell division, and production of mRNA specific to SV 40 virus. These features clearly indicate that the viral genome has been integrated into the fibroblast genome.

C.R. Merril, M.R.Gleir and J.C. Petricciani (1971) have attempted to treat human galactosemia, by allowing the incorporation of phage (bearing the gene *gal operon*) into the fibroblast cells grown in tissue culture. These cells were homozygous for the galactosemia. The fibrobalst cells after incorporation of gal operon produced the necessary enzyme, which persisted for at least eight cell generations. This indicates that the phage or bacterial DNA can be inserted into the human genome.

Qasba and Aposhian (1971) working at the school of medicine. Maryland University, USA, have shown that *polyoma pseudovirus* (viral particles containing its own protein coat, but DNA from other cells) carrying DNA from mouse cells can be inserted into human DNA in cells cultured on a medium *in vitro*. It is possible that, the viral genome gets incorporated into the human cells, although evidence for this is not available.

Globulin gene and insulin gene from rabbit and rat have been successfully incorporated into the bacterial cell. Thus it is possible in future to insert new genes into the human system and replace defective genes.

Genetic engineering in plants: Human beings have been trying to improve the quality and yield of their cultivated plants by several methods of breeding and selection. Tissue culture is a new field, which aims to improve and propagate plants of the desired type in a short time. For instance a bit of a tissue from the meristem may be cultured on a culture medium to produce new plants in a very short time. This type of meristem culture is now a days quite common in ornamental and exotic plants like*Anthuriums,* Orchids etc. The rate of propagation is so rapid that it is possible to produce a million plants per year from a single source.

Tissue culture

This technique refers to the culture of whole plants from pieces of tissues from a parent plant. In order to understand the phenomenon of tissue culture we must first understand as to how the cells grow.

All diploid individuals (both plants and animals) are products of a single cell called zygote. This zygoe is formed by the fusion of a sperm and egg. This single celled zygote by dividing mitoticacly produces the entire plantbody. If mitosis is an equational division then all the cells produced by zygote should develop into identical cells. How is it some cells develop into roots, some into leaves, some into flowers etc? The answer to this lies in the process called *differentiation.* After every cell division there is a gap of time so that the cells physiologically mature and develop individuality. In otherwords the cells are allowed to differentiate and eventually develop into different organs. Basically the zygote is capable of developing into all the parts of the plant and this is called *totipotency.* Mature cells or differentiated cells lack this totipotency and that is why they can develop into only certain parts - root, stem, leaf etc depending on the differentiation they have undergone. If it is possible to induce totipotency back into the mature cells, then it would be possible to make each cell grow into individual plants. It is this mechanism that is fcllowed in tissue culture.

The technique of tissue culture

Tissue culture technique involves the following steps.

(1) Selection of suitable plant

(2) Selection of suitable part of the plant

(3) Culture of the tissue

(4) Propagation of plants and transplanation

(1)Selection of suitable plant : This is the first setp. A plant which is economically important (medicinal value, food value etc) but fails to produce seeds or seed setting is low or life cycle is very long (as in bamboos) can be selected for tissue culture micropropagation. Even tree species - teak, sandal wood also may be selected.

(2) Selection of suitable part of the plant: For the purpose of culture a suitable part of the plant to be propagated may be selected. This may be a piece of tissue of the leaf, stem, axillary bud, anther, embryo, seed or even pollen grain.

(3) Culture of the tissue : The tissue selected for the purpose must be separated from the plant. This is now called the *explant.* The explant has to be washed thoroughly to remove contaminants and placed on a culture medium. The culture medium is very important for the success of tissue culture. It may be a solid medium or liquid medium contianing all the essential nutrients and specific chemicals. Culture mediums are varied and the composition is selected depending on the individual requirement.

The first step in tissue culture is to make the tissue to undergo *dedifferentiation* ie., to give the cell back its totipotency. This has to be done by careful mainpulation of the components of the culture medium (both qualitatively and quantitatively). Under proper conditions the tissue undergoes dedifferentiation ie., the cells start dividing and form an amorphous mass called *callus.* Callus formation indicates initial success of the tissue culture experiment. However culture of an amorphous mass is not the purpose. It should produce plants ie. it should undergo redifferentiation. By careful mainpulation of the medium once again differentiation can be reintroduced and the callus produces a number of tiny shoots (plantlets). At this stage if necessary, the callus with plantlets may be cut into pieces and introduced into separate culture flasks to multiply this number. In this way a number of plantlets may be obtained.

(4) Propagation of plants and transplantation : Plantlets growing in tissue culture tubes or flasks are very delicate and can be compared to premature babies kept in incubators. These first have to be transfered to tiny pots and kept in protected green houses. This process is called *hardening*. After the plantlets are several weeks old in the green hence, they can be transplanted to the normal pots or to the field for further growth.

Advantages of Tissue culture

1. Large number of plants may be obtained from a single piece of tissue.

2. Time of propagation may be shortened. Within a short period of time a number of plants may be obtained.

3. Exact replica of the parent plants may be obtained. If a plant has got ideal characters, propagation by seeds may alther those characters. By tissue culture original traits remain unaltered.

4. In plants where seed setting is difficult, and they cannot be propagated vegetatively tissue culture propagation is ideal.

5. If a plant is disease resistant, the trait can be continued in tissue cultured plants.

Tissue culture has now been practised in our country on a large scale as a commercial proposition. Rare plants, endangered plant and medicinally important plants can be propagated and even exported to other countries. Many rare orchids, Anthuriums and other valuable plants are micropropagated by tissue culture.

Apart from the conventional tissue culture, there is a new method of culture which is still in its experimental stage. The technique of isolation, culture and fusion of protoplasts falls into this category. This is also called somatic hybridization. Protoplast culture is attempted in order to circumvent the cell wall in plant cells, which normally do not allow hybridization between cells.

Applications of Biotechnology : The possibilities are immense and sky seems to be the limit for future biotechnologists. The following are some of the applications of genetic engineering.

1. Gene therapy : A new branch of curative medicine is opened up due to genetic engineering. Many genetic disorders like hemophilia, alkaptoneuria, galactosemia etc., which are due to defective genes and can be cured by replacing them (defective genes) with correct genes. Plasmid or virus mediated genes may be incorporated into human systems.

2. Cheap production of drugs : Many of the drugs like vitamins, antibiotics, hormones are now being produced chemically or produced from higher organisms which is a slow and costly process. If the genes for the production of these are introduced into rapidly multiplying bacterial cells, we have virtually myriads for tiny living factories turning out the medicines at a fraction of their original production cost.

Genetically engineered Insulin

Insulin is an important hormone that plays a vital role in the metabolism of carbohydrates. In patients suffering from diabetes, insulin has to be injected subcutaneously to regulate the blood sugar level. Insulin producing genes can be transfered to bacteria by techniques of genetic engineering and large quantities of insulin may be extracted. This process is cheaper and convenient than the conventional one. We will briefly discuss below the method of procuring insuling by genetic engineering techniques.

Structure of human insulin : Insulin molecule is composed of two chains - A and B of aminoacids. Initially it is produced as a single long chain called *Pre-proinsulin* which is later cut into a shorter chain called *proinsulin* (composed of 84 aminoacid units). Proinsulin then breaks into A and B chains when a piece from the middle is cut off. Chain A is composed of 21 aminoacids and chain B has 30 amino acids - The remaining 33 aminoacids act as a connecting peptide chain.

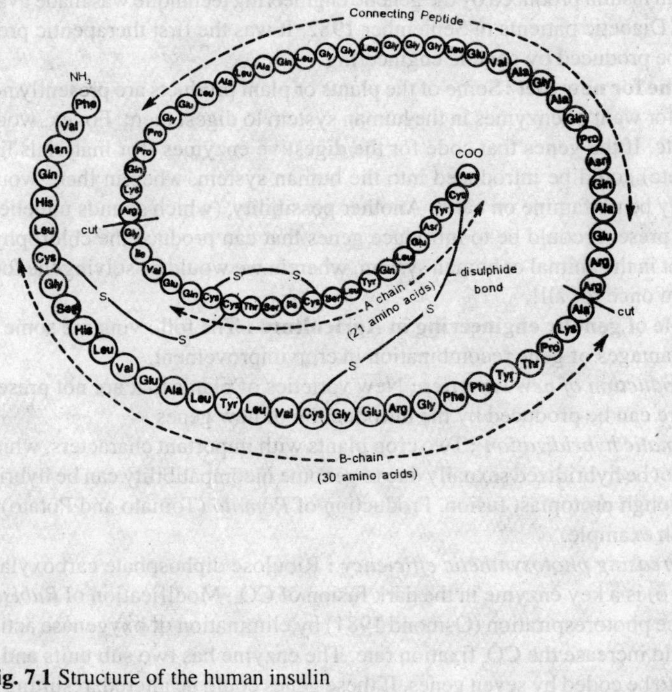

Fig. 7.1 Structure of the human insulin

Production of Insulin in E. coli : The genes for insulin chain A and B are indentified and located. This may be done first by identifying the codons and then the corresponding mRNAs can be reconstructed. Using these mRNA molecules and with the help of the enzyme *reverse transcriptase* the corresponding segments of double stranded DNA molecules are produced. Then with the help of the enzyme *terminal transferase,* the ends of the insulin genes (DNA strands) are extended with short sequence of identical bases.

Plasmids obtained from *E.coli* are opened up at specific portions by the use of restriction enzymes. The insulin genes (DNA strands) of A and B chains are added to the plasmids separately forming recombinant DNA plasmids. Gaps if any in the circular plasmids of A and B chains are filled by lac operon gene for switching on the synthetic process. The plasmids with recombinant DNA for A and B chains are inserted into *E. coli* finally.

The recombinant DNA plasmids multiply in *E. coli* and trigger the synthesis of A and B chains of insulin as a genetic expression. Following large scale production cultures, the insulin chains are obtained following specific purification processes. Finally the A and B chains are chemically bonded by disulphide bonds to produce the complete molecule of Insulin.

Human insulin produced by the genetic engineering technique was made available to Diabetic patients in September 1982. It was the first therapentic product to be produced by genetic engineering.

3. Gene for new diet : Some of the plants or plant products are presently non edible for want of enzymes in the human system to digest them. For ex. wood, grass etc. If the genes that code for the digestive enzymes (for materials like wood etc) could be introduced into the human system, wherein there would virtually be no famine on earth. Another possibility, (which sounds unbelievable at present) could be to introduce genes that can produce the chlorophyll pigment in the animal or human system, wherein we would be solving the food problem once for all!

4. Role of genetic engineering in Agriculture : The following are some of the advantages of gene recombination in crop improvement.

a) *Production of new varieties* **:** New varieties of plants that are not present in nature can be produced by the introduction of new genes.

b) *Somatic hybridization* **:** Two crop plants with important characters, which could not be hybridized sexually due to genome incompatibility can be hybridized through protoplast fusion. Production of *Pomato* (Tomato and Potato) is one such example.

c) *Increasing photosynthetic efficiency* **:** Ribulose diphosphate carboxylase (Rubisco) is a key enzyme in the dark fusion of CO_2. Modification of *Rubisco* to reduce photorespiration (Osmond 1981) by elimination of oxygenase activity would increase the CO_2 fixation rate. The enzyme has two sub units and is known to be coded by seven genes. If these genes could be altered, resulting in the changed *Rubisco,* it would increase the photosynthetic efficiency.

d) *Nitrogen fixation* **:** The roots of leguminous plants normally possess nitrogen fixing nodules, induced by various strains of *Rhizobium.* As a result, leguminous plants require little or no nitrogen fertilizer. Arable crops require about 100 kg of nitrogen per hectare. If this can be replaced by N_2 fixing genes, it would lead to a lot of saving in terms of fertilizer cost. Nitrogen fixation is catalyzed by the ATP dependent electron reduction of dinitrogen to ammonia by nitrogenase. If it is possible to introduce these genes in higher plants, it would bring about a revolution in agriculture. This may be achieved by one of the following ways.

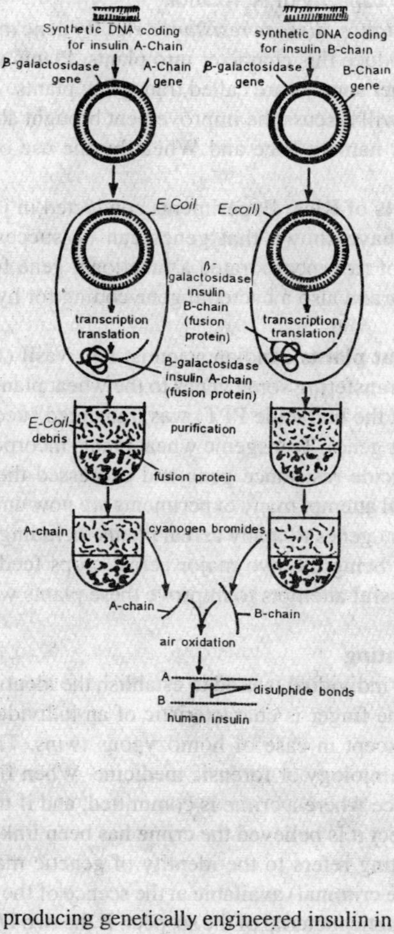

Fig. 7.2 Technique of producing genetically engineered insulin in E.Coli

(i) Transfer of nitrogen fixing gene (*nif*) from free living bacteria like *Klebsiella, Azatobacter* etc., to higher plants. This would enable the plants to fix the N_2 directly. This is still in the experimental stage, as the *nitrogenase,* produced by the *nif* gene is highly oxygen labile and gets degraded in higher organisms.

(ii) Expansion of host range of symbiosis to crops other than legumes. At present non leguminous plants do not have symbiotic association with *Rhizobium.* If it is possible to expand the host range for the bacterium, a lot more plants will have the capacity of N_2 fixation.

e) *Herbicide resistance, disease resistance etc*: By gene manipulation, it will be possible to introduce this character into plants. Plants which have genese transfered from other sources are called transgenic plants.

In this context we will discuss the improvement brought abut in two very important crop plants namely Rice and Wheat by the use of biotechnological principles.

Transgenic plants of Rice: Experiments conducted in japonica and indica vaireties of rice have shown that genes can be successfully transfered. Transgenic plants of rice incorporated a functional gene for *aminoglycoside phosphotransferase* and also a bacterial gene coding for hygromycin B resistance.

Transgenic wheat plants : Indian scientist I.K. Vasil (1992) has done remarkable work in transfering some genes to the wheat plants. Herbicide resistance gene (against the herbicide PPT) was transfered sucessfully along with *gus* gene (a marker gene). Transgenic wheat plants incorporated both the *gus* gene and the herbicide resistance gene and expressed them phenotypically. After this successful attempt many experiments are now underway to incorporate disease resistant genes to many cereal plants including wheat.

Rice and Wheat being the two major cereal crops feeding practically the entire world successful attempts to improve these plants would certainly be a boon to mankind.

Genetic fingerprinting

Finger print of an individual is used to establish the identity of a person. The lines or loops on the finger is characteristic of an individual and it is not repeated in others except in case of homozygous twins. This is made use of particularly in criminjology or forensic medicine. When finger prints are obtained from the place where a crime is committed, and if these match the fingerprints of a suspect it is believed the crime has been linked to the suspect.

Genetic fingerprinting refers to the identity of genetic material (DNA) (left inadvertently by the criminal) available at the scence of the crime and its comparison with the genetic material of the suspect. If the two DNAs match in their base pair sequence, then the crime is linked to the suspect.

Genetic or DNA fingerprinting may be defined as *comparison of two sets of DNA for their identity* ie. whether they belong to same source or not. Genetic fingerprinting has wide applications in forensic science, instances of disputed parentage, identification of rapists etc. The technique of DNA fingerprinting was first developed by Alec Jeffreys and his colleagues (1985,86) at Leicester university in UK.

In this technique, DNA will be extracted from the blood stains, hairs skin fragments, semen etc and subjected to southern blotting and DNA hybridization using specific DNA probes.

In southern blotting (named after E.M. Southern) DNA sample is separated on gel electrophoresis into single strands. It is then transfered on to a nitrocellulose membrane. The membrnae is heated to 80°c to permanently fix the single stranded DNA on to the membrane. This can now be used for hybridization with radioactive labelled DNA strand and an X-ray film is exposed which shows if any binding is there. Usually the DNA extract from the sample is used for hybridization with labelled DNA obtained from the suspected person in case of a crime. A complete match or hybridization is a proof that both DNA samples belong to the same person. This is the technique of genetic fingerprinting.

In conclusion it may be said, that biotechnology is a highly potential tool which man can use either for good or for bad. He has to use it carefully so as not to alter the balance of nature.

Eight

Histology

PLANT TISSUES

Cells are grouped together to form tissues. Generally cells which are alike in origin, form and function constitute a tissue. In higher plants like, gymnosperms and angiosperms and to a certain extent in pteridophytes cells are of diverse types. They vary in their origin, shape, size, structure etc., Consequently the assemblage of these different types of cells form various types of tissues.

Basically tissues are classified into meristematic and permanent tissues. The basis for this classification is the stage of development of the cells. In a meristematic tissue the cells are actively dividing and have not attained maturity. The cells of a permanent tissue on the other hand have lost the capacity of cell division and have attained their mature form.

MERISTEMATIC TISSUES

The cells of a meristematic tissue are immature and are in an active state of division. The cells are usually closely packed with no intercellular spaces. The shape of the meristematic cells varies. They may be oval, polygonal rounded or rectangular in shape. The cell inclusions will be in their precursor stage. The nuclei are very prominent. The plane of division in the cells varies. It may be periclinal (parallel to the surface) or anticlinal (perpendicular to the surface).

Classification of meristems: There have been different systems of classification based on different criteria, such as origin, location, function etc. A few important types of classifications are given below:

A. Meristems based on stage of method of development: Based on their stage of development meristems may be classified into promeristems and marginal meristems.

Promeristems constitute the region of new growth in a plant body, where the foundation of new organ is initiated. Promeristem is also known as embryonic meristem, primordial meristem or Urmeristem. Structurally the promeristem region consists of initials and their derivatives. The cells have thin cell walls, prominent nucleus, dense vacuolate or non vacuolate cytoplasm. Pits are in their early stages of development. When the cells of the promeristem begin to differentiate they may be said to have lost the characters of a promeristem.

Marginal meristems generally give rise to tissues of cortex, epidermis etc. Based on the plane of division they may be further classified into mass meristems, plate meristems and rib meristems. In mass meristems, growth is by division in all the planes and the result is increase in mass. In plate meristems division is in two planes, and in rib meristems division is in only one plane. Plate meristems increase the area of a tissue while rib meristems produce rows or columns of cells increasing the length of organs. Examples of mass meristems are early stages of development of sporangia, embryo, endosperm etc. Plate meristems are seen in the leaves. Rib meristems play an important part in the development of young roots and the pith and cortex of young stems.

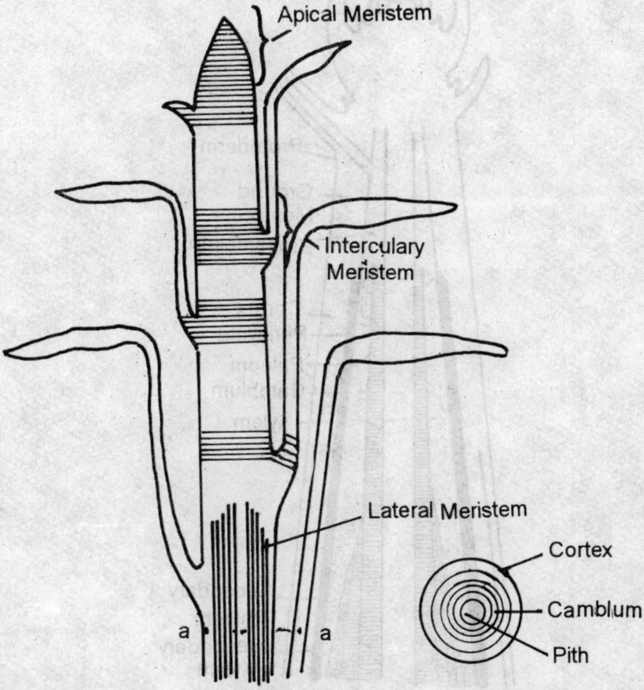

Fig. 8.1 A. Diagrammatic L.S. of shoot apex showing the positions of meristems, B. T.S. of shoot at A-A.

B. Meristems based on their origin: Based on the history of the initiating cells meristems are classified into primary and secondary meristems. Primary meristems make up the fundamental parts. They consist of a part of the promeristem. Mainly primary meristems are located at the apices of plant organs.

Secondary meristems always arise late in the ontogeny of the plant. They take their origin in permanent tissues. In position, they are always lateral. Sometimes permanent tissues acquire the capacity of cell division. They are also known as secondary meristems. Typical example of a secondary meristem is the cork cambium.

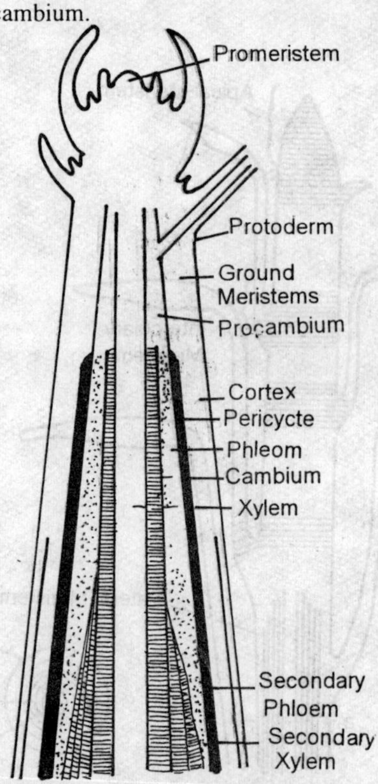

Fig. 8.2 L.S. of shoot apex showing meristems based on their function

C. Meristems based on their location in the plant body: According to the location, there are three types of meristems, viz., apical, intercalary and lateral.

Apical meristems as the very name suggests are located at the apices of shoot, root and other organs. They constitute the growing points of plants. The growth in length of the plant and plant organs is due to apical meristems. Growth is initiated by one or more cells of the apical meristems. Apical meristem consists of two zones, promeristem and meristem. Promeristem consists of actively dividing cells called apical initials. The meristematic zone is classified into protoderm, procambium and ground meristems.

Intercalary meristems represent the detached portions of the apical meristems seperated due to the growth of an organ. Intercalary meristems are always internodal. In the early stages of development of the plant, whole or part of the internode is meristematic. Subsequently a protion of it matures, as a result, patches of meristems are found embedded in permanent tissues.

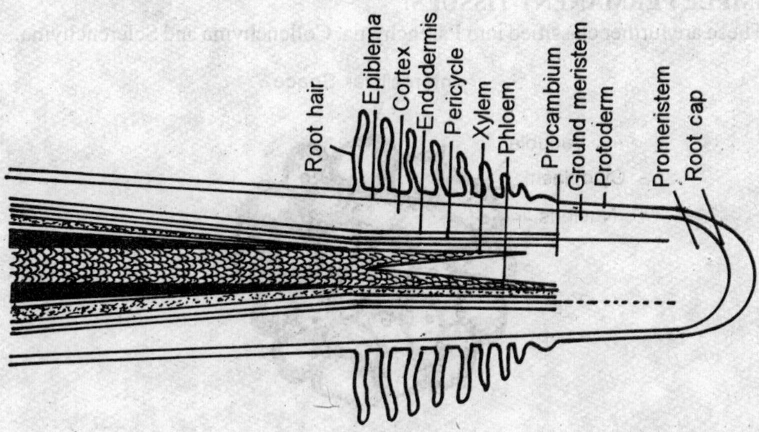

Fig. 8.3 L.S. of root apex showing meristems

Lateral meristems are present along the side of the stem. The cells divide in only one plane and add to the girth of the tissues. Typical example of a lateral meristem is the vascular cambium. Vascular cambium produces secondary Xylem and Phloem thus bringing about secondary growth.

D. ***Meristems based on their function:*** According to the function performed by the derivatives of the meristems, they may be classified into protoderm, procambium and ground meristem. This classification was proposed by Haberlandt at the end of the 19th century. Protoderm develops into epidermis. Procambium gives rise to the primary vascular tissues and the ground meristem are thin and isodiametric. They give rise to hypodermis, endodermis and pericycle.

PERMANENT TISSUES

As a contrast to the meristematic tissues, the cells of permanent tissues have no capacity of division. They have attained their mature form and shape. Very rarely (injury etc.) permanent tissues may become meristematic. This is called dedifferentiation.

The cells of the permanent tissues may be living or dead, thin or thick walled. Permanent tissues may be classified into simple and complex, based on their composition. A simple tissue is homogenous i.e., it is made up of only one type of cells. In a complex tissue there will be at least two different types of cells. But these cells coordinate and function like a single unit. Generally one or more types of simple tissues combine together to form a complex tissues.

SIMPLE PERMANENT TISSUES:

These are further classified into Parenchyma, Collenchyma and Sclerenchyma.

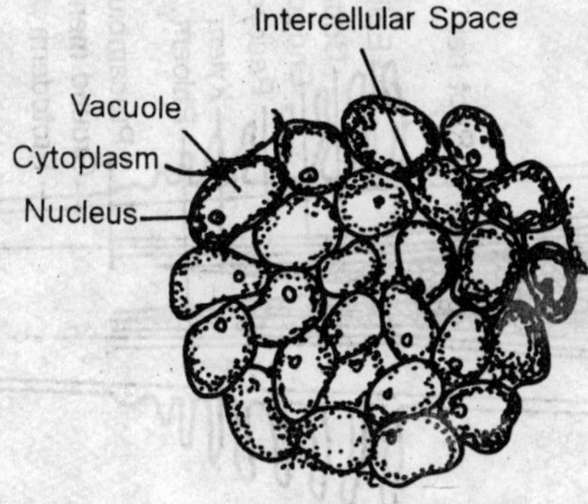

Fig. 8.4 Parenchymz

PARENCHYMA: Parenchyma refers to a tissue composed of living cells variable in their morphology and physiology. The word parenchyma is derived from the Greek **para**, beside and **en-chein**, to pour, meaning a semi-liquid substance poured beside other tissues.

Parenchyma is often considered as the fundamental or ground tissue in which other tissues, notably the vascular are embedded. Ontogenetically parenchyma is the precursor of other tissues. Phylogenetically also it is more ancient than other tissues as seen buy the fact that the body of primitive organisms is always parenchymatous. Parenchyma tissue is the principle seat of such essential activities of the plant like photosynthesis, assimilation, respiration, storage, secretion etc.

Developmentally, parenchyma cells are relatively undifferentiated and are unspecialised morphologically and physiologically when compared with other complex tissues.

Origin: The ground meristem gives rise to parenchyma. But the parenchyma in the vascular tissues is formed by the procambium and the vascular cambium. Parenchyma may also arise from phellogen in the form of phelloderm.

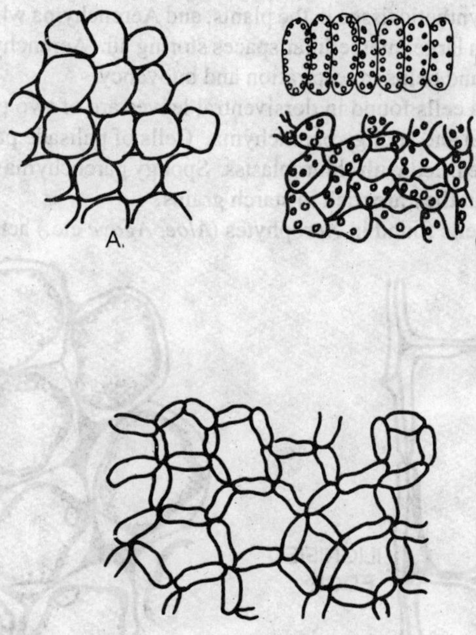

Fig. 8.5 Types of Parenchyma, A. Simple, B. Palisade, C. Spongy, D. Acrenchyma.

Distribution: Parenchyma cells are present in almost all parts of the plant, like the pith and cortex of roots and stems, mesophyll of leaves, the flesh of succulent roots, the endosperm of seeds etc. As components of heterogeneous tissues parenchyma cells form vascular rays.

Cell shape: Parenchyma cells are spherical or polygonal in shape (Fig). Sometimes they may be elongated and columnar as in the leaves. Studies based on geometric models have indicated that closely packed parenchyma cells have the shape of a polyhedron with approximately fourteen faces.

Parenchyma cells are closely packed or arranged loosely with large number of intercellular spaces.

Cell structure: The cells have thin walls and have dense cytoplasm with a prominent nucleus. Parenchyma cells in storage organs have abundant reserve food materials like proteins, fats and carbohyderates. Parenchyma cells found in leaves have number of choloroplasts . These help in photosynthesis.

Cell types: Based on their structure and function parenchyma cells are classified into the following types

Simple parenchyma which forms the ground tissue, Chlorenchyma which is the chief photosynthetic tissue in the plants, and Aerenchyma which has parenchyma cells with large intercellular spaces storing air. Aerenchyma is present in hydrophytes and helps in respiration and buoyancy.

Chlorenchyma cells found in dorsiventral leaves are of two types viz. Palisade parenchyma and spongy parenchyma. Cells of palisade parenchyma are columnar and are filled with chloroplastss. Spongy parenchyma cells also have chloroplastss, but they are rich in starch grains.

Parenchyma cells found in Xerophytes (*Aloe, Agave* etc.) act as water storage tissues.

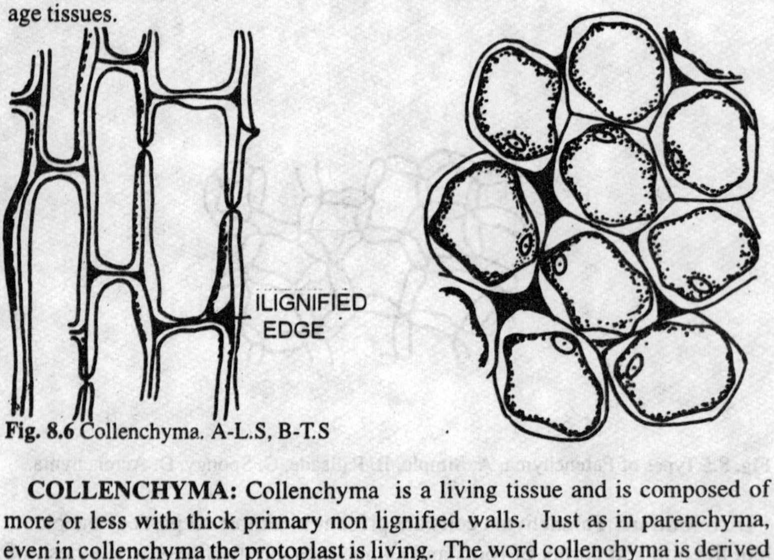

Fig. 8.6 Collenchyma. A-L.S, B-T.S

COLLENCHYMA: Collenchyma is a living tissue and is composed of more or less with thick primary non lignified walls. Just as in parenchyma, even in collenchyma the protoplast is living. The word collenchyma is derived from the Greek word colla, meaning glue, which refers to the thick glistening wall of collenchyma .

Origin: According to Ambronn (1881). Haberlandt (1914) and Majumdar (1941), collenchyma originates jointly with the vascular tissues from the procambium. According to Esau (1936) however, collenchyma independently originates from the procambium.

Distribution: Collenchyma characterstically occurs in the pripheral position of stems, leaves, floral parts etc. It may also occur in root cortex particularly if it is exposed to light. In the stems and leaves of dicotyledons, collenchyma is hypodermal situated immediately below the epidermis or seperated from it by one or two layers of parenchyma. Collenchyma cells are considerably elongated. Sometimes, cells are 2 mm, long resembling fibres. The shorter collenchyma cells are prismatic and appear polygonal in a transection (Fig) In a long section the cells appear rectangular.

Cell structure and function: The structure of the cell wall is the most characterstic feature of the collenchyma cells. The thickenings are deposited unevenly, mainly deposited at the corners where several cells join together. In long section, collenchyma cells show thin and thick wall portions dependng on the direction of the cut. The walls containing mainly of cellulosic and pectic compounds and consist of much water. Ultrastructurally the thickenings of collenchyma show an alternation of layers of longitudinally oriented microfibrils and non cellulosic material.

Collenchyma cells have living protoplasts and may have chloroplastss. Chloroplastss are less in number in long and narrow cells and more in cells which approach parenchyma in form. Tannin may be present in collenchyma cells.

Collenchyma is primarily a mechanical tissue adapted for support of growing organs. The thick walls and close packing, make collenchyma a strong tissue. At the same time, the peculiarities of growth and the structure of its walls impart a considerable tensile strength with flexibility and plasticity.

SCLERENCHYMA: This tissue consists of thick walled cells, (lignified), their main function being mechanical. The word is derived from the Greek **Sclerous**, hard and **Enchyma** an infusion which emphasizes the hardness of the cell walls. Haberlandt (1914) and Foster (1949) combine collenchyma and sclerenchyma under a common name *stereome* which includes the entire mechanical system of the plant.

Sclerenchyma cells show variety in form, structure, origin and development with the different types often integrating. Generally, sclerenchyma cells are classified into *fibres* and *sclereids*. Fibres are considered as long cells while sclereids, are relatively short cells.

Fibres: Fibres are long narrow, thick, lignified cells with tapering ends (Fig. 2.7). The wall thickenings sometimes even block the lumen of the cells. The pits of fibres are simple, small round and often oblique. The middle lamella is conspicuous.

Fibres are classified into three types viz. surface fibres, wood fibres and bast fibres.

Surface fibres are found in the seed coats or the pericarp of the fruits (eg cotton fibres, fibres present in the mesocarp of *Cocos nucifera*)

Wood fibres are present in xylem. These are also called xylary fibres (For details see xylem)

Bast fibres are obtained from the pericycle and phloem of various plants. In *Cannabis sativa*, the pericycle yields fibres, which is used in the making of carpets, ropes, sacs etc. Phloem fibres are seen in many plants like *Linum* (for details see phloem).

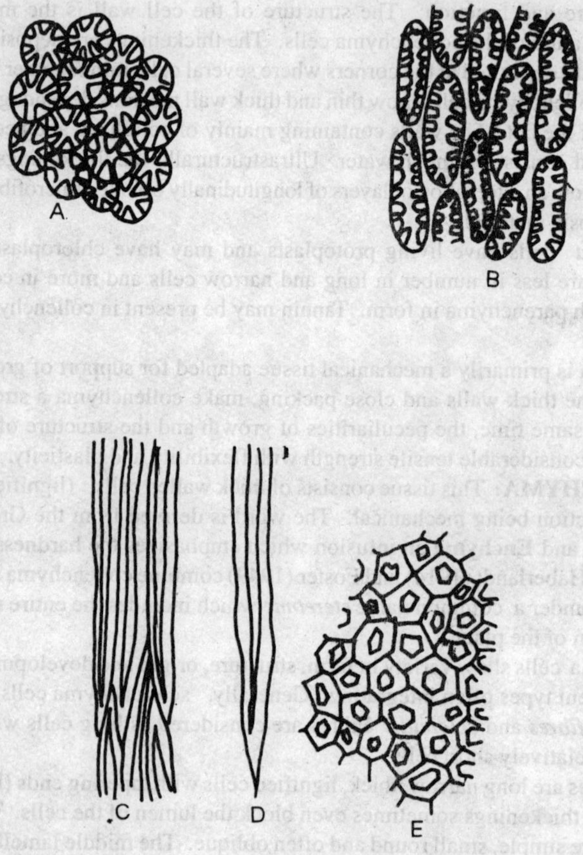

Fig. 8.7 Sclerenchyma, A. T.S. of sclereids, B. L.S. of the same, C and D. L.S. of fibres, E. T.S. of fibres

Fibres are of great economic value. They find a place of pride in textile industry. To the plants, the fibres are of great use as they offer mechanical support.

Besides, xylem and phloem, fibres also occur as bundle caps in *Zea, Saccharum* etc. In *Zea mays* (Murdy 1960) a hypodermis containing fibres as long as 1mm, has been recorded. In palms, central cylinder is demarcated by a sclerotic zone that may be several cm., wide. Fibres may be present in the leaves of monocotyledons, where they form sheaths enclosing the vascular bundles of strands extending between epidermis and vascular bundles.

Sclereids: These are short isodiametric cells (Fig. 2.8) but sometimes may be elongated also. The cell walls are very thick and lignified but the cells do not have tapering ends. The cells are dead and have a very narrow lumen. The secondary walls show prominent pits which are simple. Sometimes there is slight overarching of the secondary wall over the pit chamber. The lignin deposition is uniform, but may be uneven as in the seed coats of some Leguminosae.

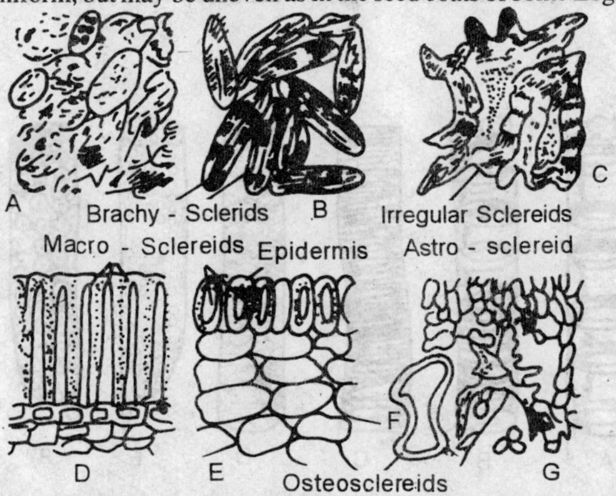

Fig. 8.8 Types of Sclereids, A and B. Brachysclereide, C. Irregular sclecrids, D and E. Macrosclereids, F. Osteosclereid, G. Astrosclereid

Sclereids arise through a belated sclerosis of ordinary parenchyma cells or directly from cells called sclereid primordia.

Sclereids are widely distributed in the plant body. The cortex and the pith of gymnosperms and dicotyledons very often contain sclereids arranged singly or in groups. Xylem and phloem may often contain sclereids besides fibres. Sclereids may be present in leaves, seeds and fruits. In fruits they are sometimes dispersed in the soft tissues.

Sclereids are classified into the following types viz., brachysclereids (fig 25A) or stone cells (short isodiametric cells widely distributed in the flesh of fruits), macrosclereids long rod like cells found in the epidermis of leguminous seeds (Fig 25B), Osteosclereids (bone like cells), Astrosclereids (star like cells) Bizarre sclereids (Irregular sclereids), Filiform sclereids (long cells resembling fibres) and trichosclereids (branched, thin walled sclereids).

COMPLEX PERMANENT TISSUES

The vascular tissues are the most important among the complex tissues. There are mainly two components in the vascular tissue viz., xylem and phloem.

X Y L E M: This is mainly a conducting tissue of water and mineral nutrients. The term xylem was introduced by Nageli (1858) and is derived from the Greek *Xylos* meaning wood. Structurally the xylem is most complex, because it consists of various types of living and nonliving cells.

The xylem is mainly classified into primary and secondary xylem. The former differentiates in the primary plant body and is derived from the procambium, while the latter originates from a fascicular cambium during secondary growth.

Elements of xylem: The xylem is composed of different kinds of elements viz., tracheids, fibre tracheids, vessels or tracheae, wood fibres and wood parenchyma.

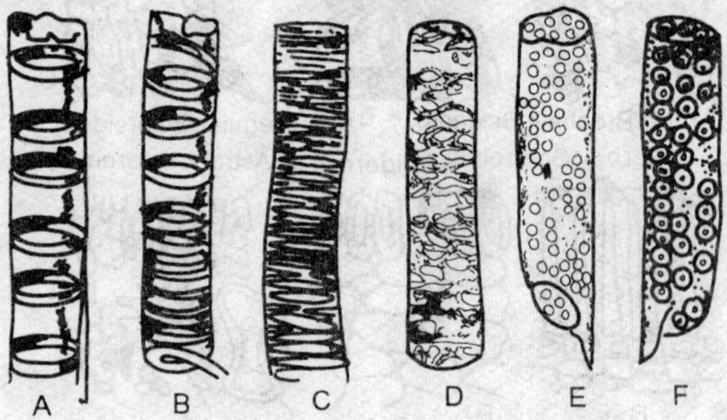

Fig. 8.9 A. Annular, B. Spirat, C. Scalariform,
D. Reticulate, E. Simple pitted, F. Bordered pitted

Tracheids: They are the fundamental type of xylem elements present in pteridophytes and gymnosperms where they constitute the chief conducting elements. Occassionally, they occur in Angiosperms also (Magnoliaceae). A tracheid is an elongated tube like cell (fig) with tapering rounded oval ends. The tip of the tracheid of a secondary xylem may be chisel-like. The walls are thickened due to the deposition of lignin, which are of various types.

Thickening of the cell wall: The secondary wall acquires different types of thickenings. This is specially seen in cells which are to differentiate into tracheary elements i.e, vessels and tracheids. In these cells, the thickening is due to the deposition of lignin. The deposition of lignin is not uniform, hence we find different types of thickenings. Chiefly, the following types of thickening are seen in xylem.

Annular: Lignin deposition is in form of annuli or rings. These rings are unconnected. The space between the rings is unthickened. (Fig)

Spiral: Lignification occurs in the form of a spiral band (fig)

Scalariform (ladder like): Wall thickening is in the form of transverse rods, very much resembling the rungs of a ladder (Fig).

Reticulate: Lignin is deposited in the form of network or reticulum. The unthickened areas of the cell wall are irregular in shape (Fig 26D)

Pitted: Lignin deposition is almost uniform over the wall, except for small pores here and there. These pores are called pits. (Fig 26E)

Structure of Pit: Pits are formed on either side of the wall. In any area there will be two pits lying opposite to each other. These are usually termed as "pit pairs" . There is a space inside each pit called pit cavity or pit chamnber. There is a thin membrane which seperates the pit cavities of the opposite pits. It is called pit membrane. The opening to a pit is called the pit aperture.

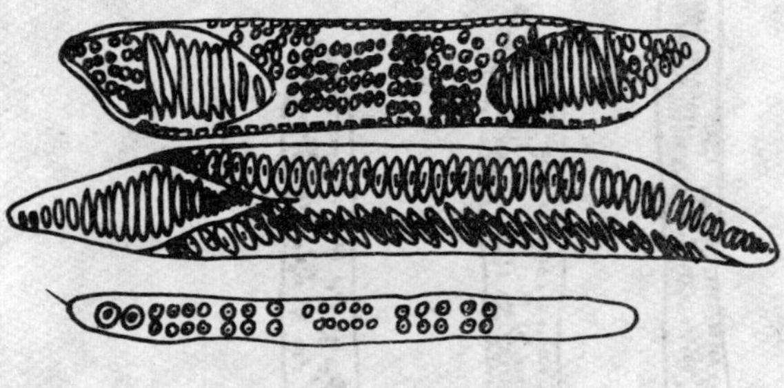

Fig. 8.10 Vessels

A pit membrane is common to both the pits. It has a primary wall on either side seperated by middle lamalla.

There are two types of pits viz. *simple pits* and *bordered pits*. A simple pit is just an opening in the wall. Whereas in a bordered pit, the pit aperture is overarched by the projection of the secondary wall. In a pit pair both may be simple, both may be bordered or one may be simple and the other one bordered. The last mentioned one is generally called a half bordered pit pair. A pit that occurs opposite to an intercellular space will not have its complementary and hence known as a blind pit.

Simple pits generally occur in parenchyma cells, medullary rays, phloem fibres, companion cells, and also in tracheids.Pit cavity is uniform in its diameter and the pit membrane is simple. Simple pits may vary in shape (oval, polygonal, irregular etc.)

In a bordered pit, the pit membrane becomes thick at the centre. The thickening is known as the *torus*. The pit membrane may also change its position. The torus instead of remaining at the centre may shift towards a pit aperture. In bordered pits, the aperture is narrow (pit canal)_ and it widens towards the interior (pit chamber). Bordered pit occurs in the vessels of Angiosperms and in the tracheids of many Conifers.

Elements of xylem:

(a) Tracheids: These are dead cells and like sclerenchyma have neither the nucleus nor the protoplast. The lumen of the cell is large. The wall of the tracheids is provided with a number of bordered pits.

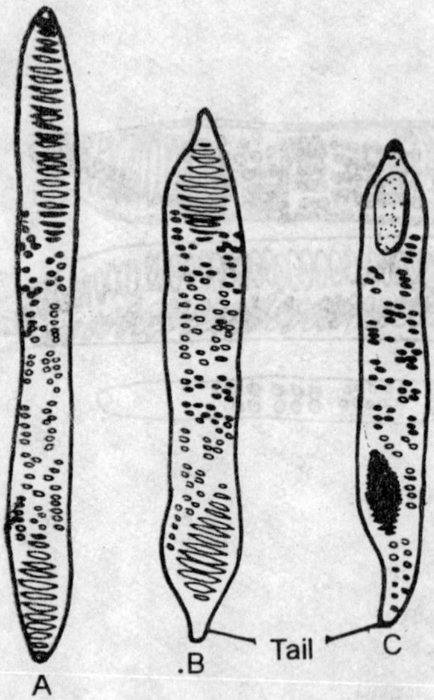

Fig. 8.11 Types of vessels. Simple C and Scalariform A-B

In a cross section, tracheids appear polygonal or rectangular. Primary xylem tracheids develop from the procambial strands.

Besides helping in conduction, tracheids also give mechanical support to the plants. Tracheids rarely exceed 1 mm in length, though, there are reports of tracheids more than 10mm long (Conifers).

(b) Fibre tracheids : During the development of fibres the wall thickening increases whereas the lumen decreases. In many cells, the lumen becomes so narrow that it is altogether obliterated. In such cells there is little or no conduction and they may be said to have reached the stage of true fibres.

(c) Vessels: These are evolved from tracheids and are present mainly in Angiosperms. However., there are several instances of their presence in pteridophytes (*Selaginella, Pteris* etc.,) and Gymnosperms. Vessels arises through the dissolution of the end walls of tracheids . Thus sometimes a tracheid may represent a developmental stage of a vessel. Vessels are composed of rows of cells placed one above the other with the transverse walls being eliminated. This facilitates the easy flow of water. Vessels may arise from tracheids either by the dissolution of the pit membranes which makes the end wall preforated or by the dissolution of the entire wall.

Vessels rarely exceed a length of 10cm. Occassionally however they may be two meters long (*Quercus*) or even 3-6 meters long (*Eucalyptus*). Like tracheids vessels also have different types of walls thickenings due to the uneven deposition of lignin. Vessels also have a wide lumen.

Vessels (tracheids also) may be of two types protoxylem and metaxylem. The former matures a little early in the ontogeny and has a narrow lumen while the latter matures later and has a broad lumen.

(d) Wood fibres: They are also called xylary fibres. Sclerenchyma fibres associated with xylem are termed wood fibres. They possess very thick walls and few small pits. They originate from the derivatives of the cambium and have varied shapes. In wood fibres, known as libriform fibres, the wall is extremely thick and lumen is almost absent. Like the tracheids , wood fibres are also dead and their main function is mechanical support.

(e) Wood Parenchyma: Parenchyms cells assicoated with xylem are called wood parenchyma . These are living cells and remain so, as long as the tissue in which they lie conducts water. The cells of wood parenchyma may be thin or thick walled have plenty of storage food and may help in conduction. In secondary xylem wood parenchyma forms the rays which may help in radical conduction.

PHLOEM: The phloem is the principal food conducting tissue in vascular plants. Unlike xylem all the elements of phloem (except for phloem fibres) are living. Like the xylem, phloem is also of two types viz., primary phloem (derived from procambium) and secondary phloem (derived from fascicular cambium). The basic components of phloem are sieve elements, companion cells, phloem parenchyma and phloem fibres. Occasionally sclereids, laticiferous cells and idioblasts are also found in phloem.

(a) Sieve elements: What the tracheids are to xylem sieve elements are to phloem. Sieve elements may be seggregated into relatively less specialized sieve cells (found in pteridophytes) and more specalized sieve tubes. Sieve tubes are living cells and have a thin cellulose wall. In angiosperms they are enucleate, while they are nucleate in gymnosperms.

The protoplast of a sieve tube has a big central vacuole which is filled with viscous and albuminous proteins. Leucoplasts are also reported in the vacuole.

The cross walls between the seieve tubes are porus and are called sieve plates as they resemble a sieve (fig). Sieve plates are usually of two types viz., simple and compound. In simple sieve plates, the pores are large and the plate is horizontal or transverse in position. The compound sieve plate is somewhat oblique. The sieve pores are large and the plate is horizontal or transverse in position. The sieve pores are very often plugged with a shining substance called callus pad which is usually deposited at the end of the spring season. Chemically callus pads are made up of a carbohydrate called callose. At the approach of the spring, the callus dissolves. Through the sieve pores, thin cytoplasmic strands come out and enter into other cells.

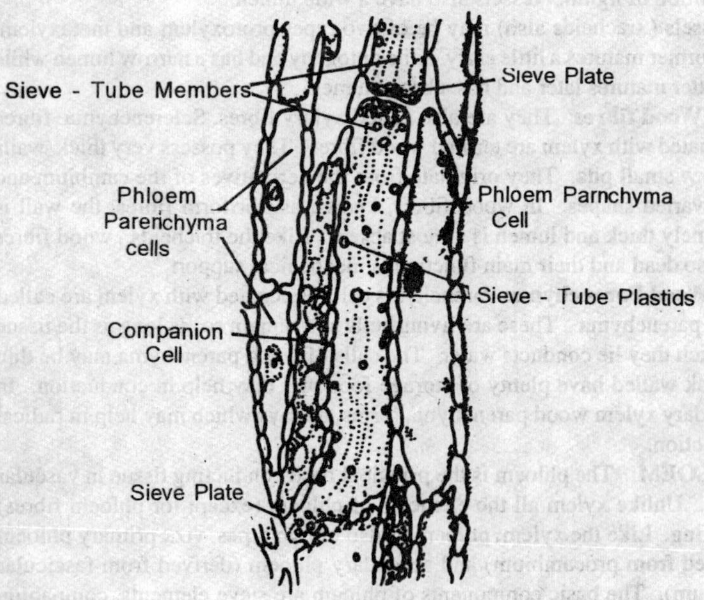

Fig. 8.12 Phloem. Long sectional view

(b) Companion cells: These are specialised small cells usually associated with the sieve tubes. These are living, nucleate cells and are supposed to compensate the absence of nuclei in sieve tubes. Companion cells are absent in gymnosperms, where albuminous cells are present. The cytoplasm of companion cell is dense and granular with a prominent nucleus. Ontogenetically both sieve tubes and companion cells are derived from the same mother cell.

(c) Phloem parenchyma: Parenchyma cells in association with phloem are called phloem parenchyma. They are no different from other ordinary parenchyma cells. In shape, the cells may be elongated, pointed, broadly cylindrical or even subspherical. The cells contain oil, starch mucilage, latex etc. Phloem parenchyma is absent in most of the monocotyledons.

(d) Phloem fibres: These are also called bast fibres. These are found both in primary and secondary phloem. The walls have simple pits. The bast fibres may be lignified or non lignified. The bast fibres of *Linum usitatissimum* are cellulosic and are of great value in commerce. The fibres of *Linum* are derived from protophloem and are arranged in irregular bands.

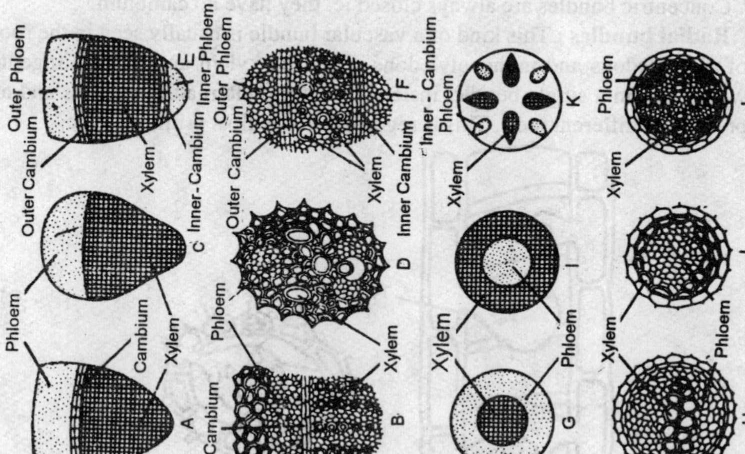

Fig. 8.13 Phloem. A.T.S. of sieve plate, B. L.S. of sieve tube

Types of Vascular bundles

A vascular bundle may be defined as grouping together of vascular elements namely xylem and phloem to form a functional unit. Within a vascular bundle the two major components - xylem and phloem are arranged in various ways. Depending on their relative position vascular bundles are categorized into the following types - cambium which brings about secondary growth may or may not be present.

Open vascular bundle: This is seen in dicotyledons. Here between xylem and phloem, cambium (secondary meristem) is present. As a result of this the vascular bundle is open to secondary growth. There is no bundle sheath surrounding the vascular bundle.

Closed vascular bundle: This is seen in monocotyldeons. Here there is no cambium in the vascular bundle. Hence there is no secondary growth. Further a bundle sheath surrounds the vascular bundle.

Based on the relative positions of xylem and phloem vascular bundles are classified into - *concentric, radial and conjoint types.*

Concentric vascular bundle: In concentric vascular bundles xylem and phloem form concentric rings; either xylem surrounds the phloem or phloem surrounds the xylem. Concentric vascular bundles are of two types-

Amphivasal (Leptocentric): Here phloem is in the centre surrounded by xylem. e.g. *Dracena, Yucca* etc.

Amphicribal (Hadrocentric) : Here xylem is in the centre surrounded by phloem eg. Ferns.

Concentric bundles are always closed ie. they have no cambium.

Radial bundles : This kind of a vascular bundle is usually seen in the roots of dicotyledons and monocotyledons. Here the xylem and phloem together will not form a single bundle, instead they form alternating bundles and are present on different radii. This is regarded as the most primitive type.

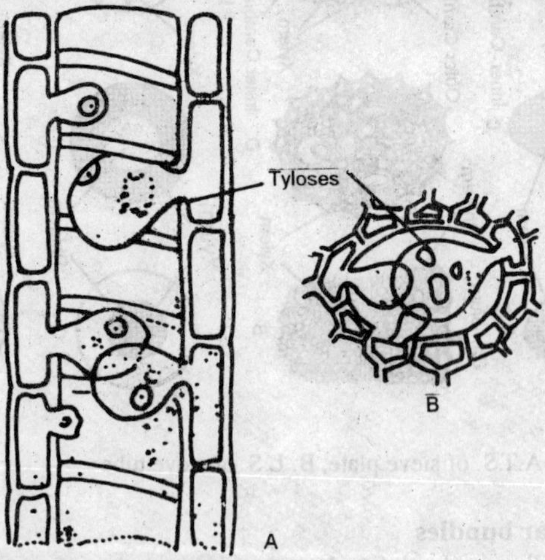

Fig. 8.14 Types of vascular bundles- A and B. conjoint, collateral and open C. and C, conjoint, collateral and closed, E and F. conjoint, bicollateral and open, G and H. concentric and amphicribral, I and J. concentric and amphivasal, K. and L. radial

Conjoint bundles : In this type xylem and phloem together form compact bundles and they are present on the same radius. Conjoint bundles may be open or closed and are usually present in stems. Conjoint vascular bundles of monocots are always closed (no cambium) while those of dicots are always open (cambium present).

Conjoint vascular bundles are of two types.

Collateral : Here xylem and phloem lie on the same radius with xylem towards the centre and phloem towards the periphery. In otherwords there is one group of phloem and one group of xylem. In dicotyledons cambium is present in between xylem and phloem while it is absent in monocotyledons. Collateral vascular bundles are found in stems of Angiosperms.

Bicollateral : In these bundles there is one group of xylem surrounded by two groups of phloem- outer and inner. Similarly two strips of cambia are also present. Here the arrangement of various elements are as follows - Inner phloem, inner cambium, xylem, outer cambium and outer phloem. Bicollateral vascular bundles are usually seen in the members of the family cucurbitaceae.

Nine

Angiosperms

Internal structure of plants organs like roots, stems and leaves may be studied by taking either a long section or a transverse section. In a transverse section all the tissues will be cut crosswise and they give us an idea of their structure, distribution etc. In this chapter we will discuss the primary and secondary structure of roots, stems and leaves.

Internal structure of stems

STEM OF HELIANTHUS (Sunflower):

A transverse section of a young sunflower stem shows an outermost epidermis, a middle cortex and a central stele (Fig).

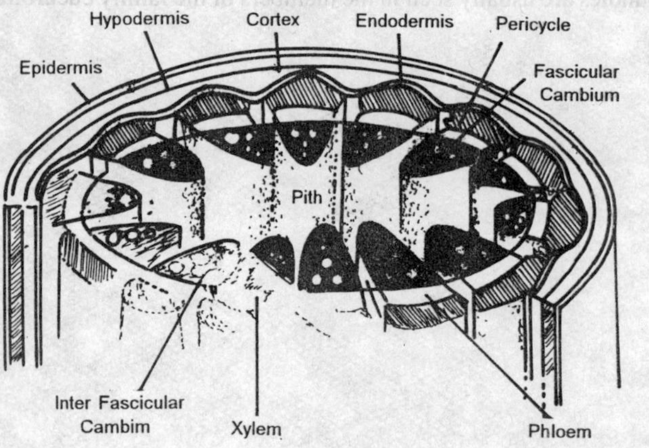

Fig. 9.1 T.S. of Sunflower stem (Ground plan)

Epidermis: This is single layered with closely packed rectangular cells. Above the epidermis there is a well defined cuticle (Fig.). Some of the epidermal cells produce multicellular hairs. Occasionally stomata may be seen in the epidermis.

Cortex: There are two regions in the cortex. The outer cortex (immediately below the epidermis) consists of a few layers of collenchyma whose corner walls are thickened by the deposition of pectin and cellulose. Next to the collenchyma is parenchyma constituting the inner cortex. This consists of sev eral layers of parenchyma , some of the outer layers of which are chlorophyllous.

Stele: The outermost layer is the endodermis whose cells are rich in starch grains; hence also known as the starch sheath. It has the thickening on radial walls called casparian thickenings.

Next to the endodermis is the pericycle. It is represented by patches (cap like) of sclerenchyma present immediately above each vascular bundle.

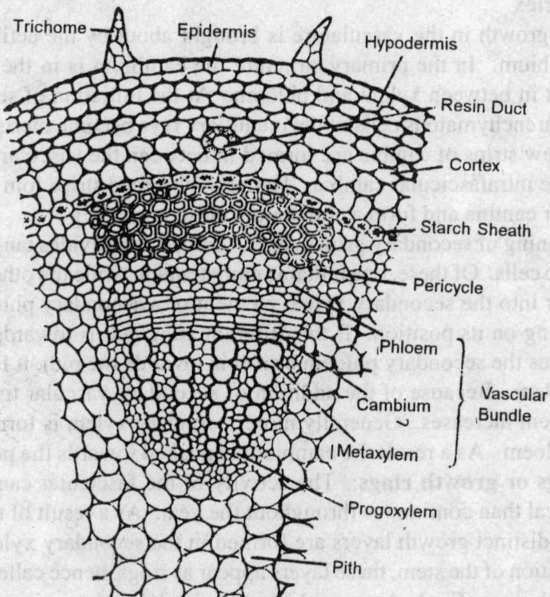

Fig. 9.2 T.S. of Sunflower stem (a portion enlarged)

The vascular bundles are many in number, arranged in the form of a ring surrounding a central pith. Each vascular bundle is conjoint (xylem and phloem combining to form a single bundle), collateral (xylem and pholem on the same radius with x/p arrangement), open (open to secondary growth i.e. cabium is present) with endarch (protoxylem pointing towards the pith) xylem. Phloem consists of seive tubes companion cells and phloem parenchyma .In between phloem and xylem is present cambium which is said to be intrafascicular (present within a vascular bundle). The cambial cells are thin walled, rectangular and actively dividing. Next to the cambium, internally is the xylem, with the protoxylem facing the pith. Xylem consists of vessels, trachieds, wood fibres and xylem parenchyma . Between the vessels are found the wood parenchyma.

Between the vascular bundles are present radially elongated parenchyma cells constituting the primary medullary rays. Occupying the central region of the stem is a wide parenchymatous pith.

SECONDARY GROWTH

In sunflower stem, after a period of growth of one year secondary development takes place. Secondary growth is seen both in the vascular cylinder as well in the cortex.

Secondary growth in the vasculature is brought about by the activity of a fasicular cambium. In the primary structure the cambium is in the form of strips present in between xylem and phloem. At the initiation of secondary growth, the parenchymatous cells of the medullary rays become meristematic. As a result, new strips of cambia are formed in between the vascular bundles adjacent to the intrafascicular cambia. The newly formed strips join with the intrafascicular cambia and form a completely closed ring (Fig)

At the beginning of secondary growth each cambial cell divides tangentially and forms two cells. Of these, one remains meristematic while the other differentiates either into the secondary xylem element or a secondary phloem element depending on its position. If the differentiating cell is towards the periphery it forms the secondary phloem or if it is towards the pith, it forms the secondary xylem. Because of the addition of secondary vascular tissues the grith of the stem increases. Generally more secondary xylem is formed than secondary phloem. As a result the cambium is pushed towards the periphery.

Annual rings or growth rings: The activity of the fascicular cambium is more periodical than continuous throughout the year. As a result of this periodic activity, distinct growth layers are formed in the secondary xylem. In a transverse section of the stem, these layers appear as rings, hence called annual rings or growth rings (Fig.). An annual ring may be defined as a ring of vascular tissue produced by the fascicular cambium during the course of one year. In each annual ring the xylem consists of two types of wood elements namely *spring wood* or early wood and *autumn wood* or late wood. During spring season growth is maximum and hence additional vessels are needed to supply the nutrition. Hence spring wood is more in extent and has vessels with a wider lumen. During autumn season, there is less growth and hence less amount of vessels are formed with narrow lumen. The spring and autumn woods of successive rings alternate.

Since each annual ring corresponds to the growth during a year, the number of rings corresponds to the age of a particular plant after its initial growth of one year.

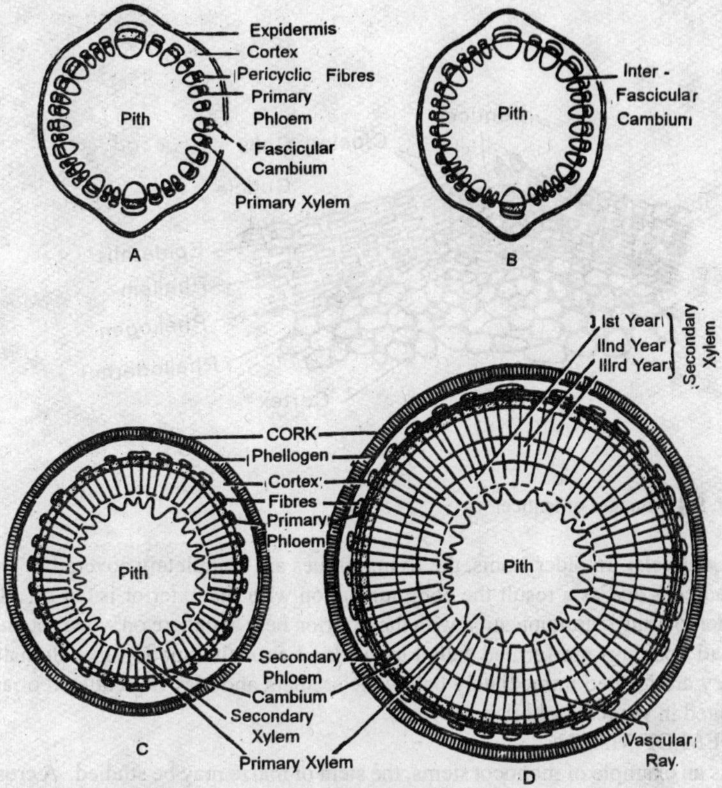

Fig. 9.3 Stages of secondary growth in a Dicotyledonous stem

Sap wood and Heart wood: After a considerable period of secondary growth, the secondary xylem gets differentiated into two regions. The central region consisting of old wood is called the heart wood or duramen. The peripheral zone of wood or the recently formed wood is called the sap wood or alburnum. While the heart wood has mostly dead cells, sap wood has some living cells in association with the vessels.

Secondary growth in the cortex takes place simultaneously with that of stele. In the Cortex, a cork cambium or phellogen arises in the outer layers and cuts off derivatives towards both the faces. The outer derivatives form the cork or the phellem while the inner derivatives form the phelloderm or the secondary cortex. These three together constitute the periderm tissue which replaces the epidermis.

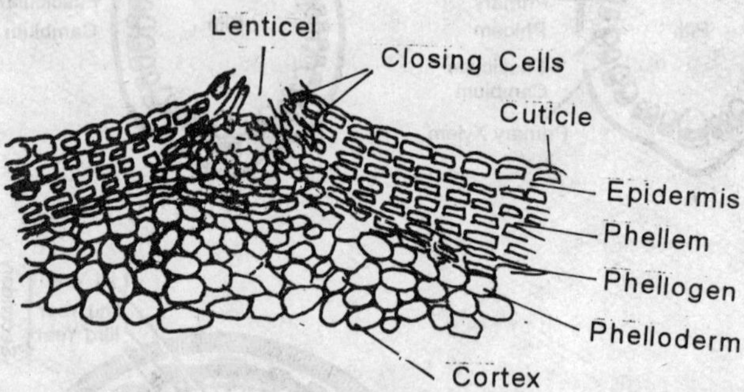

Fig. 9.4 Structure of lenticel

Lenticels: In older stems, the living tissues are completely covered by the dead cork and as a result the communication with the exterior is cut off. In order to bring communication with the exterior, here and there on woody stems, small openings are formed called lenticels. Lenticels function like stomata. They are formed immediately below the stomata and may be scattered or arranged in vertical or horizontal rows.

STEM OF MAIZE

As an example of monocot stems, the stem of maize may be studied. A cross section of young stem shows an epidermis, a hypodermis and a ground tissue in which the vascular bundles are embedded (Fig) . There is no clear cut distinction into stele , cortex and pith. Endodermis and pericycle are absent.

Epidermis is made up of a single layer of compactly arranged cells. It is covered with a cuticle. There are no epidermal hairs.

Hypodermis is internal to the epidermis. It consists of few layers of sclerenchyma cells.

Ground tissue is represented by loosely arranged, thin walled parenchyma cells. Embedded in this tissue are the vascular bundles. Vascular bundles are scattered in the ground tissue. They are crowded towards the margin but smaller in size, while towards the centre they are bit in size and sparsely distributed.

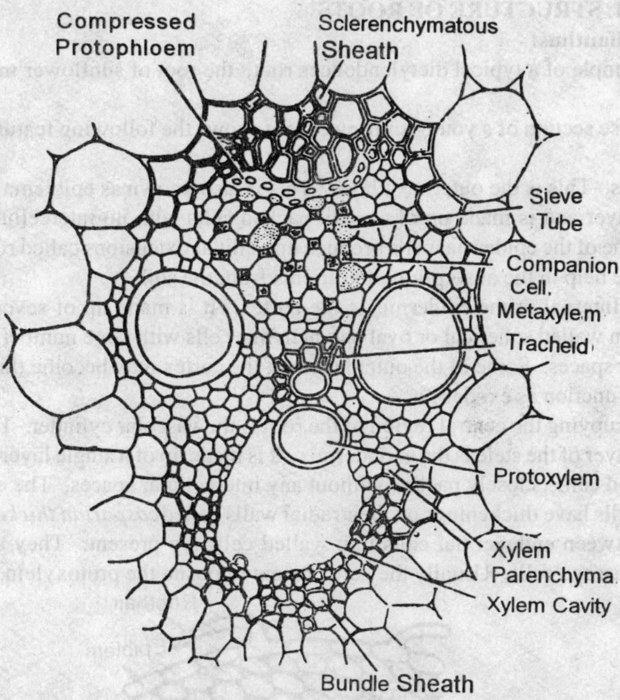

Fig. 9.5 A vascular bundle of Maize stem (magnified)

Structure of a vascular bundle: Each bundle is oval in shape, conjoint, collateral endarch and closed (cambium is absent hence no secondary growth). Surrounding the bundle there is a sclernchymatous bundle sheath (Fig). The phloem is above the xylem. It consists of sieve tubes and companion cells. There is no phloem parenchyma . The outer cells of phloem (protophloem) are crushed while the inner cells (metaphloem) are healthy. Xylem is reduced; it consists of four (sometimes three) vessels (two meta and two proto) arranged in the form of 'Y'. Metaxylem vessels are pitted, while protoxylem vessels are annular and spiral. Sometimes in between the metaxylem vessels, small, pitted tracheae are present. Below the protoxylem a lysigneous cavity is present. It consists water and is known as the water containing cavity. Surrounding this are present cells of wood parenchyma .

INTERNAL STRUCTURE OF ROOTS

Root of Helianthus:

As an example of a typical dictylendonces root, the root of sunflower may be studied.

A transverse section of a young sunflower root shows the following features (Fig).

Epidermis: This is the outermost layer. This is also known as epiblema or piliferous layer and is made up of closely packed cells with no intercellular spaces. Some of the epidermal cells produce unicellular extensions called root hairs. These help in the absorption of nutrients from the soil.

Cortex: Internal to the epidermis is the cortex. It is made up of several layers of thin walled spherical or oval parenchyma cells with large number of intercellular spaces. Some of the outer layers of the cortex may become thick walled and function as exodermis.

Stele: Occupying the central region of the root is the vascular cylinder. The outermost layer of the stele is the endodermis. It is made up of a single layer of barrel shaped cells., closely packed without any intercellular spaces. The endodermal cells have thickenings on their radial walls called *casparian thickenings*. In between endodermal cells thin walled cells are present. They are called the passage cells. Usually they are present opposite the protoxylem.

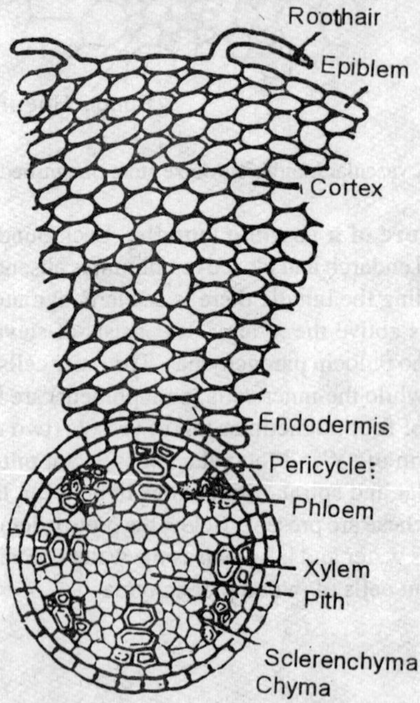

Fig. 9.6 T.S. of sunflower root

Internal to the endodermis is the pericycle. It is made up of a single layer of rough, thin walled parenchyma cells.

Surrounded by the pericycle are the vascular bundles. They are radial (Xylem and Phloem form separate alternating bundles) with exarch (protoxylem pointing towards the periphery) and tetrarch (four protoxylem points) xylem. Xylem consists of annular and sprial vessels with metaxylem being made up of reticulate and pitted vessels.

Occupying the central region of the stele is a narrow parenchymatous pith. Sometimes the metaxylem groups join together so that pith is obliterated.

Between the radial vascular bundles is present parenchyma constituting the conjuctive tissue.

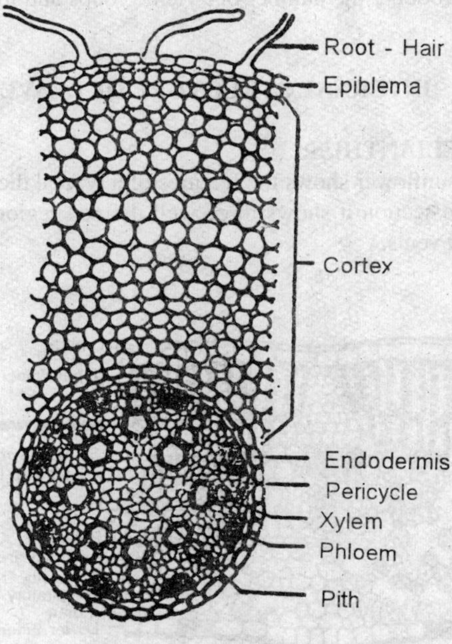

Fig. 9.7 T.S. of Maize, root

ROOT OF MAIZE: As an example of a typical monocotyledonous root, the root of maize plant may be taken. A transverse section of a young root shows the following features. Mainly three regions may be distinguished. These are epiderms, cortex and stele.

Epidermis: This is the outermost layer, and is also known as epiblema or piliferous layer. The epidermis is made of a single layer of cells which are closely packed without any intercellular spaces.

Cortex: Immediately internal to the epidermis is the cortex which is made up of many layers of loosely arranged parenchyma cells. Some of the cells in the outer region of the cortex may be closely packed constituting the hypodermis.

Stele: The stele is bounded by an endodermis and pericycle. Endodermal cells are barrel shaped and have casparian thickenings on their radical and tangential walls. Pericycle is represented by a single layer of thin walled parenchyma cells.

The vascular bundles are many in number (polyarch) and radial with exarch xylem. In the centre of the stele is a parenchymatous pith.

The main points of difference between the dicotyledonous root and the monocotyledonous root are the number of xylem groups and the presence or absence of pith.

INTERNAL STRUCTURE OF LEAVES

LEAF OF HELIANTHUS:

The leaf of sunflower shows the features of a typical dicotyledonous leaf. In a transverse section it shows three well defined regions viz. epidermis, mesophyll and veins.

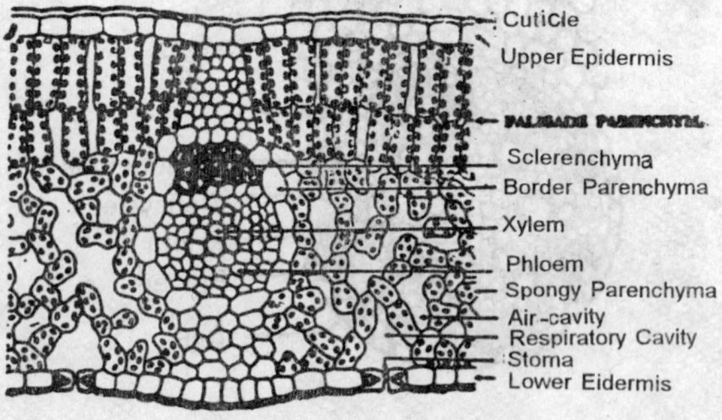

Fig. 9.8 T.S. of a dicot leaf

Epidermis: There are two epidermal layers covering either surfaces of the leaf. The epidermis on the upper surface is called the upper epidermis and the one on the lower surface is called the lower epidermis. In some cases, covering the epidermal layers will be a layer of cuticle. In the leaf of sunflower the upper epidermis is continous while the continuity of the lower epidermis is interrupted by the presence of stomata. These are small pores helping in gaseous exchange and the loss of water during transpiration. Each stoma is covered by two guard cells which regulate the opening and closing of the stomatal aperture. The guard cells are crescent shaped with thick inner walls. Internally each stoma leads into an air space.

Mesophyll: In between the two epidermal layers is the mesophyll. It is differentiated and has well developed palisade tissue below the upper epidermis. Above the lower epidermis and below the palisade tissue is the spongy parenchyma which has loosely arranged cells with large number of air chambers. Palisade tissue is chlorophyllous and it is the chief photosynthetic tissue of the plant. There are a few chloroplasts in the spongy parenchyma cells also but their main function is storage. There are rich in starch grains.

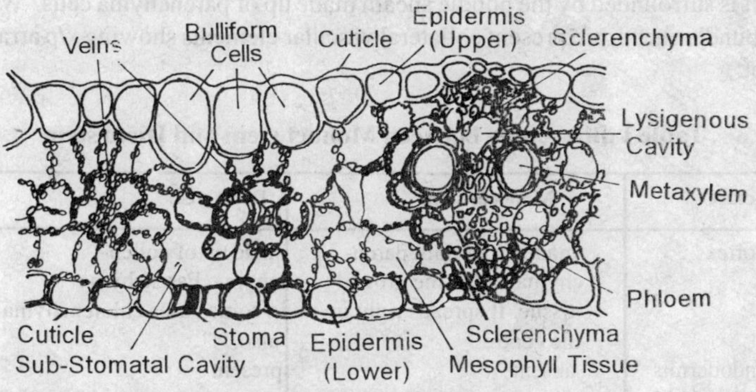

Fig. 9.9 T.S. of monocot leaf (paddy)

Vein: The veins are arranged in the form of a network and in a cross section they show the vascular bundles. Each vein representing a vascular bundle is surrounded by a parenchymatous envelope called the bundle sheath. Within the bundle sheath are the vascular elements viz, the xylem and the phloem. Unlike in stems the xylem is above and the phloem is below.

LEAF OF MAIZE:

As an example of a typical isobilateral leaf, the leaf of maize may be studied. A cross section shows the following (Fig. 35).

Epidermis : There will be two epidermal layers upper and lower. The epidermal layers are uniseriate and composed of more or less oval cells arranged compactly. There is the deposition of a cuticle both on the upper epidermis as well on the lower epidermis . Besides the normal cells, the upper epidermis consists of special cells called bulliform cells or motor cells. These cells are larger than the other epidermal cells and are thin walled. These help in the rolling of the leaves during dry weather . Leaves are amphistomatic (stomata are present on both the epidermal layers). Each stoma internally leads into a large substomatal air chamber.

Mesophyll: It is undifferentiated. But the mesophyll cells below the upper epidermis may show a compact arrangement.

Vascular bundles: The veins are arranged parallely and are not interconnected. There are two pads of sclerenchyma supporting the vein. The vein itself is surrounded by the bundle sheath made up of parenchyma cells. Within the bundle sheath are present collateral vascular elements showing x/p arrangement.

Table I differences between Monoct stem and Dicto stem

Structure	Monocot stem	Dicot stem
1. Cortex	made up of only parenchyma called the ground tissue. It spreads right up to the centre.	made up of collenchyma, Parenchyma in some cases chlorenchyma also.
2. Endodermis	absent	present
3. Pericycle	absent	present
4. Vascular bundles	scattered, and closed	arranged in the form of a ring open.
5. Cambium	absent	present
6. Pith	absent	present

Table II Difference between Moncot root and Dicot root

Structure	Monocot root	Dicot root
1.Xylem	many bundles, exarch	2 to 6 in number exarch.
2.Pith	wide pith is present	either very narrow or absent.
3.Pericycle	gives rise to lateral roots and cambium	gives rise to lateral roots only.
4.Cambium	absent	present

Table III difference between Dicot leaf and Monocot leaf

Structure	Monocot leaf	Dicot leaf
1. Mesophyll	undifferentiated. Has loosely arranged parenchyma cell.	differentiated into palisade and spongy parenchyma.
2. Sclerenchyma	two pads of sclerenchyma present attaching the vascular bundle of the upper and lower epidermis.	a group of parenchyma cells present above the vascular bundle.
3. Upper epidermis	motor cells (bulliform cells) present.	motor cells are absent.

Ten

Movement of Substances

SIGNIFICANCE OF WATER

Water is one of the most important compounds that profoundly influences life. In fact it is impossible to visualize life in its myriads of forms in a dehydrated state.It is widely believed that life itself originated in some quiet corner of the primordial ocean (on the young earth) in the remote past. There is no wonder then water is aptly called the liquid of life or the elixir of life. From the point of view of plant life, their very sustenance is dependent on water. The following is a list of the role played by water in the life of plants.

1. Water is an esséntial and major constituent of protoplasm. In fact bulk of protoplasm is water. All the cell organelles maintain their structure and function in the presence of adequate quantity of water.
2. Distribution of various dissolved substances from one cell to another and from one part to another can take place only in the medium of water.
3. Water is an excellent solvent. It also acts as a reagent in various reactions in plant cells. In photosynthesis and respiration water plays an important role.
4. Breakdown of various chemical compounds takes place only in the medium of water. During germination, complex food substances stored in the seed breakdown into simple constituents. This is possible only when water is present.
5. During metabolic reactions, the inner temperature of the plant rises. Water stream flowing throughout the plant body acts as a coolant and maintains the temperature.
6. Adequate amount of water is necessary in cells and tissues to maintain the turgidity of the plant body.
7. Enzymes, without whose mediation, virtually, no biological reaction can take place, function only in a hydrated medium.
8. Plants take in their nutrients from soil only in a solution form. Soil water thus is very essential for the survival of the plants.
9. A network of thin layer of water surrounding each cell plays an important role in the movement of dissolved substances.
10. In lower plants water is essential for the mobility of gametes.
11. Dehiscense of sporangia and dispersal of spores (hygroscopic reaction) is aided by atmospheric water.
12. Water also helps in pollination and dispersal of fruits in many of the higher plants.

MOVEMENT OF SUBSTANCES

Various types of substances move into the cell from the external medium. A few also move in the reverse direction. Mainly two kinds of forces are responsible for the movement of particles to and from the cell. These are: 1) Plant forces or external forces and 2) Forces inherent in the particles. Here we shall discuss the latter. There are three types of movements of substances or particles by the force or kinetic energy which is inherent in them. These are Imbibition, Diffusion and Osmosis.

IMBIBITION: Imbibition may be defined as the absorption of liquid by the particles of a solid without forming any solution. The substances which absorb the liquid may be termed as imbibants. In plants, the solid particles are present in the form of colliods and they absorb water, (The absorption of ether by rubber is also an example of imbibition).Due to imbibition the solid particles increase their size.

In plants, the imbibants consist of many hydrophilic compounds like proteins, pectic compounds, starch and cellulose. Of all the hydrophilic colloids proteins have maximum imbibitional capacity.

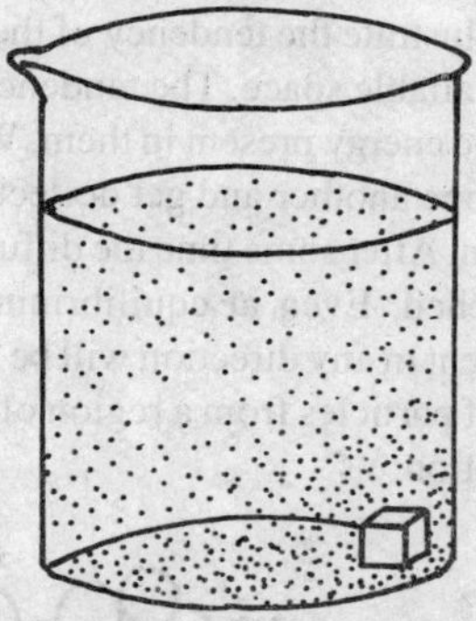

Fig. 10.1 Diffusion of solid particles

During imbibition, water molecules are not only absorbed but also adsorbed (held over the surface). Due to adsorption water molecules are immobilised and they lose their kinetic energy. This kinetic energy appears in the form of heat. If the imbibant is not allowed to swell in volume, it will bring about a force or pressure called imbibitional force. considerable force is generated in this way. In olden times this force was used in breaking rocks. Large wooden pegs used to be inserted in the rocks. They were then allowed to absorb water. The resulting imbibitional pressure would break the rock.

The swelling of a bean or gram seed is a familiar example of imbibition. It is because of the imbibitional force only, the seed coats break open and the radicle comes out.

Imbibition is influenced by a number of factors such as temperature, pH, texture of the imbibant etc.,

In plants, imbibition is of great importance in the process of germination.

DIFFUSION: The tendency of the molecules of a liquid solid or gas to get evenly distributed throughout the available space is called diffusion. We will study some examples.

Place a copper sulphate crystal in a beaker containing water. The crystal gradually dissolves and one can see particles of copper sulphate migrating in all directions in water.

Fix a porous partition at the base of a 'V' tube. Fill one arm with water and the other with sugar solution. Slowly sugar solution migrates to the arm containing water. After sometime the concentration of solution in both the arms will be same.

Burn an agarbathi or open a bottle of ammonia in one corner of the room. After sometime it could be smelt even in the fartherest corner of the room.

The three examples clearly illustrate the tendency of the molecules to evenly distribute themselves in the available space. The tendency of the molecules to migrate is because of the kinetic energy present in them. When more molecules are present, they collide with one another and get deflected in the direction of least resistance or concentration. After some time the diffusion rate slows down, because equilibrium has reached. Even at equilibrium there will be some movement, but the net movement in any direction will be zero. Hence diffusion may be said to be movement of particles from a region of higher concentration to a region of lower concentration.

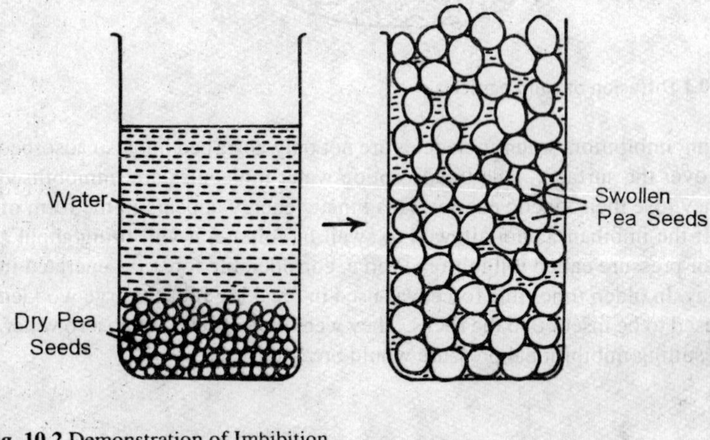

Fig. 10.2 Demonstration of Imbibition

Diffusion pressure: When more number of particles are present in a given space than what it possibly can accommodate, there will be pressure of particles to migrate. This is called diffusion pressure and is proportional to the concentration of particles. In a region of less concentration there will be more space and less particles, hence there will be a deficit of pressure. This is called "Diffusion pressure deficit (DPD). In terms of diffusion pressure, diffusion pressure (lesser DPD) to a region of lesser diffusion pressure (higher DPD). Several factors influence the rate of diffusion. There are density, concentration of the medium, temperature etc.

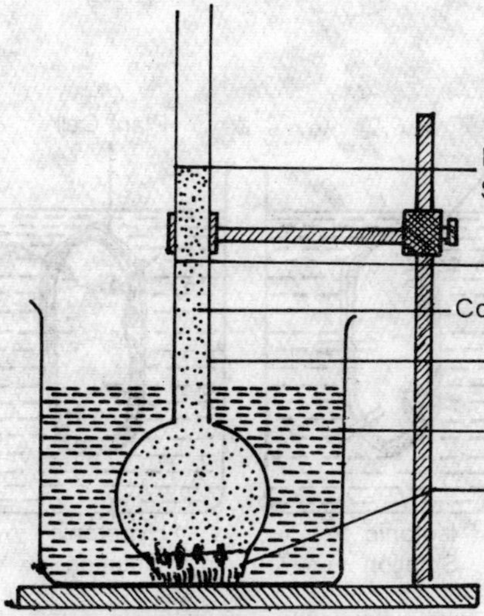

Fig. 10.3 Physical demonstration of osmosis

OSMOSIS: While diffusion deals with the movement of particles of solid, liquid and gas, Osmosis deals with the movement of only the particles of a solvent (liquid). We will study osmosis with the help of an experiment (fig3.3) Take a beaker which is half filled with water. Take a thistle funnel and tie its mouth with the help of parchment paper, sheep bladder or egg membrane. These membranes are called selectively permeable or differentially permeable, since they allow only certain particles to pass through and not all. In this case the selectively permeable membrane is permeable to only molecules of water but impermeable to others. Fix the thistle funnel to a stand and immerse it in the water contained in the beaker. Fill the thistle funnel with 5% sugar solution and note the level. After some time it may be observed that the level of the sugar solution in the thistle funnel rises. Why ? Water from the beaker must have entered into the thistle funnel. Now taste the water in the beaker. It does not taste sweet, because the membrane has allowed only the water molecules to enter into the thistle funnel but has not allowed the sugar molecules into the beaker. Water migrates into the thistle funnel because it is in higher concentration in the beaker. Hence Osmosis may be defined as the movement of the molecules of a solvent (water) from the region of their higher concentration through a selectively permeable membrane. In this, there will be no movement of the solute particles.

Osmosis was first discovered by Abbe Nollet in the year 1748. Subsequent work on osmosis was carried out by Pfeffer (1887), Traube (1826), Dutrochet (1827) etc.,

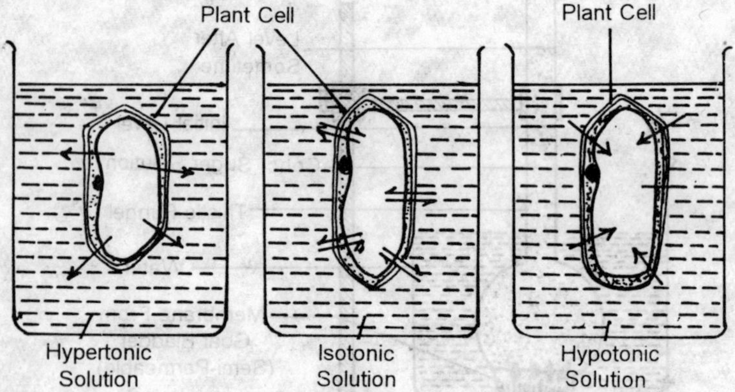

Fig. 10.4 Concentration of solution in relation to cell sap

Osmotic pressure (osmotic potential):

When two solvent systems of different concentrations are separated by means of a selectively permeable membrane, solvent migrates from the region of its higher concentration to the region of its lower concentration due to osmosis. Both these solvent systems may be termed osmotic systems. A strong solvent migrates to a weak solvent. This means that the weak solvent has the capacity or attraction or force to draw water from a strong solvent (a weak solution).

Since a weak solvent is a strong solution, it has less solvent and so can attract water from a weak solution (strong solvent).

This force or pressure that is there in a strong solution is called its osmotic force or pressure. It is due to this osmotic force that a strong solution draws water from a weak solution. However a more correct definition of osmotic pressure (OP) should be explained in negative terms, because OP cannot be seen or demonstrated directly. When the strong solution draws water from a weak solution it does so with some force or pressure. It we apply a counter pressure on an osmotic system so that it cannot attract water, that counter pressure will be equivalent to OP. Hence OP may be defined as the pressure that is applied on an active osmotic system in order to prevent the entry of water into it. Initially when the pressure is applied, entry of water into an osmotic system may continue. But at a particular magnitude of the pressure osmosis would stop and that is the OP of that system. Strong solutions (weak solvents) have higher OP ie. more amount of pressure is needed to prevent the entry of water into that system. OP is directly proportional to the number of solute molecules per solvent molecules. Hence OP may also be defined as "excess hydrostatic pressure which must be applied on a solution in order to make it's solvent potential equal to that of a pure solvent"

OP is usually expressed in terms of bars (atmosphere) -1 atm= 1.01 bar). For example, a molar solution of sucrose separated by pure water by means of a selectively permeable membrane has a pressure of 22.4 atms at $O^{O}C$. OP varies with the temperature of the medium. At lower temperatures tends it to be low.

OP is usually denoted by the symbol p. Vant Hoff formulated an equation to relate OP to concentration of the solute in the solution. This is as follows-

$p\,V = n_S\,R\,T$

Here p = is OP,V =volume of the solvent, ns=moles of the solute, R=gas constant (0.0820 litre atm/degree mole), T= absolute temperature.

Turgor pressure and wall pressure: Plant cells can survive in a wide range of osmotic concentrations in comparison with the animal cell, because of the presence of cell wall which is rigid. When the cells are placed in water or hypotonic solutions, they absorb water and increase their size. As a result, the elastic plasma membrane presses the cell wall. The pressure that is built up by the cytoplasmic membrane is called the turgor pressure. Turgor pressure will be maximum when the cell is not flaccid ie. when solvent concentration is maximum inside the cell. When the cell has less solvent (as it happens when it is placed in a medium of hypertonic solution), the membrane will not press the cell wall and turgor pressure is lowest at that time. When the cell membrane is applying a pressure, the cell wall being rigid, (in order not to burst and retain the shape), also applies an equal amount of pressure in the opposite direction. This is known as the wall pressure. As a result of these two pressures the cells will be turgid. At any given time turgor pressure and wall pressure are equal (TP = WP)

Water potential: Potential may be defined as the capacity to do the work. When a chemical system does some work at a constant temperature, the free energy falls down. Thus the free energy represents the potential of the system to do the work. The free energy per molecule in any chemical system is called its chemical potential. The chemical potential of water is called water potential. Water potential in any solvent system actually represents the difference between the chemical potential of water in that system and pure water. Water potential is always represented by a negative sign. Thus the water potential of pure water is zero. The greek letter psi(Y) designates the water potential in a system. It is expressed in terms of bars or megapascals (Mpa = 10 bars). Water potential in a system can be calculated as follows.

$Y = (RT \text{ in } e/e^{o})/V$

Here Y = water potential in the system in bars, R=ideal gas constant (8.3x 10^{-1} ergs mol^{-1} deg^{-1}), T=absolute temperature in K,e^{o} = vapour pressure of water in the system at T,e^{o}=vapour pressure of pure water at T and V=molal volume of water (18.0 cm^{3} mol^{-1}).

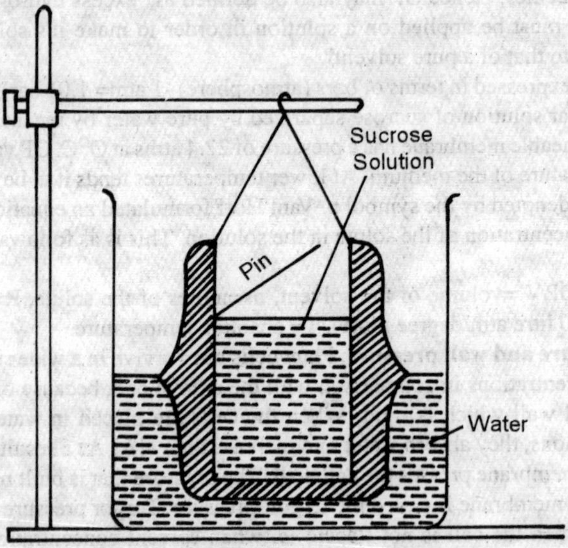

Fig. 10.5 Physiological demonstration of osmosis

Water potential is a diagnostic character of the status of water in a cell or tissue. The lower the potential, the greater is the capacity to absorb water. Water potential is made use of in measuring water deficit and water stress. In plant tissues water potential is always less than 0 bar. Leaves and aerial parts of plants in soils with adequate moisture supply will normally have a water potential between - 2 and -8 bars. With decreasing supply of water, the values become more negative and at -15 bars plant tissues cease to grow.

There are three major components of water potential in a plant cell. These are-

a) matric potential,

b) solute potential and

c) pressure potential.

Water potential in a plant is the sum total of all these three.

Matric potential represents the surface of the cell, soil etc., where hydrophilic colloids are present and water molecules get absorbed to them. The matric potential (Ym) is generally not significant in osmotic movement because it does not allow the free movement of water molecules.

Solute potential (Ys)is also known as the osmotic potential. It may be defined as the amount by which water potential is reduced as a result of the presence of solute particles. Solute potential is always represented in negative numbers (ie. so much quantity of solvent is absent in comparison with a pure solvent)

Pressure potential (Yp) is represented by the cell wall and cell membrane. The cell wall exerts a pressure (WP) on the cell contents. Similarly the cell membrane also exerts a pressure (TP) on the cell wall by virtue of its solvent content. The pressure potential is positive in that it denotes the amount of moisture already present. Hence water potential may be represented symbolically as follows.

$Y = Ym + Ys + Yp.$

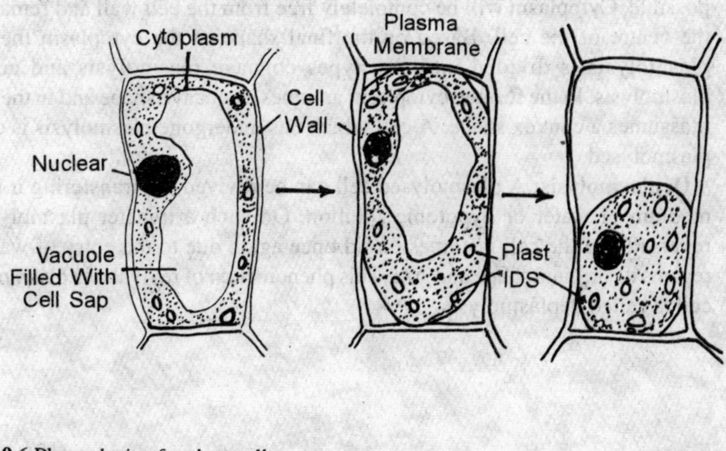

Fig. 10.6 Plasmolysis of a plant cell

PLASMOLYSIS (*Plasma* = cytoplasm, *lysis* = breakdown): When a plant or animal cell is kept in a hypertonic solution, water from cell sap flows out due to a reverse concentration gradient (concentration gradient pointing out of the cell). As the solution in the external medium is more concentrated, it has more OP and hence water moves out of the cell resulting in shrinkage of the cell volume. If the conditions persist for a long time the cell will die. This phenomenon of water loss from cells which is akin to exosmosis is called plasmolysis. Wilting of plants noticed under conditions of water scarcity is an indication of plasmolysis of cells. There are three stages in plasmolysis - viz, Incipient plasmolysis, evident plasmolysis and final plasmolysis.

Incipient plasmolysis: In a turgid cell, with adequate amount of water, the plasma membrane closely appresses the cell wall and is in contact with it alround. When this cell is kept in a hypertonic solution, water starts moving out of the cell. Initially there will be no effect on the cell wall. But as water continues to go out, the cell shrinks in volume. But the plasma membrane maintains its contact with the cell wall alround due to its elastic capacity. As the exit of water continues, the plasma membrane reaches its limit of elasticity and gets torn from the cell wall at the corners, while maintaining contact at other regions. This is the first stage of plasmolysis called incipient plasmolysis.

Evident plasmolysis: The cell under hypertonic conditions continues to lose water to the external medium and shrinks further in volume. The plasma membrane gets torn completely from the cell wall and goes on contracting. This is the second stage of plasmolysis called evident plasmolysis.

Final plasmolysis: As exosmosis continues, the shrinkage of the cell and cytoplasm reaches the minimum limit and no further shrinkage in volume is possible. Cytoplasm will be completely free from the cell wall and remains in the centre of the cell. Based on the final shape of the cytoplasm the final plasmolysis is divided into two types concave plasmolysis and convex plasmolysis. In the former, cytoplasm assumes a concave shape and in the latter it assumes a convex shape. A cell which has undergone plasmolysis is called plasmolysed.

Deplasmolysis: A plasmolysed cell can be revived, by transfering it into a medium of water or hypotonic solution. On such a transfer plasmolysis is reversed and the cell becomes turgid once again due to the entry of water. It regains its original shape and size. This phenomenon of revival of a plasmolysed cell is called deplasmolysis.

Eleven

Absorption of water

Water absorbing system in plants: Different groups of plants have different methods of absorbing water from the enviornment. In lower plants like algae, fungi etc, many of which are aquatic, there is no special organ to absorb water. Water is absorbed by the general surface of the plants as they are present in the medium of water. In Bryophytes, which are partially terrestrial, separate organs meant to absorb water have been evolved. These are the rhizoids. Rhizoids are unicellular organs, hair like in appearance. In many of the liverworts (Ricia, *Marchantia* etc), two types of rhizoids are seen -simple and pegged. The simple rhizoids have a smooth inner wall, while the pegged rhizoids, have their inner walls infloded. The latter type of rhizoids, have their inner walls infolded. The latter type of rhizoids form an anastamose on the ventral surface of the thallus assuring uniform distribution of water.

In terrestrial plants, soil is the only source of water. As all the parts of the plant body are not in direct contact with the soil, special organs (roots) and special tissues (vasculature) evolve to provide anchorage, absorption and transportation of water to all the parts of the plant body. Roots are seen in the plant kingdom from ptreridophytes onwards. Roots are generally underground. But in some exceptional cases as in *Selaginella rupestris* (Pteridophyte) the leaf surface may absorb water. In some of the flowering plants like epihytic orchids, the roots are aerial. These roots have a special tissue called velamen to absorb water from the atmosphere. These roots are chlorophyllous and carry out photosynthesis.

In a large majority of plants the roots are underground. Apart from the conventional classification of the root system into tap root system and adventitions root system seen characteristically in dicotyledons and monocotyledons, it (root) may be classified into two categories based on their absorption of water in the soil. These are *deep feeders* and *surface feeders*. In the former, the root system grows very deep in search of water, while in the latter, the root system is not very deep, but spreads very wide a little below the surface of the soil. Both the types of root systems are advantageous to the plants depending on the availability of water in soil. In conditions when the surface temp., is high as in deserts, the rain water may get evaporated or it may percolate the lower reaches of the soil very quickly due to loose particles of sand. In such a situation, a spreading root system can catch as much amount of surface water as possible before it becomes unavialable. In situations, where water is available, but deep within the soil, deep feeders are of advantage.

The root system is branched and the ultimate branches are called root hairs. The root hairs are unicellular, delicate and vary in length from 1-10 mm. They are extensions of the epidermal cells. The root hairs are present a little above the root tip. This region is called the root hair zone. The root hairs remain alive and functional from few days to several months. According to Kramer (1959), maximum water absorbtion takes place in the root hair zone, while the epidermal cells at the root tip, where there is no suberization or cutinization might also absorb some quantity of water.

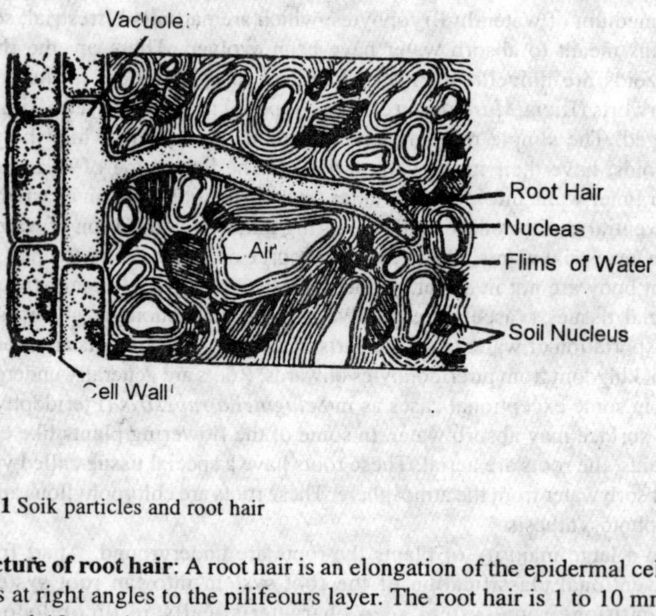

Fig. 11.1 Soik particles and root hair

Structure of root hair: A root hair is an elongation of the epidermal cell. It extends at right angles to the pilifeours layer. The root hair is 1 to 10 mm in length and 10 mm in diameter. The number of root hairs per unit area is a measure of the absorptive capacity of the root. The cell wall of root hair has two layers. The outer layer is mainly composed of pectic substances, while the inner layer is made up of cellulose. The wall composition varies in some instances as in *Taxus baccata*, where the outer wall is gelantinized and soil particles get integrated with it. In Corn, the outer cell wall of root hair has no pectin, while *Tradescantia*, callose is present at the tip of the root hairs. Inside the cell is a large vacuole, which is continuous with that of the epidermal cell. There is no separate nucleus in the root hair. The nucleus of the epidermal cell controls the activity of the root hair also. The cell wall of root hair is permeable to water. The cytoplasmic layer functions as a differentially permeable membrane.

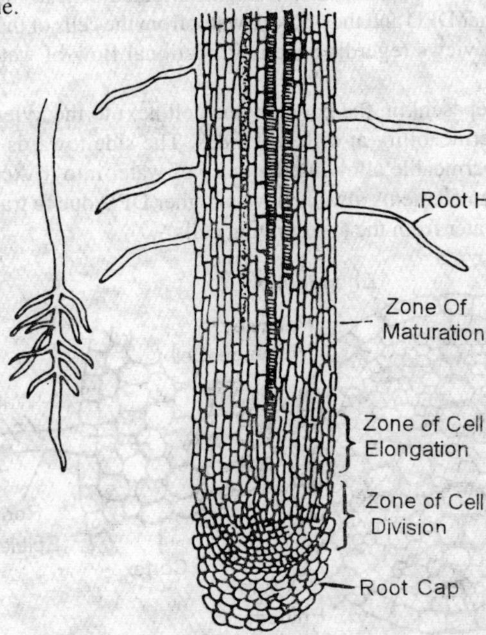

Fig. 11.2 Structure of a root tip. A- root tip, B- long sectional view of a root tip

The path of water in the root: Water is first absorbed by the root hair cell from the soil due to a pressure gradient. As a result, the root hair cell becomes turgid and its DPD decreases. The adjacent cortical cells have a higher DPD as they have less water. As a result of this, water moves into the cortical cells increasing its TP. The next cortical cell draws water from this cell because it has less TP. In this way, through a series of concentration gradient water flows from one cortical cell to the other and finally reaches the endodermis. According to Sutcliffle (1969), the movement of the water through cortex is in response to water potential gradient.

There are two ways of radial movement of water- (i) Cell to cell movement across the root and (ii) movement of water along the water filled spaces in the cell wall forming a continuous system from soil water upto the endodermis. The flow of water meets with an impregnable barrier in endodermis as its cells have thick strips on the radial and tangential walls called casparian strips. However there are certain thin walled cells in between the endodermal cells called passage cells. It is through these cells, water enters into the stele. The passage cells are usually positioned opposite the protoxylem elements. The passage cells allow the entry of water into the cells of the pericycle, which in turn become turgid. The protoxylem elements situated next to the pericycle always have a higher DPD and they absorb water from the cells of the pericycle. There are various views regarding the unidirectional flow of water into the xylem elements.

According to Lepeschkin, the parenchyma cell next to the xylem element has differential permeability at different sides. The side towards the xylem element is more permeable allowing the entry of water into it. According to Priestley, the xylem elements always have a higher DPD due to transpiration and hence draw water from the surrounding cells.

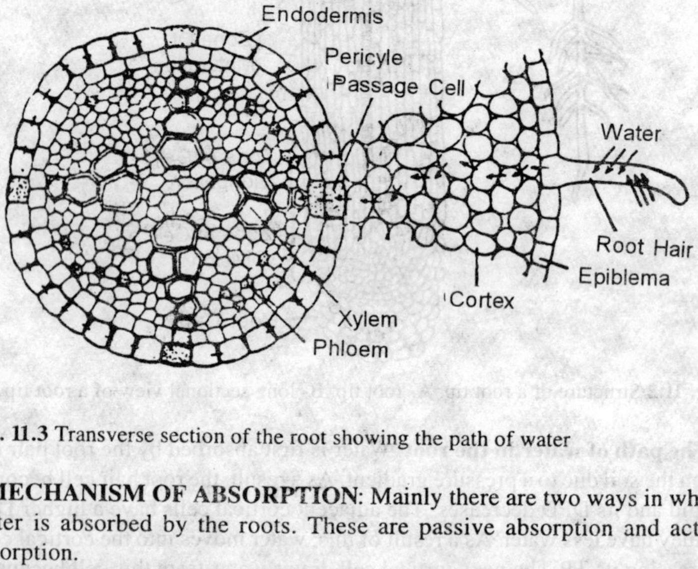

Fig. 11.3 Transverse section of the root showing the path of water

MECHANISM OF ABSORPTION: Mainly there are two ways in which water is absorbed by the roots. These are passive absorption and active absorption.

Passive absorption:The roots do not play any role in the absorption of water in the sense, the force or forces responsible for water absorption are generated elsewhere in the plant body. It is widely beleived that transpiration has a direct bearing on the rate of water absorption. That is why in rapidly transpiring plants, rate of absorption of water is very high. In a rapidly transpiring plant, a continuous gradient of DPD is built up due to evaporation of water. The mesophyll cells have a higher DPD, and as a result, water in the xylem elements are attracted towards them. This creates a tension(DPD) in the xylem elements which is transmitted down due to the cohesion of water molecules. The tension or DPD reaching the lower tip of the xylem is carried across the root cortical cells and ultimately to the root hair cell. As a result of this continuous chain of DPD, water migrates radially across the root cortex and travelling through the xylem finally reaches the leaf mesophyll cells, from where it escapes. Hence transpiration force or pull or suction force generated away from the roots, is responsible for the absorption of water by the roots. It would be more appropirate to use the term absorption of water through the roots rather than by the roots because roots are passive and do not influence the water uptake in any way. No event taking place in the root will influence the absorption of water (either way).

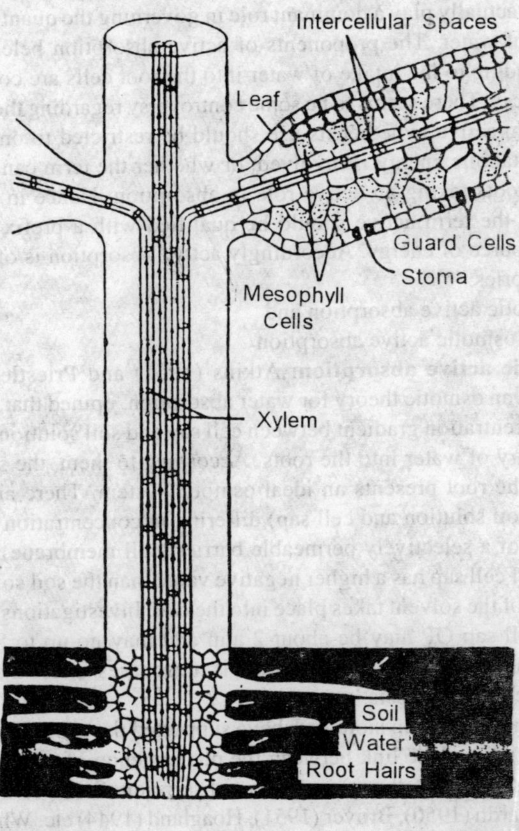

Fig. 11.4 Diagrammatic representation of passive absorption of water

When the plants are rapidly transpiring, it has been estimated, that the DPD in root hair cells increased by -3 to -5 bars, whereas the DPD of the soil solution is less than -1 bar. It is this higher DPD that forces large quantity of water into the roots, ultimately to be lifted up to the leaves.

Evidences for passive absorption is obtained from the fact,that the rate of transpiration is approximately equal to absorption of water, and any inhibition of transpiration inhibits water absorption also (Kramer 1937, Lachenmeir, 1932). To sum up, it may be said that in passive absorption roots provide merely a conduit for the entry of water into the plants.

Active absorption: The mechanism of water absorption due to the forces generated in the root itself is called active absorption. According to this theory, roots are not just passive surfaces, witnessing the passage of water through them, but actually play a dominant role in governing the quantum and the rate of entry of water. The proponents of active absorption beleive that all the events leading to the intake of water into the root cells are controlled by the root surfaces. There seems to be some controversy regarding the actual purport of the term active, whether its use should be restricted to only in instances where metabolic energy is involved, or whether the term can be used in the sense of roots taking an active role in absorption. Hence in order to avoid confusion the term active should be qualified with a prefix indicating the specific source of energy. Accordingly active absorption is of the following tow categories.

(i) Osmotic active absorption and

(ii) Non osmotic active absorption

Osmotic active absorption: Atkins (1961) and Priestley (1922) who postulated an osmotic theory for water absorption, opined that, the difference in the concentration gradient between cell sap and soil solution is responsible for the entry of water into the roots. According to them, the situation in the soil near the root presents an ideal osmotic system. There are two solvent systmes (soil solution and cell sap) differing in concentration and separated by means of a selectively permeable barrier (cell membrane). As the water potential of cell sap has a higher negative value than the soil solution osmotic migration of the solvent takes place into the cell. Investigations have revealed that the cell sap OP may be about 2 atm and may go up to 8 atm in some circumstances. The OP of soil solution is most cases is less than 1 atm. As a result there is endosmosis.

A number of plant physiologists have supported the osmotic absorption, by demonstrating a close link between the phenomenon of sap exudation (root pressure - see ascent of sap) and water absorption. Some of these scientists are Lundegardh (1950), Broyer (1951), Hoagland (1944) etc. When an actively water absorbing plant is decapitated, sap exudes out of the cut end of the stem indicating active absorption of water. According to Atkins, the xylem sap always has a high concentration of solutes due to the release of sugars into them by the adjoining parenchyma cells. Anderson and House (1967) have reported that salts are absorbed by the xylem elements containing living protoplasts. Priestly (1922) on the other hand opined that a cell differentiating into xylem gets higher salt concentration because of its contents.

Objections to osmotic theory: The main objection comes from the fact, that the sap concentration in xylem is not always high, as a result a gradient may not exist between cortical cells and xylem. (Xylem sap has 2 atm; cortical cell have 3-4). Secondly, root pressure is not universal in plants.

Non osmotic active absorption: Many physiologists beleive that water absorption is driven through the expenditure of metabolic (respiratory) energy. Absorption of water has been observed against the concentration gradient and even when there is a gradient, the rate of absorption far exceeds what is possible due to a concentration gradient

(Bennet Clark 1936, Thimann 1951, Bogen and Preil, 1953 etc). The following are the evidences to suggest, the involvement of respiratory energy in water absorption.

a) Many of the respiratory poisons (Malonate etc) also inhibit water absorption.

b) KCN which reduces metabolic activities also reduces water absorption.

c) Respiratory promoters increase the rate of absorption (e.g. hormones)

d) Low temperature, which retards the rate of energy release during repsiration also, reduces water intake.

e) Reinders (1942), Kelley (1947) etc., have discovered that aerobic conditions are necessary for water intake.

f) Wilting of roots occurs in water logged soils as they have poor aeration.

g) Only living cells in roots absorb water(?)

h) Bennett Clark et al (1936) showed that OP of cell sap determined by plasmolytic method yields a more negative value than the cryoscopic method. According to them, this indicates a non osmotic component in water absorption. They suggested a secretion force produced at the expense of metabolic energy, pumping water into the vacuole. Kramer (1956) on the other hand opines that the descrepancy in OP is due to experimental errors.

i) House and Findaly (1966) found that the OP of xylem sap was equal to that of external solution and hence opined the involvement of a respiratory energy in water intake.

j) A diurnal peridicity has been observed in water absorption indicating fluctuations in the available energy from metalobic processes.

ABSORPTION OF MINERAL SALTS

Plants require different types of mineral salts besides water for their healthy growth. Most of these (mineral salts) are taken up from soil in the form of solution. Many of the mineral elements are electrolytes and are present in the soil in the form of charged particles called 'Ions'. Ions which are positively charged are called 'cations' (ions which move towards cathode) and negatively charged ones are called 'anions' (ions which move towards anode). The ions will be usually attached to clay micelles. It is in these ionic forms that the mineral elements enter into the plants.

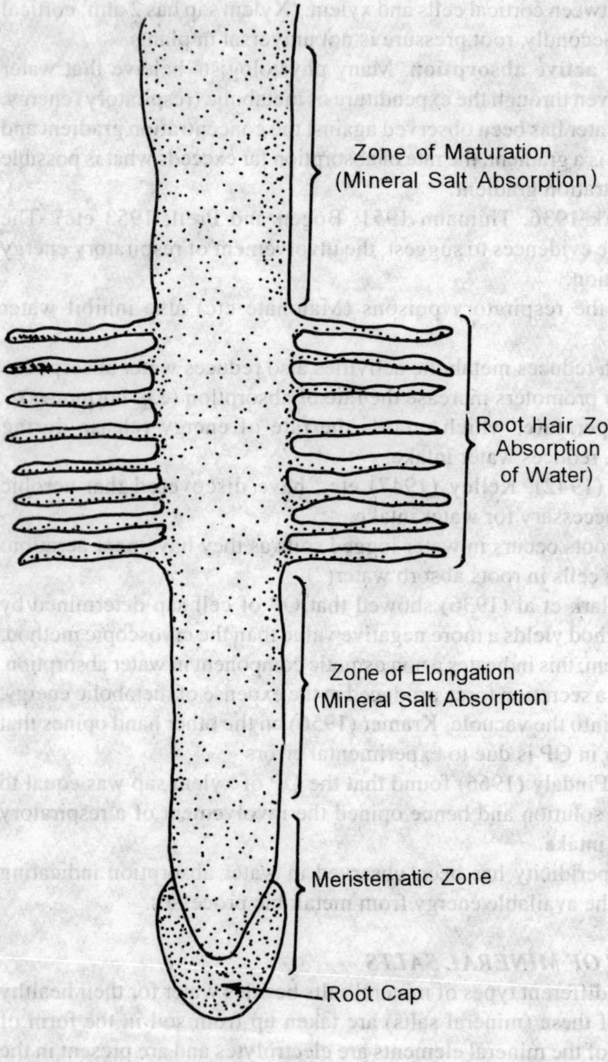

Fig. 11.5 Diagram of root tip showing the site of mineral salt absorption

MECHANISM OF ABSORPTION OF SALTS

Previously it was thought that together with the water, salts also get into the plants and that no special force governs their entry. But some of the recent works have clearly shown the existence of different forces that govern the entry of salts into the plants. Broadly categorizing, the mechanism of mineral salt absorption may be classified into two viz., Passive absortpion and Active absorption. This classification is based on the assumption that no metabolic energy is expended in the passive absorption.

Passive absorption:

1. Ion exchange: The exchange of ions (anions or cations) from within the plant cells to the ions of soil is called ion exchange. By this method, the plant cell can exchange the unwanted ions for the required ones. Ion exchange is of two types. *Contact exchange* and *Carbonic acid exchange*. In contact exchange, the ions that oscillate around the root hair overlap with the ions that oscillate around the soil particle and there will be mutual exchange of ions. According to carbonic acid exchange theory CO_2 released during respiration combines with water to form carbonic acid. This acid disassociates into H^+ and HCO_3^- in soil solution. These H^+ ions may be exchanged for K^+ ions held by the soil particles. Since the exchange of ions here is mediated by carbonic acid, it is called carbonic acid exchange theory.

2. *Mass flow*: According to this theory, when water enters into the plant passively through the influence of transpiration, salts also get in. Kramer (1956) is one of the supporters of this theory.

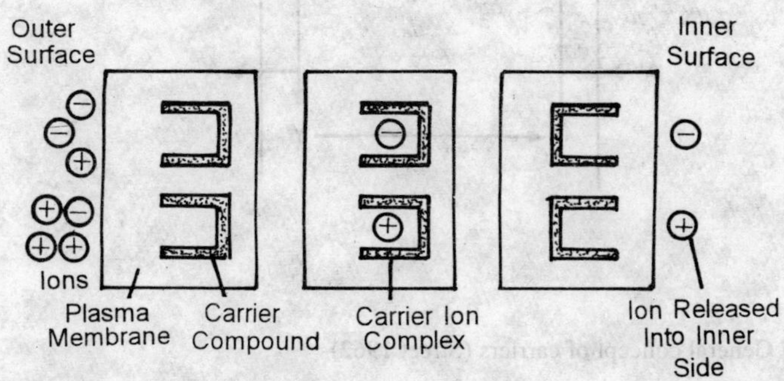

Fig. 11.6 Model of carrier concept

3. Diffusion: Initially when there is difference in concentration gradient (when the concentration of salts inside is less than outside) salts enter the plants through the simple process of diffusion.

4. Donan equilibrium: Normally the total amount of cations and anions present inside the cell are balanced with the ions outside and they would be in constant motion. But there are certain fixed ions in the plant cell which are immobile. In order to balance this more ions have to enter the plant cell to maintain the equilibrium. This is called Donan's equilibrium.

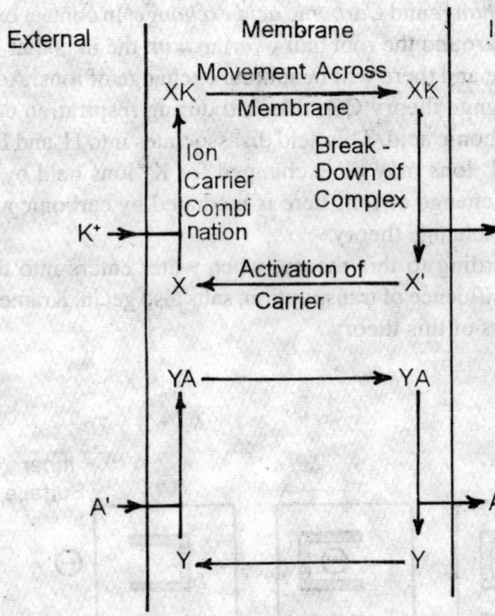

Fig. 11.7 General concept of carriers (Street 1962)

Active Absorption: The continued absorption of salts against the concentration gradient sometimes to the tune of concentrations several hundred times greater than the concentration outside gives credence to the ivew that some other energy, most likely metabolic energy is invovled in the uptake of salts. Indication of a direct relationship between salt uptake and respiration enhancement of salt uptake when rate of respiration increases and reduction of salt uptake when respiratory inhibitors are introduced prove the existence of active mechanism. Active absorption of salts is of the following types:

Carrier Concept: This theory assumes that the membrane is impermeable to ions alone. But there are certain compounds called 'Cytochromes' which act as carriers. The ions combine with the carriers to form an "ion carrier complex". This ion carrier complex can pass through the membrane. Within the cell, the ion carrier complex breaks up releasing the free ions. The carrier now is free to receive one more ion. Metabolic energy is supposed to be involved in the activation of the carrier. There are two hypotheses based on the carrier concept. These are Lundegardh's cytochrome pump theory and Bennet Clark's protein lecithin theory Cytochrome pump hypothesis viauslises the existence of a direct relationship between anion absorption and salt respiration. (The increase in the rate of respiration of a plant organ when transferred to a salt solution is called salt respiration). Lundegardh postulated that anions are actively absorbed via a cytochrome pump (series of cytochromes); here cytochromes act as carriers.

Bennet Clark's protein lecithin theory postulates that the carrier could be a protein combined with a phospholipid called lecithin instead of the cytochrome as visualised by Lundegardh. Their view is based on the prevalence of proteins and lipids in the membrane.

Twelve

Ascent of sap

In terrestrial plants whether they are prostrate plants, or herbs of about 1 or 2 feet high or tall trees reaching a height of more than 350 feet, the only source of water for all the cells, whether it is a cell of the root or a cell of the apical bud (sitting pretty on a pedestral of 350 feet high), is soil. The herb or a tree absorbs water from the soil and after reaching the lower reaches of xylem in the root, it (water) travels up. In the case of small herbs the distance to be travelled by water to reach the top is not much; it may be 2-3 feet at the maximum. But if we consider the situation in trees, where even the apical bud has to be served by water taken from the soil, water has to travel up a considerable distance. In other words, water has to be lifted up to a height of 300-350 feet. Indeed an enormous amount of force or energy is required to lift the water to such dizzy heights, and that too against the force of gravity. What type of energy is involved in this task? Does it involve biological energy or is it merely physical? Conclusive answers to these questions are still wanting. In this chapter we shall study the various theories that account for the upward movement of water (ascent of sap).

Path of ascent of sap: There is no doubt that, the water travels up along the vascular tissue, as it (vasculature) is ideally suited, both in structure and location to perform this task. But the vascular tissue has two components viz. xylem and phloem. Between these two which tissue is the path for translocation of water? Is it only xylem or phloem or both? The following experiment conducted with the balsam plant will show that xylem is the only element throught which water moves up. The balsam plant has a transparent stem so that the interior is visible.

Cut a few branches of the balsam plant and place them in a beaker containing coloured water. After some time coloured streaks appear in the transparent stems indicating the course of water. Cur a transverse section of the stem and observe it under the microscope; only the xylem elements are coloured indicating that they consitute the path for ascent of sap. Phloem is not coloured indicating it has no role in this. The fact that only xylem is the path for ascent of sap may further be confirmed by ringing experiment. In ringing experiment, a girdle of bark external to xylem is removed and it has been observed that this has no effect on ascent of sap.

Another point that needs confirmation in the path of ascent of sap is, whether the water moves in the xylem along the lumen or along the cell wall. Because, the cell wall of xylem having hydrophilic colloids can imbibe water and transport it to upper regions. This can be ascertained by dipping the balsam twig in molten paraffin for some time and then placing it in coloured water. On dipping in paraffin, the lumen of the xylem element gets blocked. No streaks appear in such a twig indicating that there is no upward movement of water, because of the blockade of lumen by paraffin. Hence it can be ascertained that water moves up to the plant along the lumen of xylem elements.

MECHANISM OF ASCENT OF SAP: The problem of the forces involved in ascent of sap has attracted the atention of plant physiologists, since a long time. A number of theories have been proposed from time to time. All these can be grouped under three headings. a) Vital force theories, b) Root pressure theory and c) Physical force theories.

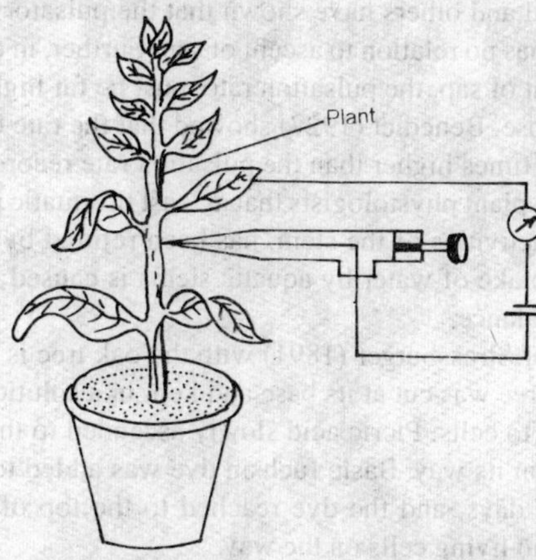

Fig. 12.1 Cresograph used by J.C. Bose

VITAL THEORIES: There are various opinions in the vital theories of ascent of sap as to the exact source of energy required for the movement of water. But all the proponents of vital theories are unanimous in that, the xylem, though it happens to be the path of ascent of sap does not have a role in the process. Instead, the living cells of the stem, with their vital energy play a dominant part in the same. Some of the proponents of vital theories are - Westermeir (1883-1884), Godlewski (1884), Janse (1887), Jagdish Chandra Bose (1932), Mollisch (1928-1929) etc.

Jagdish Chandra Bose (1923) an eminent physicist and a plant physioligist strongly supported the vital theory of ascent of sap. He was of the opinion that the innermost layer of cortical cells abutting the endodermis are of a special category as they are endodwed with rhythmic pulsations, which is electrical in nature. These cells are responsible for the pumping up of water into the xylem. Bose's theory, called *pulsation theory* envisages that the alternating contraction and expansion of these cells forces water into .he xylem. In order to prove the electrical pulsation of the innermost cortical cells, Bose invented an instrument called the *cresograph,* which consisted of an electric probe was attached to a galvanometre. When the fine needle of the probe was pierced into the cortex there were minor oscillations. When the probe reached the inner cortical cells however, the oscillations in the galvanometre indicator became violent and continuous. This showed the high electrical activity of the inner corticai cells. Bose beleived that these cells expand when they absorbed water from the outer cortical cells and after pushing water into the xylem they contracted. Each contraction and expansion took 14 secs to several minutes.

Molisch (1929, who supported Bose's thoery, experimentally showed, the electrical activity of the inner layer of cortex could be increased when treated with certain chemicals.

Shull, MacDougal and others have shown that the pulsatory activity of the inner cortical cells has no relation to ascent of sap. Further, in order to account for the rate of ascent of sap, the pulsation rate must be far higher than what is demonstrated by Bose. Benedict (1927) showed that the rate of ascent of water, is 8,000-30,000 times higher than the pulsation rate recorded by Bose.

The claim of some plant physiologists that at least in aquatic plants, ascent of sap is due to vital activities of the stem, has been refuted by Wilson (1947). Who argued that uptake of water by aquatic stems is caused by the readjustment of the water balance.

The experiment of Strassburger (1891) with the oak tree is worth mentioning. A 75 year old tree was cut at its base and kept in a solution of picric acid which is poisonous to cells. Picric acid slowly ascended to the top and killed all the living cells on its way. Basic fuchsin dye was added to the picric acid solution after three days, and the dye reached to the top of the cells, even though there were no living cells on the way.

ROOT PRESSURE THEORY: If a well watered plant is cut a few inches above the ground level, a sap exudes out with some force. This is called *sap exudation* or *bleeding* and is found commonly in most of the plants. Joseph Priestly was the first one to have observed this phenomenon. According to him a sort of hydrostatic pressure builds up in in the roots due to continuous absorption of water, and this is responsible for the upward flow of water. Stephen Hales (1727), who also observed this coined the term *root pressure* to explain this phenomenon. He fixed a tube of 8 mm diametre to the cut end of the stump and noticed that in a few days time, water can rise up to 6 metres. This is due to the pressure built up by the accumulated water. The water absorbed by the roots reaches endodermis, after travelling across cortex, where it meets with resistance for its smooth flow. As water continues to be absorbed, a hydrostatic pressure is built up, which forces water into the xylem resulting in its upward flow. This is called root pressure. Root pressure may be defined as the *hydrostatic pressure built up by water in the outer boundary of the stele due to continuous absorption.* Stephen Hales (1827) recorded a pressure of over one atm, in the roots. However it has been observed that root pressure is largely an osmotic phenomenon and disapperas, when isotonic or hypertonic solutions come into contact with roots, when there is a possibility of no gradient or reverse gradient away from the interior of the root. Root pressure may be demonstrated by attaching a mercury manometre to the cut end of the stem. Within a short time, the level of mercury rises up indicating upward movement of sap due to root pressure.

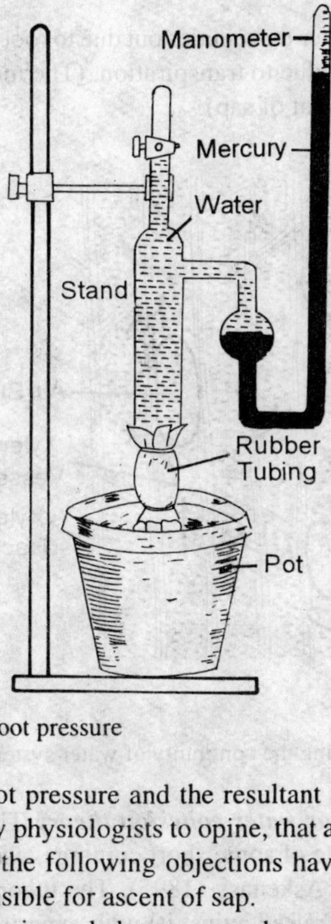

Fig. 12.2 Demonstration of root pressure

The phenomenon of root pressure and the resultant upward movement of water in xylem made many physiologists to opine, that ascent of sap is largely due to root pressure. But the following objections have been raised against root pressure being responsible for ascent of sap.

1. Root pressure is not a universal phenomenon, but ascent of sap is. In gymnosperms, which includes some of the tallest plants, with a high rate of ascent of water, little or no root pressure is observed.
2. There is no relationship between root pressure and ascent of sap in terms of rate of the process. During summer, when the demand for water is more and the rate of ascent of sap is high, a very low root pressure has been recorded in plants. On the other hand in autumn, when the rate of ascent of sap is low, a high root pressure has been recorded.
3. Ascent of sap continues even in the absence of roots.

4. The magnitude of root pressure is not sufficient to account for ascent of sap in such tall trees as *Eucalyptus* and *Sequoia.*

5. In plants which are actively transpiring, and require a regular supply of water to the distal regions, very low root pressure has been recorded. On the other hand, when the rate of transpiration is low a high root pressure has been observed.

6. The quantity of water that exudes out due to root pressure constitutes barely 5% of the amount lost due to transpiration. (The main source of water supply for transpiration is ascent of sap).

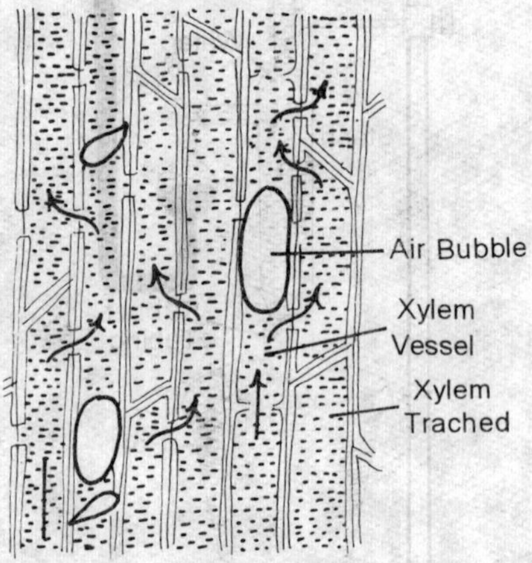

Fig. 12.3 Diagram showing the continuity of water system within the xylem

***Transpiration pull or water cohesion theory*:**This theory which has been widely accepted inspite of some short comings, was first proposed by Dixon and Joly in 1894 and Askenasy (1895). The theory was proposed again (to answer the objections raised aginst it) with experimental detail with the support of several plant physiologists like Renner (1911), Curtis and Clark (1951), Bonner and Galston (1952), Kramer and Kozlowski (1960), Milburn and Johnson (1966), Hammel (1967) and Levitt (1969). There are three main aspects in this theory. These are -

a) Cohesion of water molecules: Attraction and adhesion to one another is called cohesion. Water molecules have a strong cohesive force due to which they are held tightly to one another and cannot be easily separated. In addition to this, water molecules have an adhesive property also which makes them have a strong attraction to the inner wall of xylem.Hence water molecules cannot be easily torn away from the wall of the xylem. Some calculations have been made about the strength of cohesion and many physiologists agree that it is in the range of 350 atm far in excess of what is required for ascent of sap. However there are certain forces within the plant itself that work against the cohesive force of water these are - a) the weight of the column, b) resistance encountered by water column. All these add up to about 50 atm. But as has been pointed out already, the observed cohesive force far exceeds the forces that can break it. In order to answer some objections raised against the maintainability of water column, Dixon (1914) measured the cohesive force of water in the xylem sap of *Ilex aquifolium.* He took a J shaped glass tube and partially filled with xylem sap. The sap was heated until it almost filled the long arm. The tube was then sealed and gently inclined, so that the long arm was completely filled by the sap. Afterwards the tube was inverted to see, whether the water column would break in the long arm. The water column did not break due to the cohesive and adhesive force among its molecules. The water column would break, on cooling. Dixon tested several samples of the sap and observed the extent of cooling required to break the column. From this he calculated, that the sap had a cohesive strength of 133-207 atm. The works of Gremidge (1957), Loomis (1960) etc., have supported the existence of cohesive force among water molecules, but not as high are reported by Dixon. They opine that the maximum cohesive strength is about 30 atm. Considering the fact that even a force of one atm can maintain a height of 30 feet, 30 atms can account for about 900 feet. The water column can be streched also without breaking. According to Fischer, a force of 13,000 bars is required to break the water column.

b) Continuity of water column within the plant: According to Dixon and Joly, the entire water column in the plant, starting from root hair cells, cortical cells, xylem, to mesophyll cells of the leaf around a stoma is a continuous chain, so that any influence in the upper part or the lower part is felt by the entire column. The water columns present in different xylem vessels are not independent. They are laterally connected. In otherwords, the whole of the water present in the plant from root tip to stem tip is a single continuous system.

c) Transpiration pull or suction force: The water network has two terminals, one at the root tip near the absorbing regions and the other at the sub stomatal cavities in the leaf mesophyll. These two *menisci* of water represent the two extremities. Water evaporates from the mesophyll cells due to transpiration. This increases their DPD and as a result water flows into them from the adjacent mesophyll cells. The DPD of these cells increases and they attract water from the mesophyll cells abutting the vasculature, which (mesophyll cells) draw water from the xylem elements. The DPD of water in xylem increases as a result of continued drawal of water by the mesophyll cells. Continued transpiration in the leaf creates a tension (negative pressure) in the xylem elements in the upper parts. Since the water column is continuous and unbroken, this tension is transported down into the root up to the area of absorption and water moves up due to this force. In other words water is pulled up due to the suction force generated by transpiration. This may be likened to the pulling up of a rope (water column) through a tube (xylem).

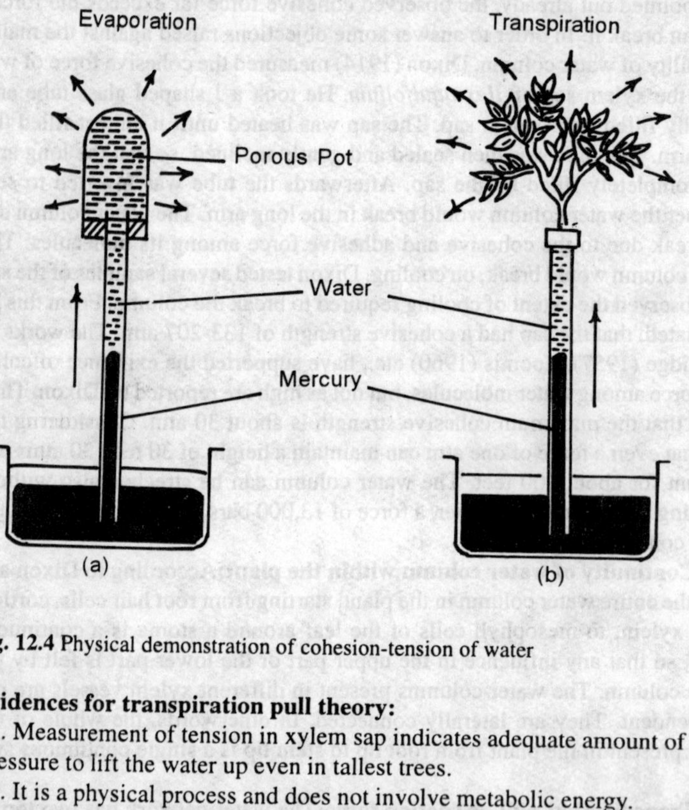

Fig. 12.4 Physical demonstration of cohesion-tension of water

Evidences for transpiration pull theory:

1. Measurement of tension in xylem sap indicates adequate amount of pressure to lift the water up even in tallest trees.
2. It is a physical process and does not involve metabolic energy.
3. The tensile strength of xylem sap is in the region of 25-300 atm which is sufficient to withstand any force that attempts to break it.
4. Kramer and Kozlowski (1960) made dendrographic measurements of tree trunks and showed a diurnal variation of the girth. There is decrease during day time as a result of shrinkage of cells. This shrinkage is due to water loss and the resultant tension in the xylem elements.

Objections: One of the serious objections raised against the continuity of water column is in view of several physical barriers in the form of tissues. In addition, a number of air bubbles present in the xylem sap disturb the continuity of the water column. Copeland (1902) stated that the air bubbles do not allow continuity of water column. Preston (1958) also expressed the same opinion. There is no doubt that bubbles do get into the xylem sap, but soon get dissolved in water column. Hammel (1967) has pointed out that the air blocks get dissolved in gymnosperms (growing in high altitudes) during melting of ice.

Another objection that is raised against the continuity of water column is that tress, whose trunks are cut at sides alternately on right and left still have an unceasing flow of sap. Postlethwait and Rogers (1958) have interperted this actually as a proof of continuity of water column in the face of disturbances. By using radio active isotopes (P^{32}) they have shown, that water molecules,whenever they face a dead end, avoid it, move internally and maintain the continuity of the water system. Scholander et al (1957), showed that whenever the stem is cut, the cut vessels may not function, but others continue to do so. Even in the cut vessels, water continued to move above and below the cut, apparently because it was in connection with water of the adjacent vessels. This proves that the water inside the plant represents a labrynthine system and any disturbances here and there are going to have only a local effect and not a general influence.

Thirteen

Loss of water

Water is one of the most important compounds that not only helps as a medium for various biological reactions, but goes into the structural framework of the cell in plants. A large quantity of water is absorbed by plants from the soil in order to perform various biological activities, but one peculiarity is, most of this water is evaporated into the atmosphere and only a fraction is retained in the plant system. It is one of the oddities of nature as to why it makes a plant absorb large quantities of water and again forces it to lose much of it into atmosphere.

The phenomenon of loss water in the form of vapour from the aerial parts of the plant body is called *transpiration.*

Magnitude (quantity) of transpiration: In order to understand the physiological impact of transpiration, it becomes necessary to know as to what is the quantity of water lost from the plant. This is not uniform for all the plants and also for different parts of the plant. There are various physical and biological parameters governing the amount of water loss from the aerial parts of the plant. The rate of transpiration is generally high during summer. It has been estimated that in trees like *Betula,* the amount of water lost due to transpiration during summer is as much as five times the fresh weight of its leaves. (This approximately amounts to about 15,000 litres of water per acre per day).

Types of transpiration: All the aerial parts of the plantbody without exception transpire; but their efficiency will vary. However the chief transpiring organ is leaf, even though stems, flowers etc., will also transpire. Hence the major type of transpiration in land plants is Foliar transpiration. In the leaf again, the epidermal cells, stomata etc., are the chief places where water loss occurs. Similarly in the stems transpiration takes place through lenticels. Based on these, transpiration is divided into three categories viz., Cuticular transpiration, Lenticular transpiration and Stomatal transpiration.

Cuticular transpiration: The surface of leaves and very often stems are covered by a waxy coating called cuticle which is actually a layer that is supposed to prevent loss of water. That is why in plants exposed to sunlight, the deposition of cuticle is heavy, while in plants of the shaded area it is thin. However, water still evaporates through the cuticle and it is known as cuticular transpiration. Cuticular water loss however is not much when compared with the overall rate of transpiration. In barely it amounts to about 3-10% of total transpiration. When compared with the overall rate of transpiration.

Lenticular transpiration: In woody plants, after the secondary growth, the epidermis is replaced by periderm whose outer layers are composed of dead cells (cork). In order to provide aeration to the inner living tissues, at certain places the cork tissue opens out and forms small slits called lenticels. Transpiration taking place through these lenticles is called lenticular transpiration. In comparison with other types of transpiration, this is very insignificant as it amounts to only 0.1% of the total.

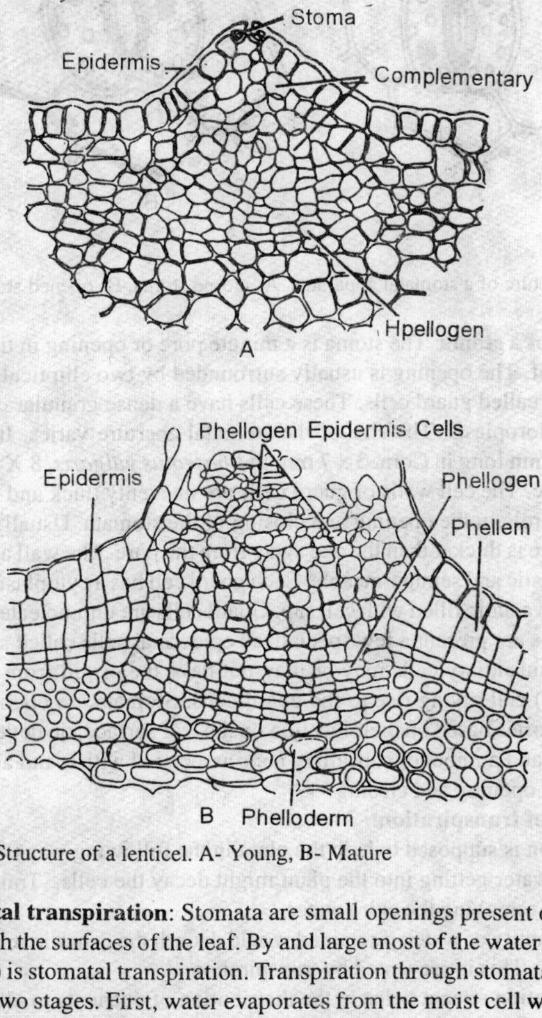

Fig. 13.1 Structure of a lenticel. A- Young, B- Mature

Stomatal transpiration: Stomata are small openings present on lower, upper or both the surfaces of the leaf. By and large most of the water loss in plants (90-95%) is stomatal transpiration. Transpiration through stomata is known to occur in two stages. First, water evaporates from the moist cell walls of mesophyll cells into the intercellular spaces and secondly from the intercellular spaces into the exterior through the stomata. Various aspects of stomatal distribution, morphology etc will affect the rate of transpiration.

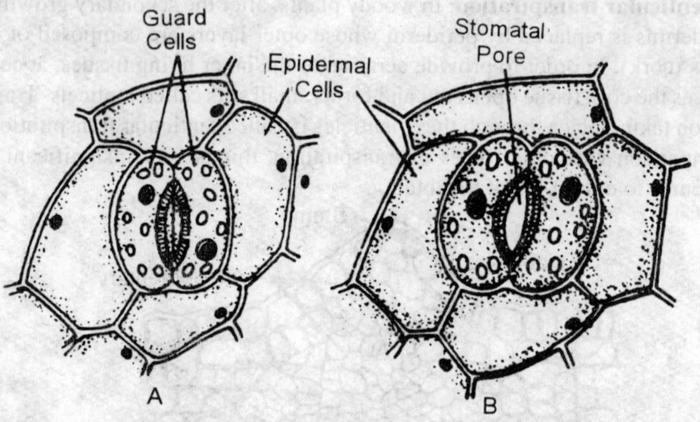

Fig. 13.2 Structure of a stomatal apparatus. A- closed stoma, B- opened stoma

Structure of a stoma: The stoma is a minute pore or opening in the epidermis of the leaf. The opening is usually surrounded by two elliptical (crescent shaped) cells called guard cells. These cells have a dense granular cytoplasm and some chloroplasts. The size of the stomatal aperture varies. It is 4 mm wide and 26 mm long in Corn. 3 x 7 mm in *Phaseolus vulgaris*, 8 X 38 mm in *Avena spp*. etc. The cell walls of guard cells are unevenly thick and this plsay an important role in the opening and closing of the stomata. Usually the wall facing the pore is thicker than the one away from the pore. The wall away from the pore is elastic and semipermeable. Each guard cell has a cytoplasmic lining and a central vacuole filled with cell sap. Guard cells are uninucleate. Close to the guard cells are present a few specialized epidermal cells called subsidiary cells. These subsidiary cells may or may not have the origin from the same (meristemoid) cell giving rise to the guard cells. Stomata are classified into various categories based on their location on the leaf surface, on development, on the nature and number of subsidiary cells associated with it and also on the periodicity of opening and closing.

Importance of transpiration:

Transpiration is supposed to help the plant in the following ways.

1. Excess water getting into the plant might decay the cells. Transpiration prevents it by removing the extra water.

2. Extra quantity of water, if retained would disturb the osmotic relationship between cells. This is prevented by transpiration.

3. Transpiration stream helps in the distribution of nutrients to all the parts of the plant body.

4. Fresh and cool water reaches all the parts of the plant body and this reduces the metabolic heat acting as a coolant.

5. Transpiration indirectly helps in the uptake of salts as the latter get into the plant together with water.

6. Upward movement of water (Ascent of sap) is due to the suction force created by transpiration.

Some of the recent researches however, have indicated that transpiration after all is not advantageous as was thought previously. The fact of water escape is a consequence of the anatomical construction of the leaf. The anatomical structure of the leaf is necessary for photosynthesis and respiration. When there is an escape route water also escapes. Transpiration does not remove excess of water. The plants in fact are forced to absorb more water because most of it will be lost by way of transpiration. That is why transpiration is often called a *necessary evil.*

Factors influencing the rate of transpiration:

Many external and internal factors influence the rate of transpiration. These are -

External Factors

Available soil water: The amount of water available in the soil governs the rate of transpiration. Less water in the soil would correspondingly retard the rate of the process.

Wind: Velocity of the wind is a very important factor in deciding the rate of transpiration. If the air is still, the water vapours that come out of the leaf remain in the vicinity. As a result of this, the exterior of the leaf gets saturated with water vapours and there is less scope for water vapours to come out. On the other hand if there is breeze the water vapours that come out are carried away from the leaf and more water vapours can come out.

Light: The intensity of the light is perhaps the most important single factor in deciding the rate of transpiration. Light has a direct influence on the opening and closing of the stomata. Lack of light closes the stomata and there will be little or no transpiration.

Atmospheric humidity: The water content in the atmosphere is an important factor, because if the atmosphere is already saturated with water, there is less scope for transpiration. Rate of transpiration is inversely proportional to the amount of atmospheric humidity in the area.

Temperature: Initially an increase in temperature would increase the rate of transpiration. But this is subject to a certain limit. Usually a temperature range that is necessary for other physiological activities is also suitable for transpiration.

Internal Factors:

Structural features of the plant: The anatomical and morphological features of the plant have a direct bearing on the rate of transpiration. Some of these are - Nature, number, size and distribution of stomata, presence or absence of cuticle etc.

Protoplasmic factors: Some of the protoplasmic factors such as enzymes have an influence on the stomatal opening and thus regulate transpiration.

MECHANISM OF TRANSPIRATION

Starch -sugar hypothesis

Lloyd (1908) discovered that the amount of starch in guard cells increases during night and decreases during day time. The reverse is true for sugar content of the guard cells. According to him the starch sugar interconversion is responsible for the opening and closing of the stomata. He believed that the change in the OP causes the interconversion of starch - sugar. Lottfield (1921) also discovered the same phenomenon in the guard cells, and he also reported that for the surrounding epidermal cells, the reverse (i.e. increased sugar and decreased starch during night) is true.

Sayre (1926) observed that stomata remain open when the medium is either alkaline or neutral even in dark and are closed in acidic pH even during day time. This means to say that the pH of the medium decides the interconversion of starch and sugar. According to Sayre (1926) utilization of CO_2 during photosynthesis causes the increase in pH resulting in the conversion of starch to sugar. Scarth (1932),Small et al (1942) supported the idea of Sayre. Scarth (1932) proposed that during day time CO_2 liberated by respiration is utilized by mesophyll cells in photosynthesis, which results in lowering of H ion concentration. Whenever the pH is high the enzymatic conversion of starch to sugar is favoured. In darkness, CO_2 accumulates in the intercellular spaces and increases the H ion concentration and decreases the pH promoting conversion of sugar into starch.

The discovery of the enzyme phosphorylase in guard cells by Hanes (1940) greatly supported the starch sugar interconversion theory. Subsequently Yung and Tun (1948), Yung and Willist (1954) have confirmed the participation of phosphorylase enzyme in the interconversion of starch and sugar in the guard cells. This can be represented as follows.

$$\text{Starch + Inorganic phosphate} \xrightarrow[\text{dark pH 5.0}]{\text{light + pH 7.0}} \text{Glucose 1. phosphate}$$

Starch is hydrolysed to form Glucose 1. phosphate at pH 7.0, but at pH 5.0 Glucose 1. phosphate condenses into starch.

Steward (1964) proposed a scheme which is slightly modified. According to him Glucose 1. phosphate should have its phosphate removed to be ultimately able to increase the concentration of the cell sap. The process requires the participation of biological energy. The process of opening and closing of stomata according of Steward's scheme is as follows-

Opening of stomata

Starch + phosphate $\xrightarrow{\text{Phosphorylase}}$ Glucose 1. phosphate

Glucose 1. phosphate $\xrightarrow{\text{Phosphoglucomutase}}$ Glucose 6. phosphate

Glucose 6. phosphate $\xrightarrow{\text{Phosphatase}}$ Glucose + phosphate

Closing of stomata

Glucose + ATP $\xrightarrow{\text{Hexokinase}}$ Glucose 1. phosphate

Glucose 1. phosphate $\xrightarrow{\text{phosphorylase}}$ Starch + phosphate

Objections:

Steward's theory has been criticised as follows.

1. Some guard cells (monocots) do not have starch, although they have other polysaccharides.
2. There is no evidence to show the presence of sugar at a time when starch disappears, when the stomata open.
3. In some instances starch is formed at a pH found during the opening of the stomata.
4. Often stomata close at mid day without any change in starch content.
5. The theory accounts for utilization of energy for closing, while opening does not require any energy.

Proton Concept

The proton concept of stomatal opening beleives in the proton transport mediated mechanism which governs both the stomatal opening as well as closure.

The proton concept may be said to be a synthesis of scarth's pH theory as well as the active K+ ion transport theory proposed by Walker and Zelitch (1963) : The proton transport mechanism is as follows.

Mechanism of stomatal opening: There are two mechanisms for the active transport of potassium ion which are supposed to take place in quick succession. These are -

1. Light induced transport of protons (H+) from the cytoplasm into the chloroplasts (during photosynthesis) creates a negative potential which eventually should stop continued movement of proton. As a result off this K+ ions from the surrounding cells move into the cytoplasm of the guard cells. This will once again increase the positive potential and the movement of proton into the chloroplasts continues. As a result of this the pH of cytoplasm (of guard cells) is raised to 8-9, while that of the chloroplast is reduced to 5. At a higher pH CO_2 changes to HCO_3 (in the cytoplasm).

2. Due to photosynthetic activity in the chloroplast, starch is produced and it is converted into phosphoenal pyruvic and (PEP) during glycolysis in the chloroplast. PEP diffuses into cytoplasm (of the guard cells) where in the presence of the enzyme PEP carboxylase it (PEP) reacts with HCO_3^- to form organic acids. These organic acids dissociate in the cytoplasm and release protons (H+). This in turn will favour the influx of K+ ions into the cytoplasm of guard cells increasing their osmotic pressure (OP). The increased OP will favour endosmosis resulting in the bulging of the guard cells and opening of the stomata. The influx of K+ ions (into cytoplasm) and efflux of protons is energized by H+, K+ ATPase pump. For every molecule of ATP hydrolyzed to ADP (near the plasma membrane) one K+ ion will enter into the cytoplasm in exchange for one H+ ion.

Mechanism of stomatal closure: During darkness H^+ ions start diffusing out of chloroplast into the cytoplasm. This continues till $(COO-)_2$ ions are available to combine with H+ to form $(CO\ OH)_2$ molecules. These organic acid molecules into chloroplast will cause a reverse influx of K+ ions. They move out of the guard cells into the subsidiary cells. This will decrease the osmotic pressure of guard cell cytoplasm. As a result, exosmosis takes place and guard cells shrink, bringing about the closure of the stomata.

Evidences for proton concept

1. Disappearence of carbohydrates from the guard cells during stomatal opening.
2. Low concentration of Co_2 (day time) favouring stomatal opening.
3. Increased pH causing stomatal opening.
4. Direct relationship between K+ accumulation and stomatal opening and efflux of K+ ions and stomatal closure.

GUTTATION

Under certain environmental conditions, water comes out through the aerial parts (leaves) in the form of drops (not in vapour form). This physical loss of water is called *Guttation.* When soil conditions favour, a high rate of water absorption and the rate of transpiration is low, guttation takes place. The term guttation was first used by Burgerstein, to describe loss of water from the leaves in liquid form.

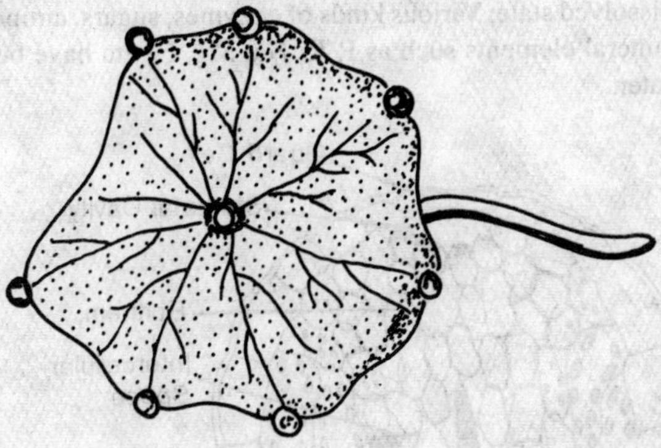

Fig. 13.3 Leaf of Nasturtium showing guttation

The magnitude of guttation in comparison with transpiration is very low, besided guttation is not universally observed. It has been reported in 333 genera belonging to 115 families. Even in these plants guttation occurs depending on the environmental conditions. *Colocasia antiquorum* is known to guttate upto 100ml of water per day. Guttatian is observed early in the morining, after a warm and humid night. Drops of water adhering to the tips or edges of leaves are often seen in grasses, tomato,*Nasturtium* etc, which are often mistaken for dew drops.

The exudation of water in guttation is due to positive root pressure. Just as transpiration takes place through stomata, guttation occurs through special structures called *hydathodes,* which are often called *water stomata.* Hydathodes are seen at the tips and margins of leaves. This is the reason why, water drops are seen at the tips and margins of leaves.

Structure of a hydathode: Each hydathode opens to the exterior by means of a pore in the leaf epidermis. Internal to the pore is the **epithem,** which is made up of loosely arranged thin walled parenchyma cells. The epithem usually occupies the tip of a vein. Surrounding the epithem is a mass of chlorenchymatous tissue.

Haberlandt, has recognised two types of hydathodes, viz. **epithem hydathode** and **active hydathode.** In the former water loss it by the force of root pressure. While in the latter, water is removed by the force developed by the cells of the epithem.

As has already been pointed out, root pressure may be the main force responsible for guttation. Conditions that discourage root pressure such as cold, dry aerated soil tec., reduce guttation also. Mineral deficiency also is known to affect guttation.

Guttation water is not pure in the sense, it is known to contain various compounds in dissolved state. Various kinds of enzymes, sugars, amino acids, vitamins and mineral elements such as P, K, Na, Mg, Fe etc have been found in guttation water.

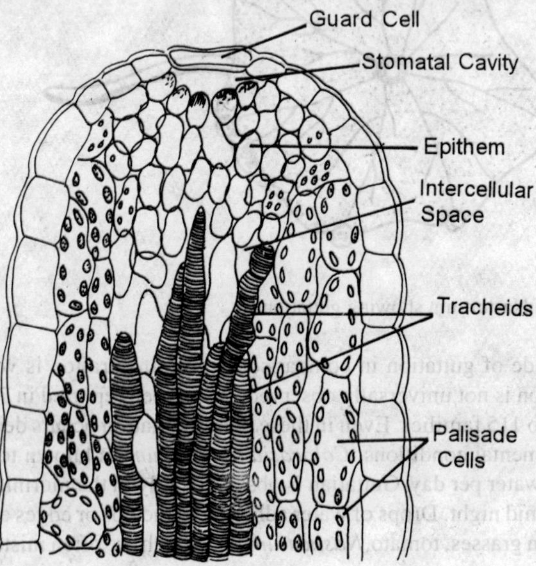

Fig. 13.4 Structure of a Hydathode

Guttation sometimes shows endogenous rhythms. It ceases during day and nearly always occurs at night. In terms of significance, guttation is of no consequence to plants. Occasionally, it may even cuase injury to the leaves, by salt deposition, which happens when the water drops dry out. This creates a favourable situation for the entry of pathogenie bacteria, fungi etc.

Fourteen

Translocation of organic solutes

In lower plants, with very little tissue differentitation, the site of food preparation is also the site of utlization. In aquatic plants like algae, the absorption of inorganic nutrients and water is carried out by the entire general surface and the whole of the general surface again is involved in photosynthesis. In higher plants however with complete tissue differentitation, the site of food preparation (leaves) and the site of its (food) utilization or storage are often separated far apart. Hence the orgainc food has to be transported to these areas. The phenomenon of food transportation from the site of tis synthesis to the site of its utilization is known as Translocation of organic solutes. The term solutes is used, because food materials move (dissolved) in a solution. In the process of translocation, the movement of solutes from one cell to another which generally can occur in any part of the plantbody is not included.

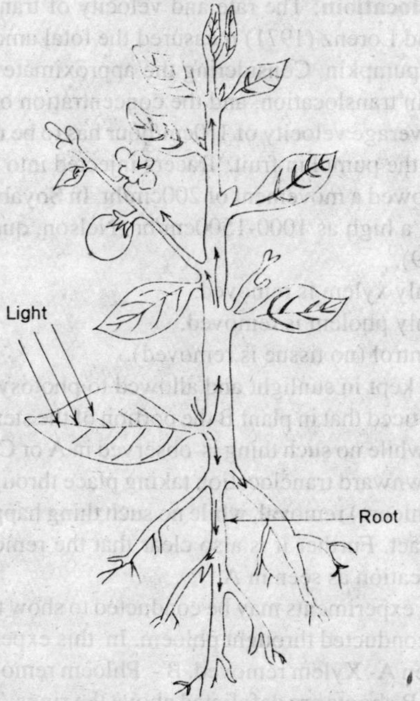

Fig. 14.1 Path of movement of organic solutes within the plants

The problem of the distance encountered by the migrating solute may not be much in herbaceous plants, but in the case of trees like, *Eucalyptus* and *Sequoia* it is indeed enormous because solutes many a time have to travel a distance of 300-350 feet.

Historically, studies on transolcation perhaps began as early as the middle of the 17th cent, even though the knowledge at that time was cursory. It was thought at that time that the 'elaborated sap' moved in the downward direction through the 'wood' to the roots.

Hartig (1837) was the first person who initiated studies on translocation. He gave the anatomical and physilogical description of the tissues involved in translocation of organic compounds. He identified seive tubes in the 'bark' tissue and attributed a role to it in nutrient transportation. It was he, who first demonstrated the collection of nutrients and the consequent bulging of the stem above a girdle (for details of girdling see later in the same chapter). Several scientists of 19th century after Hartig had engaged themselves in the study of translocation of organic solutes. These include Sachs (1891) etc. Among the modern plant physiologists who have contributed a lot for the understanding of the translocation,the names of Schmidit (1914),Curtis and Clark (1917-1950), Munch (1927), Masan and Maskel (1928), Esau (1966), Canny (1963), Fansan (1972) etc., are worth mentioning.

Rate of translocatioin: The rate and velocity of translocation varies with plants. Crafts and Lorenz (1971) measured the total amount of organic material in a mature pumpkin. Considering the approximate diametre of the stem tissue invovled in translocation, and the concentration of the solute, they calculated that an average velocity of 110cm/hour has to be maintained during the development of the pumpkin fruit. Tracers injected into the stem during peak development showed a movement of 200cm/hr. In Soyabean, the rate of translocation may be a high as 1000-1500cm/hr (Nelson, quoted in Bidwel, plant physiology, 1979).

Plant A Only xylem is removed.
Plant B Only pholem is removed.
Plant C control (no tissue is removed).

The plants are kept in sunlight and allowed to photosynthesize. After some time it will be noticed that in plant B the portion of the stem immediately above the ring bulges, while no such thing is observed in A or C. This is explained on the basis that downward tranclocation taking place through phloem is affected in B due to its (phloem) removal, while no such thing happened in A or C as the phloem was in tact. Further it is also clear that the removal of xylem has no effect on translocation as seen in A.

Similar ringing experiments may be conducted to show that the upward translocation also is conducted throught phloem. In this experiment also three sets of plants are taken A- Xylem removed, B - Phloem removed and C-control. In the plants A and B shoots are defoliated above the rings. The cut surface of the stem is enclosed in a glass cylinder filled with water to keep the tissues moist. Plant A in which only xylem is removed show elongation of the stem indicating undistrubed movement of solutes throught phloem. Plant B shows no elongation indicating stoppage of solute flow as phloem has been removed.

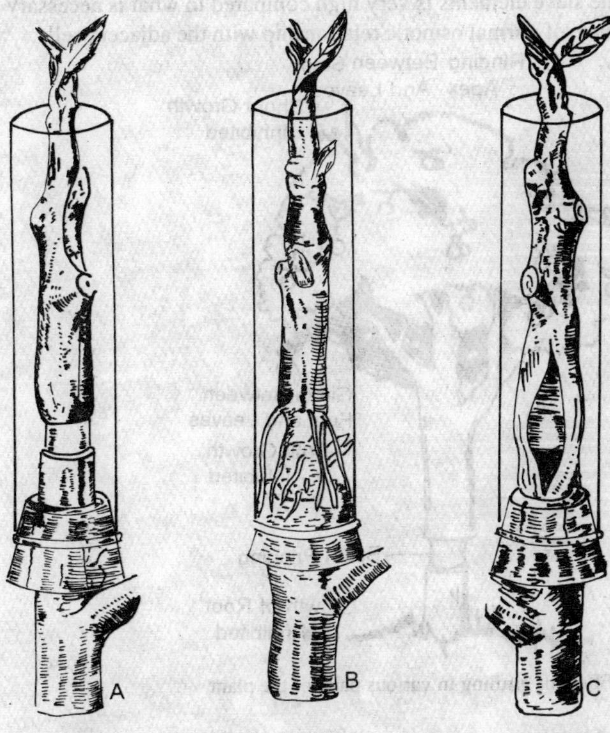

Fig. 14.2 Ringing experiment A-B Phloem tissue removed. Note the production of roots above the ring. C- xylem tissue removed.

Structure and distribution of phloem: The phloem tissue is composed of sieve tubes, companion cells and phloem parenchyma. The seive tubes are elongated like the vessels of xylem and are joined end to end forming a continuous system throughout the plant. The perforated seive plates allow for cytomplsmic connections to help in conducting the solutes. The presence of a second group of phloem as seen in bicollateral vascular bundles (cucurbitaceous plants) found in thin stems with large leaves and large sized sieve tubes seen in large fruits is an aditional evidence for phloem being involved in translocation.

Effect of blocking phloem: Occassionally the sieve proes are blocked by callose deposits resulting in stoppage of translocation.

Chemical analysis of phloem sap: Chemical analysis of the sap extracted from phloem reveals that there is a high concentration of soluble sugars. Since the carbohydrates are translocated in the form of sugars, it clearly indicates the phloem being the tissue for translocation. Further, the concentration of sugars seen in the sieve elements is very high compared to what is necessary for the maintenance of normal osmotic relationship with the adjacent cells.

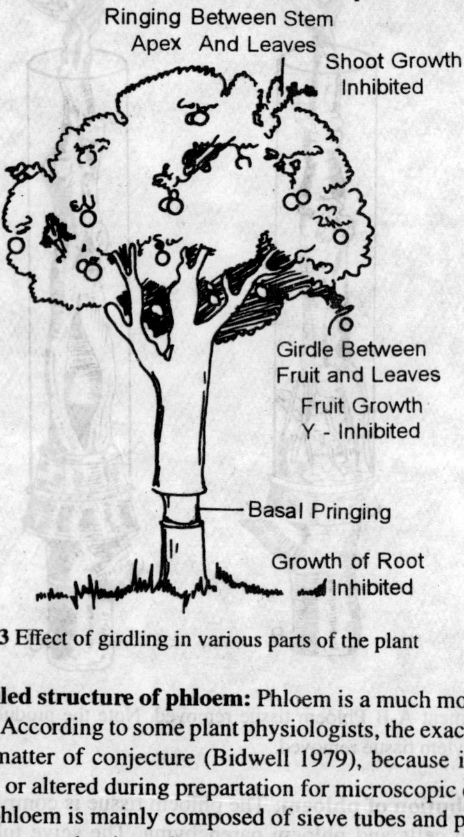

Fig. 14.3 Effect of girdling in various parts of the plant

Detailed structure of phloem: Phloem is a much more complex tissue than xylem. According to some plant physiologists, the exact structure of phloem is still a matter of conjecture (Bidwell 1979), because its fine structure is destroyed or altered during prepartation for microscopic examination.

The phloem is mainly composed of sieve tubes and phloem parenchyma. In Angiosperms, the sieve elements are always accompanied by companion cells, while in gymnosperms albuminous cells replace companion cells. Companion cells are living cells and are nucleated. The sieve elements however are enculeated, but unlike the vessel elements of xylem the sieve elements have protoplasm. They can be plasmolysed indicating that they are bound by a selectively permeable membrane, but they do not have a tonoplast. Much of the protoplasm present in sieve tubes is in the form of P-protein, a fibrillar protein.

The exact structure and funciton of this P-protein is still to be unravelled completely (Bidwell 1979). According to Esau, in mature sieve tubes, the cytoplasm and vacuole become a single system called *myctoplasm*. Ribosomes and dictyosomes are absent. Mitochondira however are present, but they differ from the normal ones in having less cristae.

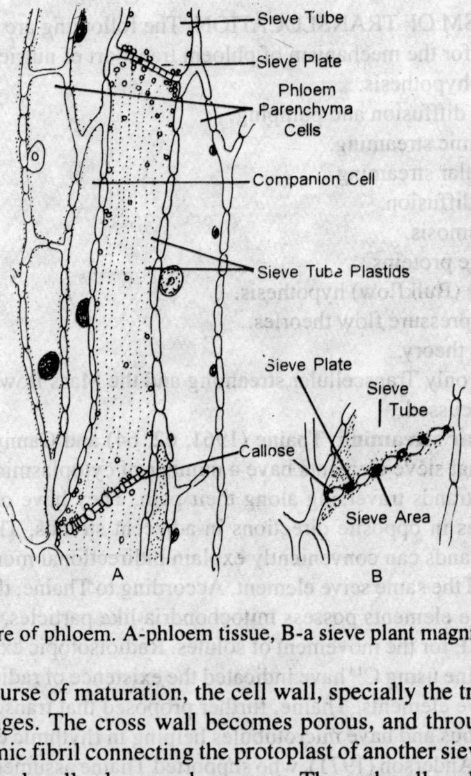

Fig. 14.4 Structure of phloem. A-phloem tissue, B-a sieve plant magnified

During the course of maturation, the cell wall, specially the transverse wall undergoes changes. The cross wall becomes porous, and through each pore runs a cytoplasmic fibril connecting the protoplast of another sieve tube. Occasionally the lateral walls also may be porous. The end walls, where the pores are many in number are called sieve plates. Occasionally the sieve pores are plugged with proteinaceous masses called slime bodies. In some sieve tubes however, the pores are partially blocked by callose (carbohydrate) deposits. It is generally beleived that in an actively translocating sieve tube the pores are largerly unplugged.

The phloem consists of phloem fibres, ray cells, sclerids etc; besides seive tubes and companion cells. These other cells however do not seem to play any role in translocation. Companion cells however are ontogenetically and physiologically related to the sieve elements.

The phloem mother cells cut off one or more companion cells before they mature into sieve tubes. Walls separating the two are thin and highly pitted. Death of companion cells renders the sieve tubes functionless. Since the sieve elements are enucleated it has been suggested that the companion cells have nuclear control over seive elements (Esau, 1939). According to Bieleski (1966), the companion cell and the sieve tube should be viewed as a single functional unit.

MECHANISM OF TRANSLOCATION: The following are the major theories to acount for the mechanism of phloem transport of nutrients.

1. Diffusion hypothesis.
2. Activated difffusion and pumping.
3. Protoplasmic streaming.
4. Transcellular streaming.
5. Interface diffusion.
6. Electro osmosis.
7. Contractile proteins.
8. Mass flow (Bulkflow) hypothesis.
9. Modified pressure flow theories.
10. Bimodial theory.

In this book only Transcellular streaming and the Mass flow hypothesis of Munch are discussed.

Transcellular streaming: Thaine (1961, 62, 64) and Canny (1962) are of the opinion, that sieve elements have a number of cytoplasmic strands called transcellular strands traversing along their path. They have observed movement of solutes in opposite directions in adjacent strands. The existence of transellular strands can conveniently explain bidirectional momement of solutes in one and the same seive element. According to Thaine, the transcellular strands in sieve elements possess mitochondria like particles, which provide energy rich ATP for the movement of solutes. Radioisotopic experiments conducted by Thaine using C^{14} have indicated the existence of radioactive strands within the seive elements. Thaine, further proposed that transcellular strands are proteinaceous and have microtubules helping in rhythmic contraction.

Aikman and Anderson (1971), who supported Thaine assumed that peristalic contractions occur as series of waves along the length of transcellular strands when there is movement of the solutes.

The theory of transcellular streaming is very appealing as it explains easily the bidirectional flow of solutes.

Several objections have been raised against this theory. These are:-

a) How biological energy is transduced into a physical force in a sieve tube?

b) What is the nature of the transcellular strands? Are they solid or tubular?

c) Whether the solute particles move in or on the strands.

d) Why the transcellular strands are not seen when sieve tubes are observed under electron microscope?

Mass flow hypothesis: Mass flow or bulk flow of solutes in sieve elements was first proposed by the German Scientist Munch (1931). Munch visualises a pumping action in mesophyll cells. According to Munch's hypothesis, the sieve tubes are connected to one another by means of cytoplasmic connections, forming a continuous system called symplast. The symplast is impermeable on the outerface.

Mass flow of solutes as per this hypothesis is as follows. A very high concentration of nutrients is present in the mesophyll cells due to two reasons - a) water is transpired continuously from the mesophyll cells and b) food is continuously synthesised. Solutes are actively loaded into the phloem thus increasing their osmotic potential. This increased potential draws water from the xylem vessels by means of osmosis. A hydrostatic pressure is built up in the sieve tubes. In the cells of the root, food material is continuously utilized reducing their osmotic potential as well turgor pressure. Thus a turgor pressure gradient is established resulting in the mass flow of solution from supply end (leaf) to consumption end (root). At the consumption end water diffuses into the xylem to be transported up to the leaves. Thus a circulatory system is established.

The principle of mass flow may be illustrated with the help of a simple experiment. Two chambers A and B with selective permeability are connected by a tube T containing water (fig 3.28) Here chamber A is comparable to mesophyll cells, chamber B to the root cells. Tube T is comparable to phloem. This forms a closed system similar to what is seen in plants.

Chamber A contains a highly concentrated sugar solution, while in chamber B its concentration is low. The system is dipped in a water filled vessel. In this experimental set up A corresponds to supply end, B to consumption end and T to longitudinal sieve tube system. The outer vessel having water corresponds to xylem. The concentration at A causes rapid diffusion of water into it resulting in a high turgor pressure. The solution from A therefore flows en masse to B until turgor pressure gradient is maintained. The flow stops, when turgor pressure both in A and B are equal. However if a high turgor pressure is maintained at A, there is a continuous flow from A to B. This is possible in the plant by the continuous synthesis of food. On reaching chamber B the water will flow into the outer vessel (xylem).

The main aspect of mass flow is that it requires a positive hydrostatic pressure in the seive tubes and a continuous supply of sugars in the leaves to generate enough pressure. Experimental evidences have clearly pointed to the presence of a positive pressure flow in phloem. An evidence in favour of this theory is that growth substances or viral particles are transported rapidly in sieve elements, when the leaves are illuminated. This suggests that sugars produced during photosyntheiss will help generate the pressure gradient.

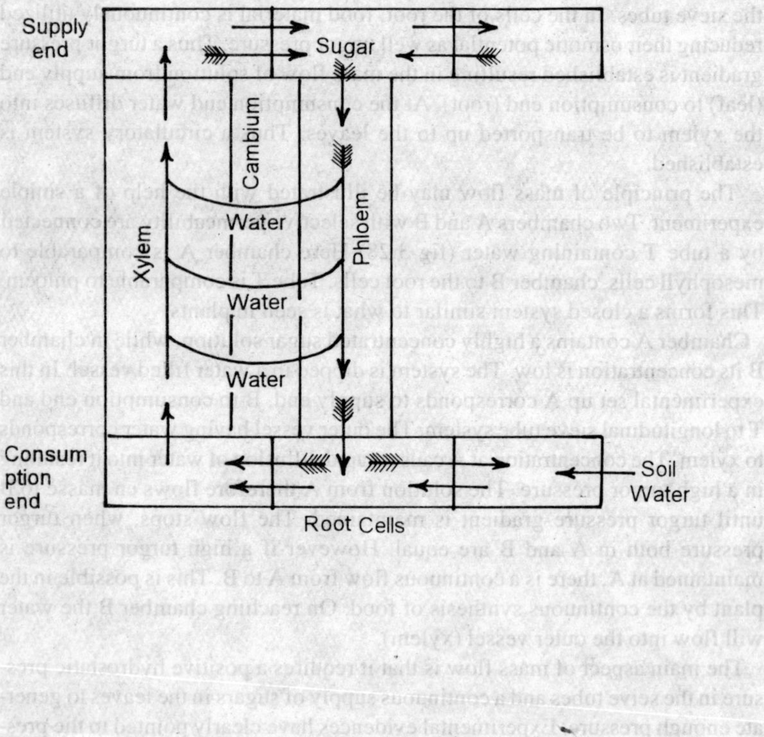

Fig. 14.5 Diagram to explain flow of solutes within the plant as per Munch's hypothesis

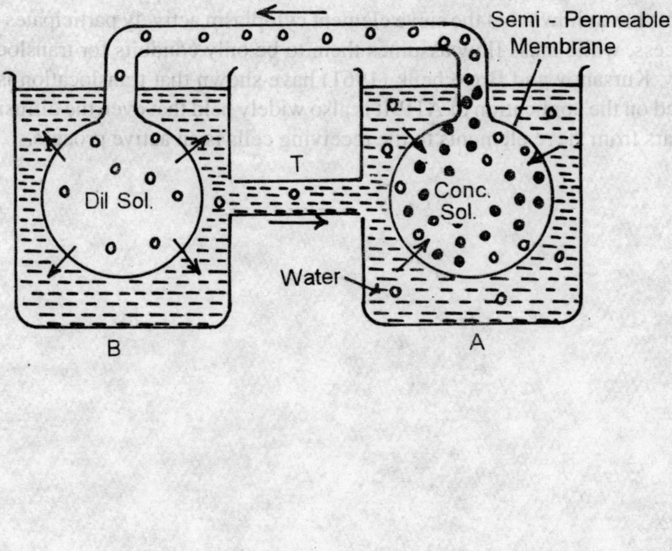

Fig. 14.6 Model to explain Munch's mass flow hypothesis

There are several objections to mass flow hypothesis. These are:-

a) Mass flow is possible only when the sieve pores are not plugged. But this is not always the case.

b) Mass flow can account for only a unidirectional movement. It is inconsistent with simultaneous bidirectional movement. But one need not always assume that bidirectional movement should take place in the same place and in the same sieve tube. It is quite possible that different sieve tubes are engaged in translocation in opposite directions simultaneously. This is compatible with mass flow (Crafts 1951).

c) Another objection against mass flow is that the observed osmotic pressure (OP) at the supply end and consumption end does not always assure a positive gradient. Many a time O.P of receiving cells is more than that of the supply cells.

d) Studies on translocation on the effect of low temperature on sieve elements have shown that the sieve element cytoplasm actively participates in the process, while mass flow assumes them to be only conduits for translocation flow. Kursanov and Brovchenk (1961) have shown that translocation is triggered on the application of ATP. It is also widely held that even the diffusion of sugars from sieve elements to the receiving cells is an active process.

Fifteen

Concept of Bioenergy

All living beings require energy to perform their vital activities. The energy for these life activities comes from solar radiation (radiant energy). The solar energy which is the basic source of energy cannot be directly utilized by all the organisms. The ability to utilize solar energy directly rests with only the green plants. All other organisms whether they be plants or animals have to obtain energy only through green plants. Thus green plants constitute the only entry point of the energy flow into the living system. What is bio energetics? The study of energy changes that occur when the energy enters into a living system is called bioenergetics. The energy first enters the living system as radiant energy, gets converted into chemical energy; eventually gets oxidized to release energy in the form of molecules of ATP which sustain all forms of biological activities including growth and development. Before we discuss further, the role of bioenergy in plant growth and development we will try to understand some of the physical principles of energy.

Energy and its forms

Energy may be defined as the capacity or the ability of a system to do work. When the system is at rest (ie. not working) also energy is there but it is not put into use. This energy is called *potential energy.* When the system is working the energy is put into use and it is called *Kinetic energy*.

Potential energy may be of various types - electrical, radiant (light), chemical, atomic etc. In the biological world potential energy of radiation is the primary source of energy which drives all the activities in the biosphere.

Free energy : The most important form of energy for the biological systems is the free energy. Free energy may be defined as the energy liberated or consumed in a reaction which can be used or supplied to another system to carry on the activities.

There are two kinds of reactions in which the free energy is either utilized or released. These are *endergonic* and *exergonic*. In endergonic reactions free energy is utilized - as in photosynthesis. In exergonic reactions - as in respiration free energy is released.

Energy conversions (Laws of Thermodynamics)

All energy conversions whether biological, or otherwise are governed by the general laws of thermodynamics. Thermodynamics literally means *dynamics of heat* as heat is the most convenient form of measurable energy.

These are two laws of thermodynamics wnich govern the behaviour of all energy in the universe.

The first *law of thermodynamics* or the law of conservation of energy states that the energy content of the universe must remain constant, since it can neither be created nor destroyed.

Biological systems absorb from their environment useful form of energy under constant temperature and pressure and return the same amount of energy in a less useful form to the environment. The useful form of energy absorbed by the biological systems is called *free energy* which is capable of doing work.

In biological systems when the energy is absorbed (by the green plants) in the form of radiation, only a little amount is put into use and gets converted into chemical energy which is used to build up the tissues or perform various metabolic activities . In this way the energy from sun changes the form i.e. in the form of biological units.

The second law of thermodynamics deals with the equilibrium since all natural processes have a tendency to reach the equilibrium. In general terms the second law states that at every stage of energy conversion the quantity of useful energy decreases; in otherwords a large amount gets converted into a form which cannot be put into use. For instance when fuel is used to heat water is a vessel, all the energy will not be used to heat water, a large amount is wasted as heat dissipated into the surroundings.

Energy that is dispersed into the environment is a wasted form because it is not available for work. This is called *entropy.* According to the second law of thermodynamics during series of conversions of energy, entropy or useless energy goes on increasing.

We can apply the second law of thermodynamics to the energy flow in an ecosystem (for details see chapter 6.2 - energy flow in an ecosystem).

Bio energy and its involvement in plant growth and development

There are two principal process in plants which are involved in energy conversions within the biological systems. These are photosynthesis and respiration. During photosynthesis which can be called an endergonic reaction free energy of the sun is utilized and converted into its chemical form and stored as potential energy in the bonds of organic molecules. In respiration which can be called an exergonic reaction the bonds of organic molecules are broken (mostly by dehydrogenation i.e. oxidation) and free energy is released.

Coupled reactions : The biological systems are dependent upon coupling of energy between exergonic and endergonic reactions. In coupled reactions there is always a molecular form which denotes energy and another molecular form which accepts energy. We will examine some exergonic reactions which provide usable energy to the biological systems to derive endergonic reactions.

The biofuels that an organism produces viz carbohydrates, proteins and fats constitute the source for starting exergonic reactions to release energy. How-

whenever their chemical bonds are broken energy is immediately stored in an intermediary. This intermediary is actually the one that runs the entire living system. The molecule of Adenosine triphosphate (ATP) is the universal energy currency. As the name indicates it has three phosphate groups attached to a nucleoside (Adenine + ribose sugar). Of the three phosphate groups the terminal one has a weak linkage. This breaks spontaneously whenever ATP forms a complex with an enzyme. This breakage of the bond releases chemical energy causing immediate shift in the bond energies giving rise to ADP (Adenosine diphosphate). During this process about 10 Kcal/mole of energy is released.

When the terminal bond of ATP is broken, the phosphate becomes inorganic phosphate (pi) and has low energy.

During the formation of ATP, pi is added to ADP utilizing the energy present in the energy rich electrons released during dehydrogenation (for details see respiration).

To sum up it can be said that whenever energy is required ATP participates donates its energy and gets converted into ADP. During the process of release of energy ATP molecules are produced from ADP by the addition of inorganic phosphate with energy.

Storage of energy ADP + pi + energy ——— ATP
utilisation of ATP ——— ADP+ pi +energy.

Growth and development in plants involve production of tissues and utilization of energy. As has already been mentioned the two process photosynthesis and respiration function as two pivotal processess which are involved in procuring, storing and releasing energy or plant growth and development.

Sixteen

Photosynthesis

The phenomenon of energy input from the abiological environment to the biological system is always through the plants-green plants. Green plants thus can be regarded as the basic fulcrum on which revolve all forms of life. In fact it is impossible to imagine what course the evolution of life would have taken but for green plants. All forms of life are dependent on green plants (directly or Indirectly) for their sustenance.

This unique phenomenon of energy input into the biological system is known as photosynthesis. The energy input has to be mediated only through the green plants for they alone are capable of absorbing the energy from solar radiation.

Source of energy: There are only two sources of energy on this planet. One is atomic energy and the other is solar energy. Of these two-atomic energy is not easily negotiable and available. Solar energy however is easily available in large quantities and living organisms have developed in the course of evolution mechanisms to trap and utilize this unique source of energy.

All forms of life however cannot directly harness solar energy. The capacity to trap solar energy is unique only for green plants. Green plants possess the green pigment viz., chlorophyll which can capture, transform, translocate and store energy which will be readily available for all activities of life.

Photosynthesis may thus be explained as the process in which the light energy will be converted into chemical energy. Except for green plants all other organisms cannot directly utilize the solar energy, hence they are dependent on green plants for their nutritional requirements. Green plants are called autotrophic (they can prepare their own food) while all other organisms are called heterotrophic (they cannot prepare their own food, there is however one group of exceptional organisms which are autotrophic in spite of being non green. These are the Chemosynthetic bacteria, which obtain energy by the oxidation of inorganic substrates such as ferrous ions and sulphur dissolved from the earth's crust or H_2S released from volcanic action. But considering the overall energy budget of life forms, the energy produced by chemosynthesis is very meagre and is of least significance quantitatively.

Definition: Photosynthesis (*Photo* = light) may be defined as the process in which CO_2 and H_2O are put together (by green plants) into an organic molecule (sugar) utilizing the solar energy.

$$CO_2 + H_2O \xrightarrow[\text{Green plants}]{\text{Light}} \text{Sugar (Carbohydrates)}$$

Photosynthesis may also be defined as the reduction of CO_2 to glucose ($C_6H_{12}O_6$) with the H necessary for reduction coming from water.

Photosynthesis involves both oxidation as well as reduction ultimately to synthesize energy rich compounds. Water is oxidized (electrons are removed) and the H released from water is used to reduce CO_2 to carbohyderates. Oxidation of water and the units of light energy (absorbed) produce energy intermediary compounds called ATP (Adenosine triphosphate) and reduced coenzymes called $NADPH^+$ —— H^+ (Nicotinamide adenosine diphosphate). These two compounds help in the fixation (reduction) of carbon dioxide into carbohydrates. These carbohydrates in turn constitute the starting point for the synthesis of various other metabolites necessary for the survival of the organism.

The definition and brief explanation of photosynthesis given above sums up the significance of the process. A few other definitions (different only in the usage of the words) of photosynthesis are given below:

1. "Biosynthesis of simple carbohydrates from CO_2 and H_2O in the presence of sunlight inside chlorophyll containing cells"

2."Production of carbon containing compounds from CO_2 and hydrogen doner by illuminaed green cells".

3. "Photosynthesis is a redox process"

4. "Photosynthesis is a sensitized photochemical oxidation and reduction mechanism between water (H_2 donor) and CO_2 taking place at biological temperature"

5. "Photosynthesis is the formation of carbon containing compounds from CO_2 and water by illuminated green cells, water and oxygen being the byproducts."

6. Photosynthesis deals with the capturing and transforming of light energy into chemical energy"

The overall reaction of photosynthesis may be represented as follows

$$6CO_2 + 6H_2O \xrightarrow[\text{Chlorophyil}]{\text{Light}} C_6H_{12}O_6 + 6O_2$$

Oxygen which is released as a byproduct originates from water (Ruben and Kamen, 1941). In order to account for this, the equation should be changed as follows

$$6CO_2 + 12H_2O \xrightarrow[\text{Chlorophyll}]{\text{Light}} C_6H_{12}O_6 + 6O_2 + 6H_2O$$

Ultra structure of the choloroplasts: The chloroplast which is the seat of photysynthesis in cells has a covering of two membranes with an inter membrane space. These membranes are smooth and there are no perforations or partticles. The membranes are differentially permeable (Mudrak 1969).

A section of the chloroplast reveals an intricate system of membranes enclosed in a granular matrix. These membranes are called *lamellae* and the surrounding matrix - the *stroma*. In a sectional view, the lamellae can be seen packed and these stacks are called *thylakoids*. In the chloroplasts of higher plants, the thylakoids themselves form highly copmact bundles called *grana*. Some thylakoids of granum extend into the stroma and maintain contact with other grana. These are called - Stroma thylakoids or stroma lamellae or intergrana.

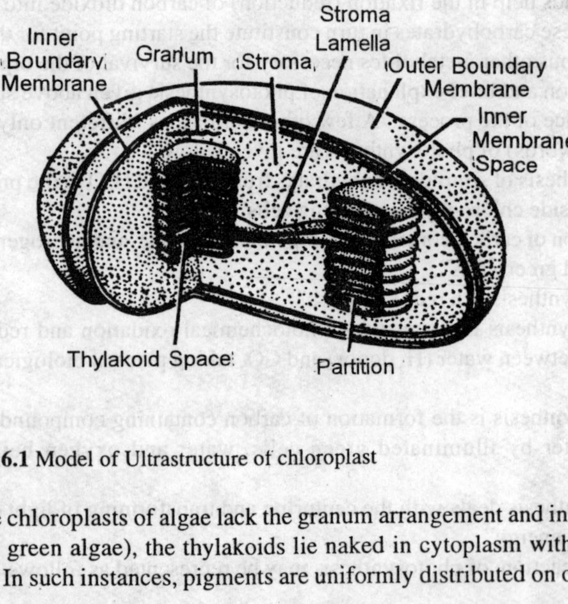

Fig. 16.1 Model of Ultrastructure of chloroplast

The chloroplasts of algae lack the granum arrangement and in cyanobacteria (blue green algae), the thylakoids lie naked in cytoplasm without any envelope. In such instances, pigments are uniformly distributed on or in the lamellae.

The stroma has in addition to the lamellae - granules (globuli), lipid droplets, starch grains and vesicles.

Strucure of the lamellae : All the photosynthetic pigments are concentrated in the chloroplast lamellae. The lamellae as well as the outer membranes are composed of lipoprotein sub units. Each sub unit according to Weir and Benson has a protein core surrounded by a lipid sheath. The chloroplast membranes have a single layer of sub units, whereas in thylakoids they are double due to the appressing of two membranes. The appressed areas are called partitions. These are hydrophobic in nature. The space bewteen the two membranes of a thylakoid (fret membrane) enclose a space called fret channel. The space between two thylakoids is called loculus while the ends of disc shaped thylakoids are called margins. These areas (loculus, fret channel) are hydrophilic. The chlorophyll molecules are concentrated in the fret channels ie. in the space between the two membranes of thylakoids. The chlorophyll molecules may be entirely in the space or the heads may be in the fret channel while the tails are buried in the sub units. Occassionally the entire chlorophyll molecule may be found in the fret membrane. Other pigments like cytochromes, carotenoids etc. are also found in the fret membrane.

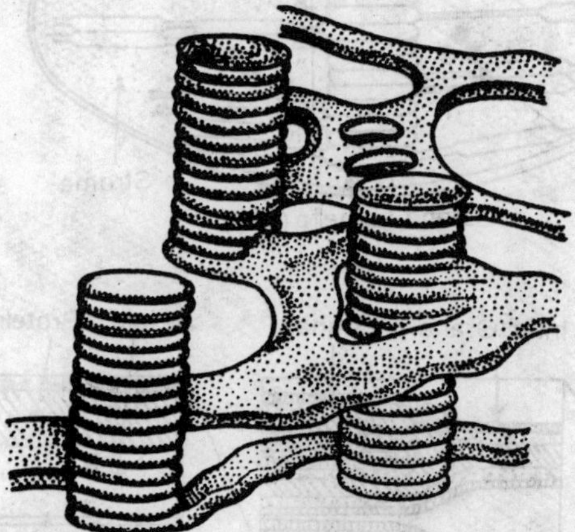

Fig. 16.2 Three dimensional view of Grana (enlarged view)

MECHANISM OF PHOTOSYNTHESIS:

Photosynthesis is an oxidation reduction process in which water is oxidised to H^+ and OH, while carbon dioxide is reduced to glucose with water and oxygen being byproducts. The mechanism has to account for the production of assimilatory power as well as its utilization to reduce CO_2 to carbohydrates. Evidences collected over the course of years have shown that there are two independent stages or phases in the mechanism of photosynthesis. These are the light phase(light reaction) and the dark phase (dark reaction). During light reaction the enrgy necessary for the reduction CO_2 is produced, while in the dark reaction CO_2 is reduced to carbohydrates utilizing the energy produced in light reaction.

Light reaction is also called photochemical decomposition of water and dark reaction is also called the thermochemical reduction of carbon dioxide. The existence of two phases in the mechanism of photosynthesis was reported as early as 1905 by F.F. Blackmann.

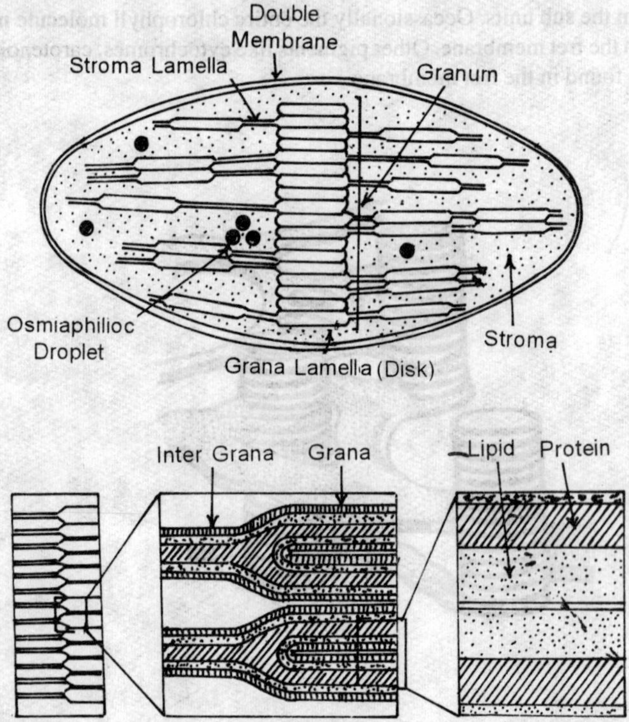

Fig. 16.3 Diagrammatic representation of ultrastructure of chloroplast showing the lamellae. A- entire chloroplast, B-D. show successive enlargement of a part of the granum.

Evidences for the existences of two phases: A number of evidences support the existence of two phases in the mechanism of photosynthesis. These are

a.*Temperature coefficient*: Studies on temperature coefficient conducted by Blackmann (for photosynthesis) showed the Q10 for light reaction was almost unity or Q10 =1. It does not show any increase in the rate with the rise in temperature. Blackmann's work further showed that Q10 is always 2 or 3 for the dark reaction. This was proved by the increase in the value of Q10 when the system was supplied with CO_2.

b.*Intermittent light*: Warburg (1920) observed that the photosynthetic yield is more when plants are illuminated by short flashes of light (followed by darkness) rather than continuous illumination. When green algae *Chlorella vulgaris* and *Scenedesmus obliquus* were subjected to intermittent flashes of about 1/16 sec, the photosynthetic yield was more rather than in a continuous supply of light of same intensity. This can be explained as follows.

There are two phases in photosynthesis-a light dependent phase and a dark phase. The two are interlinked in the sense dark phase can occur only when the products of light reaction are available. In intermittent light, flashes of light produce the products of light reaction , which are used up in the intervening dark phase to produce the photosynthetic product. Here the rate of eventual photosynthetic yield will be maximum as the products are immediately used up in dark reaction.

In continuous light there is maximum speed in light reaction and the products accumulate as they can not be used up in dark reaction as fast as they are produced. Accumulation of products of light reaction retards the rate of dark reaction (i.e production of carbohydrates)

c.*Dark reation of CO_2*: By far the best evidence for the existence of two phases comes from tracer experiments. Pre- illuminated leaves kept in darkness subsequently and supplied with C_{14} have been known to fix CO_2 eventually producing carbohydrates.

LIGHT REACTION

The light reaction or photochemical decomposition of water takes place in the grana region of the chloroplast. According to the current understanding, light reaction (photochemical excitation) takes place in about 10^{-9} seconds. Within this short span of time there will be the synthesis of molecule(s) of ATP and a molecule of NADPH. These two are utilized in the dark fixation of CO_2. In this section we will study the following aspects in detail with respect of light reaction.

a)Hill's reaction

b)Emerson effect and two pigment systems

c)Photophosphorylation

HILL'S REACTION:Hill(1937) and Hill and Scaribrick (1940) showed that the initial reaction of photosynthesis can take place even in isolated chloroplasts in the cell-free environment. When isolated chloroplasts (obtained from ground leaves) are supplied with light, water and a suitable hydrogen acceptor, release O_2 may be observed, when grana from disintegrated chloroplasts are suspended in water with ferric salts and benzoquinones acting as hydrogen acceptors. Besides these, chromates, indophenols etc also can act as H_2 acceptors. The various H_2 acceptors used by Hill are called Hill oxidants. Hill attributed the release of oxygen to the splitting of water into H and OH by light.

Hill's reaction proves the following:

a)The entire photosynthetic activity including the reduction of CO_2 takes place within the chloroplasts and

O_2 released during light reaction comes entirely from water.

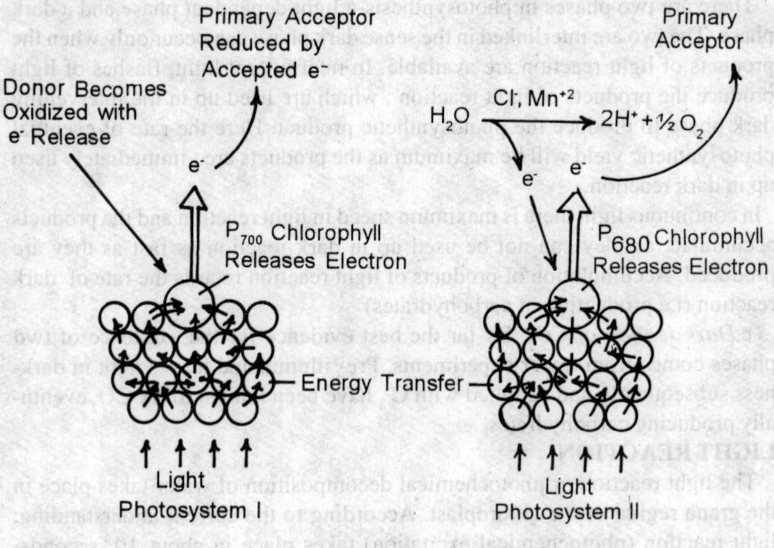

Fig. 16.4 Energy transfer in Photosystem I and Photosystem II in green plants

EMERSON EFFECT AND TWO PIGMENT SYSTMES:

Emerson and Lewis(1943) worked on different wavelenghts of light (monochromatic light) as to their quantum yield and came to the conclusion that 8 quanta of light are required for the reduction of one moelecule of CO_2 (or to release one moleucle of O_2). Accordingly the qunantum yield is 1/8 or 12 %. Since 4 electrons are required in the realeae of one molecule of O_2, Emerson opined that 2 light quanta are required to excite one electorn. Proceeding further, Emerson et al (1943) determined the quantum yield of different beams of monochromatric light and found that the average quantum yield (12%) suddenly dropped near the far red end of the spectrum. This decline in quantum yield is called red drop. Red drop begins at wavelengths greater than 680 nm in green plants and greater than 650nm in red algae. The phenomenon of red drop was also confirmed by the works of Haxo and Blinks (1950) on the red alga *Porphyra neregtis.*

The phenomenon of red drop was studied further by Emerson and Chalmers (1958), who found out that the reduction in the rate of photosynthsis due to red drop can be rectified by providing a beam of monochromatic light of shorter wavelength. In other words, by providing shorter wavelengths of light either simultaneously or quickly alternating with a longer wavelength of light, the rate of photosynthesis can be enhanced (effect of red drop is reversed).

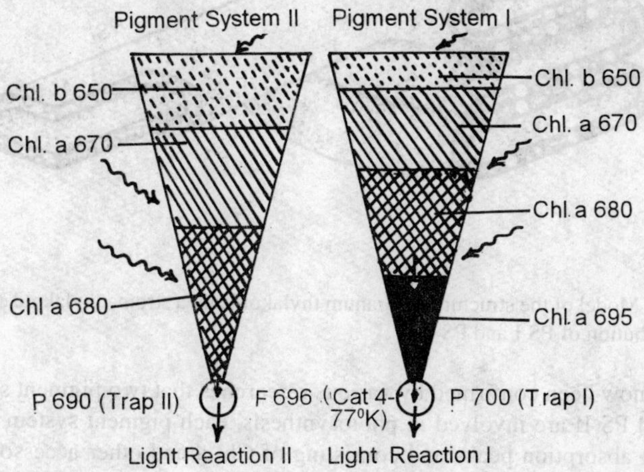

Fig. 16.5 Distribution of different pigments in PS I and PS II

This enchancement of photosynthetic rate under the influence of two wavelengths (long an dshort) of light is called Emerson effect. This may be represented as follows:

$$E = \frac{\text{Rate of evolution of } O_2 \text{ in double beam - Rate of Evolution } O_2 \text{ in red beam (short)}}{\text{Rate of evolution of } O_2 \text{ in far red beam (long)}}$$

Reason for Emerson effect

The discovery of the Emerson effect led to the idea that there must be two different pigment systems in the plants which differed in their absorption peak (of light). While one pigment system was more efficient in longer wavelength, the other was efficient in shorter wavelength.

Obviously under a specific wavelength (either short or long) the rate of photosynthesis will not be maximum as one or the other pigment system will be ineffective. Butler(1966) discovered two forms of chlorophyll viz Chl a 673 and Chl a 683 (with absorption peaks at 673nm and 683nm respectively).

Clayton (1966) discovered another form of Chl a with absorption peak at 700nm which he named P700.

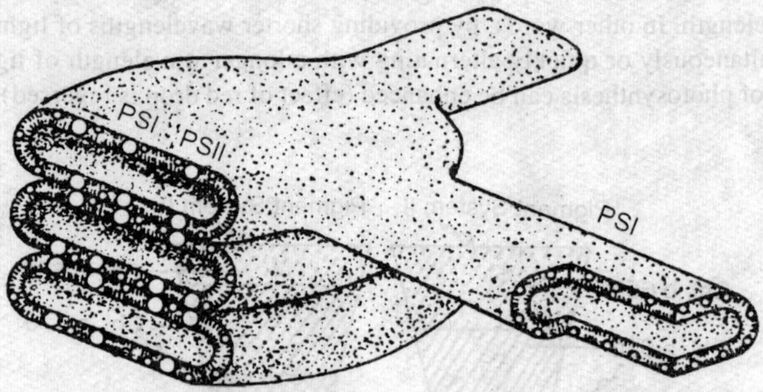

Fig. 16.7 Model of the structure of Granum thylakoid and a stroma thylakoid showing the distribution of PS I and PS II

It has now been confirmed by various researches that two pigment systems PS I and PS II are involved in photosynthesis, each pigment system having different absorption peaks and consisting of Chl a and other accessory pigments.

The two pigment systems may be removed from the thylakoids, by treating the chloroplast with mild detergents. These can be separated on polyacrylamide gel when they appear as discrete green bands. The details of the two pigments systems are as follows.

Pigment system I(PSJ)

According to Malkin (1982), Anderson(1982) and Barber(1983), PSI consists of Chl a 670, Chl a 680 and chl a 710(P 700), chl b and some b cartoenes. P700 is the reaction centre in PS I into which light energy is transferred by all other pigments. Besides, PS I also consists of iron containing proteins called Fe-S proteins and Cytochromes. PS I is weakly fluorescent.

Pigment system II (PSII):

It has several forms of Chl a with P680 (Chl a 680) being the reaction centre. A little amount of Chl b and b cartoenes are also present. There is also a colourless Chl a called Pheophyrin (Klima and Krasnovskii, 1981) which acts as the primary electron acceptor. Together with these are present quinones (Q) and manganese proteins.

PHOTOPHOSPHORYLATION: Arnon and Frankel (1954) showed that isolated chloroplasts, under illumination can produce ATP molecules. The reaction is believed to take place as follows

$$\text{ADP + Pi} \xrightarrow[\text{Chlorophyll}]{\text{Light}} \text{ATP}$$

(ADP = Adenosine diphosphate)
ATP = Adenosine triphosphate
Pi = Inorganic phosphate

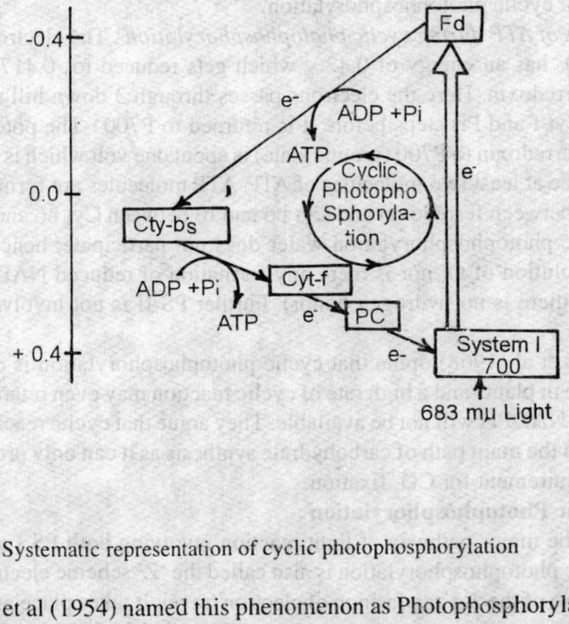

Fig. 16.7 Systematic representation of cyclic photophosphorylation

Arnon et al (1954) named this phenomenon as Photophosphorylation (addition of phosphate group to ADP under the influence of light energy). These ATP molecules are untlized to reduce PGA to PGAL during the fixation of CO_2. They are also required to phosophorylate Ribulose 5 phosphate to Ribulose 1.5 diphosphate (For details, see dark reaction). The following types of photophosphorylation have been discovered in plants.

1.Cyclic photophosphorylation
2.Non-cyclic photophosphorylation and
3.Pseudocyclic photophosphorylation

Cyclic Photophosphorylation : Here only PS I participates. The light energy absorbed by the light harvesting complex is transfered by inductive resonance to P700. P700 can accept light energy only of a wavelength longer than 680 nm. The excited electron of PQ700 is first accepted by an unknown acceptor (x) which in turn transfers its electron to FRS (Ferredoxin reducing substance) and then to ferredoxin. This causes ferredoxin to release one of its own electrons (energy rich), which shuttles through the chain consisting of Cyt b6, Cyt f, plastocyanin and finally back to P700. The P700 now returns to the ground state and is now ready to receive one more unit of radinat energy.

The electorn released from the chlorophyll (P700) is returned in a continuous chain i.e., it travels in a cyclic fashion, hence the name cyclic electron transport or cyclic photophosphorylation.

Synthesis of ATP during cyclic photophosphorylation : The electron ejected from P700 has an energy of 0.42 v which gets reduced to 0.417 when it reaches ferredoxin. Here the electrons passes through 3 down hill migration (Cyt b6, Cyt f and Pi) steps before it is returned to P700. The potential gap between Ferredoxin to P700 (ground state) is about one volt which is sufficient to reproduce at least two molecules of ATP. ATP molecules are formed at two states - a) between ferredoxin and Cyt b6 and b) between Cyt b6 and Cyt f.

In cyclic photophosphorylation water does not participate; hence there is neither evolution of O_2 nor is there any formation of reduced NADPH molecule (for there is no hydrogen donor). Further PS II is not involved in any stage.

Ramriez et al (1968) opine that cyclic photophosphorylation is of limited occurrence in plants and a high rate of cyclic reaction may even retard the CO_2 fixation as NADPH will not be available. They argue that cyclic reaction is not involved in the main path of carbohydrate synthesis as it can only provide part of ATP requirement for CO_2 fixation.

Non-Cyclic Photophosphorylation :

This is the major pathway of light reaction involving both PS I and PS II. Non-cyclic photophosphorylation is also called the 'Z' scheme electron transport because of the zig zag fashion of electron travel. It takes place as follows.

a) The radiant energy is trapped by PS I (P700) which ejects one if its electrons and is trapped by X, FRS and finally goes to ferredoxin. The reduced ferredoxin now transfers its electrons to NADPH. NADPH gets reduced. PS I however remains in the excited state as it has not got back its electron.

b) Meanwhile PS II absorbs light energy (680nm) and ejects one of its own electrons which is trapped by an unknown quinone acceptor (Q). From Q-the electron moves downhill to cyt b6, plastoquinine, cyt f, Plastocyanin (PC) and finally to PS I. PS I which is in an excited state, now returns to the ground state as it has got an electron from PS II.

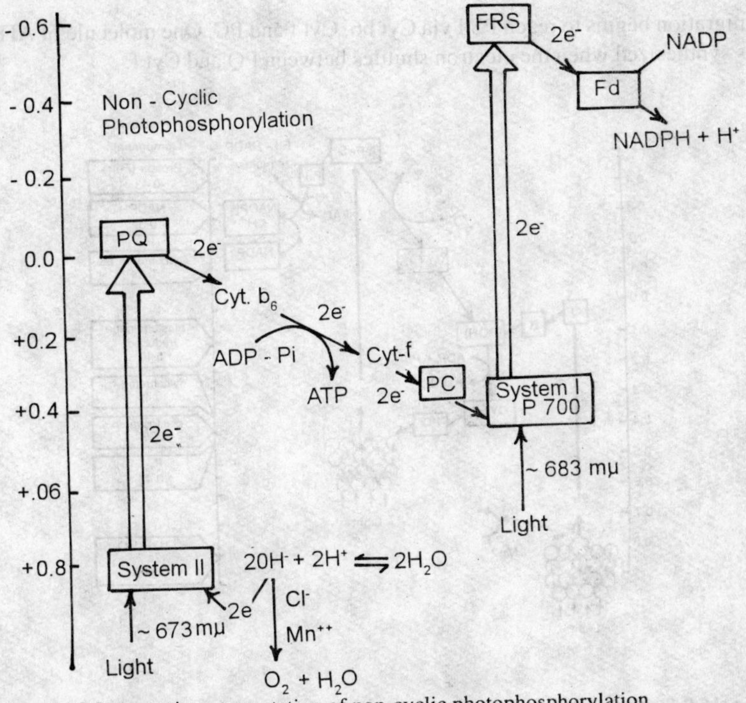

Fig. 16.8 Systematic representation of non-cyclic photophosphorylation

How does PS II come back to the ground state now that it has sent its electron to fill up the 'hole' in PS I ?. Because unless the PS II returns to the normal state, the system cannot continue functioning. The answer lies in the dissociation of water molecules. Water splits it into H^+ and OH^- ions. what is the mechanism of this splitting of water? Some physiologists call it photolysis of water, others are not sure of the mechanism but state that Mn^+ and Cl^- ions are essential for the breakdown of water. In any case the excited PS II returns to the normal state by getting an electron from the OH^- ions of water. According to Salisbury and Ross (1986), the excited PS II returns to the ground state by attracting an electron from an adjacent Mn- protein, which in turn gets an electron from the OH- of water. The H^+ ions released by water are accepted by NADPH to become NADPH $^+$ H^+. Thus a reduced NADPH is formed.

The electron transport system or photophosphorylation is called non cyclic because the excited PS I returns to the normal state by getting an electron from PS II, while the excited PS II returns to the ground state by receiving an electron from water. Hence the movement of electron is not in a cyclic fashion.

ATP synthesis during non-cyclic photophosphorylation: The electron (+ 0.8 v) released by PS II is accepted by a quinone from where the downhill migration begins to reach PS I via Cyt b6, Cyt f and PC. One molecule of ATP is synthesized when the electron shuttles between PQ and Cyt f.

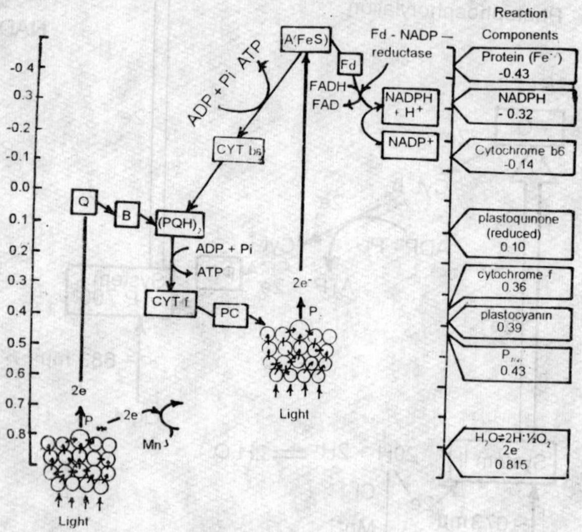

Fig. 16.9 Systematic representation of overall electron flow in light reaction combining both cyclic and non-cyclic photophosphorylation

Release of oxygen and formation of NADPH⁺ + h⁺: Four molecules of water split up for every turn of non-cyclic photophosphorylation. This is necessary (see fig 4.9) to provide four electrons to the excited PS II to bring it back to the ground state. Similarly, four H+ ions are required to reduce 2 molecules of NADPH.

These reactions take place as follows.

$$4H_2O \longrightarrow 4H^+ + 4OH^-$$

$$4OH \longrightarrow 2H_2O + O_2$$

(4 electrons from OH reach the excited PS II)

$2NADP+ \; 4H^+ \qquad 2NADPH^+ + 2H^+$

(four electrons are required to reduce 2 molecules of NADPH and these come from the excited PSI via ferredoxin)

Thus the end products of non-cyclic photophosphorylation are:

a) One molecule of ATP

b) Two molecules of reduced NADPH

c) 2 molecules of water and

d) one molecule of O_2.

DARK REACTION

The phenomenon of thermo-chemical redution of CO_2 was first established by Blackmann in 1905. He opined that light is not necessary for the reduction CO_2. In dark reaction the products of light reaction viz. ATP and $NADPH^+ H^+$ are used to reduce CO_2 into glucose. But the precise manner in which CO_2 is absorbed, and reduced resulting in the formation of a molecule of glucose was understood fairly recently thanks to the availability of radioactive isotopes. Utilizing the isotopes of carbon dioxide, one can trace the different stages of entry of CO_2 into the plant system untill it gets converted into glucose. Many physiologists, biochemists etc have contributed a great lot twoards th understanding of the path of carbon in Photosynthesis. The names of Bassham (1957-1964), Melvin Calvin (1961), Benson (1951) etc are worth mentioning in connection with the unravelling of the secrets of CO_2 fixation. Melvin Calvin was awarded the Nobel Prize in 1961 for tracing the CO_2 reduction cycle, which is also known as the Calvin cycle.

In CO_2 reduction cycle what we have to precisely understand is how the CO_2 enters the plant system, which is the compound that accepts the CO_2 molecule and how a molecule of glucose is snythesized and how the compound (that originally accepts the CO_2) is regenerated. In addition, the system should explain the acceptance of $6CO_2$ molecules, their reduction and formation of only one molecule of glucose for every turn of the cycle.

Modern researches have indicated that CO_2 is fixed in one of the following pathways in green plants.

1. C_3 pathway or Calvin Cycle
2. Hatch and Slack cycle or C_4 pathway
3. Glycollate pathway (C_2 plants)
4. Crassulacean acid metabolism (CAM plants)

CALVIN CYCLE:

This is the major pathway for the fixation of CO_2 in green plants. It is called Calvin Cycle owing to its discovery by Melvin Calvin of the university of California USA. It is also called the C_3 pathway because, the first stable compound formed after the entry of CO_2 into the plant system is a triose sugar (C_3) viz phosphoglyceric acid (PGA).

Some experimental details of Calvin's work

In order to understand the significance of the discovery of CO_2 reduction cycle, we should undetstand the experimental methods followed by Calvin. To trace the various steps of conversion of CO_2 after its entry into the plant system untill the formation of glucose, Calvin allowed the experimental plants to photosynthesize for various periods of time and analysed the various compounds formed after different periods of photosynthesis.

Calvin chose for his experimentation two green algae - *Chlorella pyrenoidosa* and *Senedesmus obliquus*. The plants have the same pigmentation as higher green plants and can be easily cultured and harvested. To make sure that the compounds to be analysed can be easily traced and not mistaken for the compounds already present in the plant, the experimental plants were starved of CO_2 and later supplied with a radioactive isotope of CO_2 (C_{14}). Only those compounds in which C_{14} was incorporated were taken into consideration because they were produced after the experiment began. By varying the time of photosynthesis (1/16 sec to 5 sec) and then killing the plant to stop photosynthesis and analysing the intermediate compounds formed, Calvin and his co-workers were able to map the whole of the CO_2 reduction cycle.

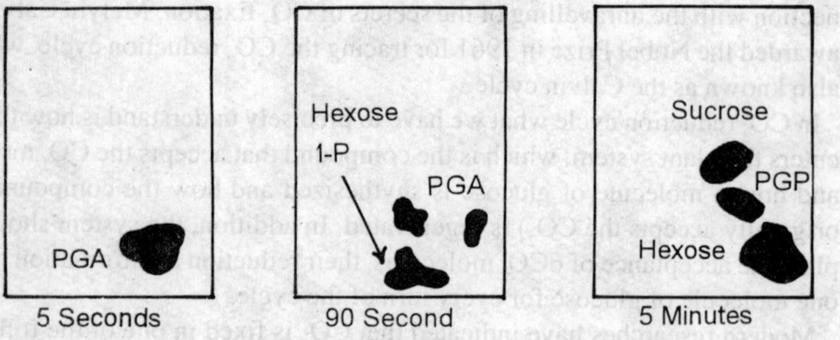

Fig. 16.10 Autoradiograph (Diagrammatic representation of chromatograma showing the fixation of C^{14} after different periods of exposure of light

Two, very simple but efficient tools namely Chromatography and Radioautography helped Calvin to trace the path of carbon in photosynthesis.

Chromatography, in particular paper chromatography is an ideal technique for the separation of compounds present in a small mixture. The extract (mixture) of the plant to be analysed is placed as a spot on a filter paper and dipped in a suitable solvent. The solvent travels on the paper due to capillary action and carries with it the different substances in the extract. The rate of migration of the different substances depends upon their solubility and specific gravity.

The filter paper (Chromatogram) should be removed from the solvent before it reaches the edge of the papte and dried. By spraying a suitable reagent (For e.g ninhydrin for amino acids) the compounds can be detected as several spots widely separated on the paper. The substances may be idetified by their relative positions on the Chromatogram or by comparing with a standard Chromatogram. (In a standard Chromatogram compounds whose identity are known are spotted and their position is marked after the solvent run).

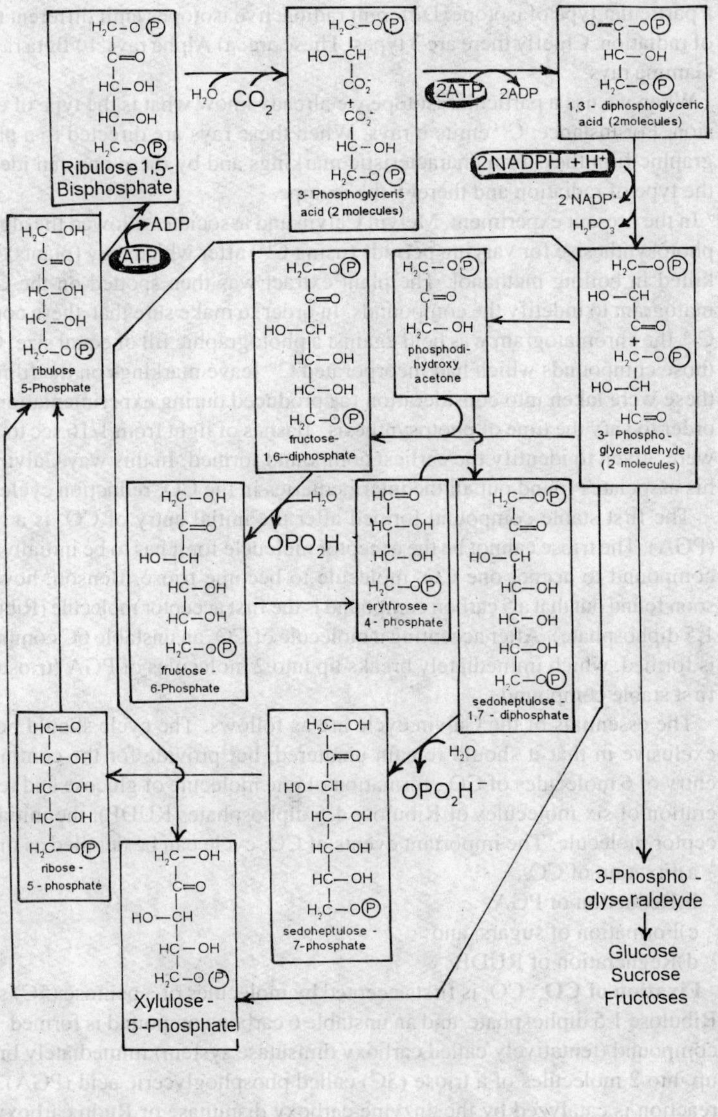

Fig. 16.11 Systematic representation of C^3 pathway (Calwin cycle)

Radioautography is a technique to find out, if a particular source consists of a particular type of isotope. Different radioactive isotopes emit different types of radiation. Chiefly there are 3 types. These are: a) Alpha rays, b) Beta rays c) Gamma rays.

When we use a particular isotope we already know what is the type of emission. For instance, C^{14} emits b rays. When these rays are directed to a photographic film they form characteristic markings and by these one can identify the type of radiation and thereby the isotope.

In the present experiment, Melvin Calvin and associates allowed the algae to photosynthesise for varying periods (using C^{14}) after which they (plants) were killed in boiling methanol. The plant extract was then spotted on the Chromatogram to indetify the compounds. In order to make sure that these contain C^{14}, the chromatogram was held against a photographic fill of equal size. Only those compounds which had incorporated C^{14} leave markings on the film and these were taken into consideration (as produced during experimentation). In order to vary the time of photosynthesis , flashes of light from 1/16 sec to 5 sec were given to identify the earliest compound formed. In this way Calvin and his associates found out all the intermediaries in the CO_2 reduction cycle.

The first stable compound formed after the initial entry of CO_2 is a triose (PGA). The triose cannot be the acceptor moleucle for it has to be initially a 2C compound to accpet one CO_2 molecule to become triose. Benson, however soon found out that a 5 carbon compound is the first acceptor molecule (Ribulose 1.5 diphosphate). After accepting a molecule of CO_2 an unstable 6C compound is formed, which immediately breaks up into 2 molecules of PGA (triose) the first stable compound.

The essentials of the Calvin cycle are as follows. The cycle should be self exclusive in that it should remain unaltered, but provide for the continuous entry of 6 molecules of CO_2, formation of one molecule of glucose and regeneration of six molecules of Ribulose 1.5 diphosphate (RUDP), the initial acceptor molecule. The important events of CO_2 cycle can be detailed as under:

a)Fixation of CO_2,

b)Reduction of PGA,

c)Formation of sugars, and

d)Regneration of RUDP.

Fixation of CO_2: CO_2 is first accepted by molecules of a pentose (5C) sugar Ribulose 1.5 diphosphate and an unstable 6 carbon compound is formed. This compound (tentatively called carboxy dimsutase system) immediately breaks up into 2 molecules of a triose (3C) called phosphoglyceric acid (PGA). The reaction is catalyzed by the enzyme carboxy dismutase or Rudp carboxylase (Rubisco). For every molecule of CO_2 ,2 mols of PGA are formed; hence 12 PGA moles for 6 CO_2 mols.

Reduction of PGA : The molecules of PGA are reduced to phosphoglyceraldehyde (PGAI) by NADPH + H. The reaction requires energy and hence ATP participates. The enzyme triose phosphate dehydrogenase catalyses this reaction.

Rudp + CO_2 + H_2O ——— PGA (2 molecules)

PGA + NADPH + H $\xrightarrow{\text{ATP}}$ PGAl

The above reaction takes place at two stages. In the first stage

PGA + ATP $\xrightarrow{\text{Phosphoglycerokinase}}$ 1,3-diphosphoglyceric acid + ADP

In the second stage

1.3 diphosphoglyceric acid + NADPH + H ——3 PGAI + NADP

Since 6 molecules of CO_2 are taken into the system (by 4 mols of Rudp) 12 PGAL mols are fomred after reduction of the 12 molecules of PGA. Only two are utilized for the formation of sugars, while the rest are channeled back for the regeneration of Rudp molecules. Several intermediate compounds are formed in the process.

Formation of sugars: The PGA molecule (one out of two) undergoes isomerization to form a mol of dihydroxyacetone phosphate. The enzyme triose phosphate isomerase catalyses this reaction. Ultimately sugars are formed as follows.

1.3 PGAL $\xrightarrow[\text{isomerase}]{\text{Triosephosphate}}$ Dihydroxyacetone phosphate

2.3 PGAL + dihydroxyacetonephosphate $\xrightarrow[\text{Aldolase}]{}$

Fructose 1,6-diphosphate

Fructose 1,6-diphosphate now gets converted into furctose 6 monophosphate due to the loss of one phosphate group under the influence of the enzyme phosphotase

3. Fructose 1.6-dip $\xrightarrow[+H_20]{\text{Phosphatase}}$ Fructose 6. Monophosphate (FMP).

Fructose 6 monophosphate isomerises into glucose monophosphate under the influence of the enzyme isomerase.

4. FMP $\xrightarrow{\text{Isomerase}}$ Glucose 6.monophosphate (GM)

Now FMP or GMP may undergo dephosphorylation to form either fructose or glucose. These monosaccharides unite to form sucrose and various other carbohydrates.

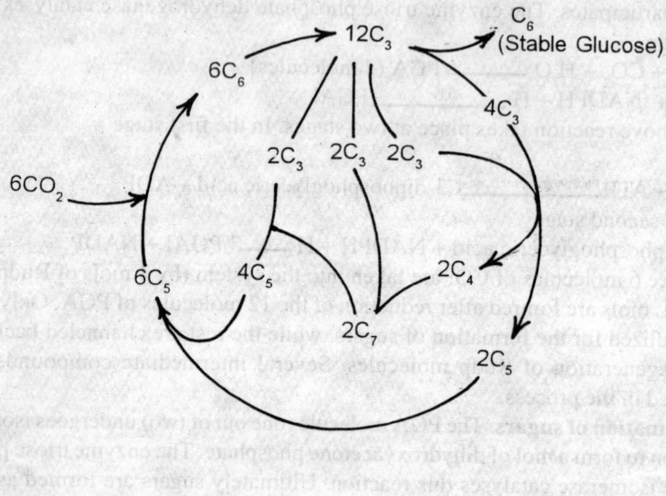

Fig. 16.12 Balance sheet of carbon in Calvin cycle

Regeneration of Rudp: In the above explanation of formation of sugars, it has been pointed out that Rudp accepts a molecule of CO_2 to ultimatley form a molecule of sguar. But the overall formula for photosynthesis accounts for the entry of 6 molecules of CO_2, hence there must be 6 molecules of Rudp to accept the CO_2 molecules and for every turn of the cycle, this should result in the formation of 6 molecules of glucose. But the output of the cycle for every 6 CO_2 molecules is only one molecule of glucose. Then what happens to the remaining 6 molecules of hexose (5 moles of 6C). Obviously they undergo transformation to regenerate 6 molecules of Rudp. This is necessary because, the cycle should be self-sufficient and there must always be 6 molecules of Rudp ready to accept 6 molecules of CO_2. If all the six molecules of Rudp combine with 6 molecules of CO_2 and form 6 molecules of glucose, the cycle will cease to operate; for the next turn of the cycle there will be no Rudp to accept CO_2 molecules. Thus for the entry of every 6 CO_2 molecules, the output is one mole of hexose and the rest go for the regeneration of Rudp.

FACTORS AFFECTING THE RATE OF PHOTOSYNTHESIS

Like any other biological process, photosynthesis is influenced by a number of enviornmental conditions. These factors or environmental conditions occur at varying concentrations and consequently the rate of the process also varies. Sachs (1860), who made a study of the influence of the concentration of factors on a biological process proposed the concept of three cardinal points. These are a minimum, an optimum and a maximum. Minimum point represents the lowest concentration of the factor when the process barely begins, and the optimum represents the highest rate of the process and a maximum (of the factor) represents the limit beyond which the process is impossible. The rate of the process beyond the maximum of the factor actually comes down. This concept of three cardinal points is based on the isloated study of a factor as to its influence on a process to the exclusion of other factors. But in actual fact when a process is conditioned by several factors, a particular factor not only influences the process but will influence other factors also mutually. For instance an optimum of a factor will fluctuate depending on the influence of other factors. Optimum concentration of CO_2 for example is not fixed, it is greater with increased light intensity and vice versa; same is true for light also. Hence the factors that influence a process are all interrelated and they condition a process not in isolation but in totality.

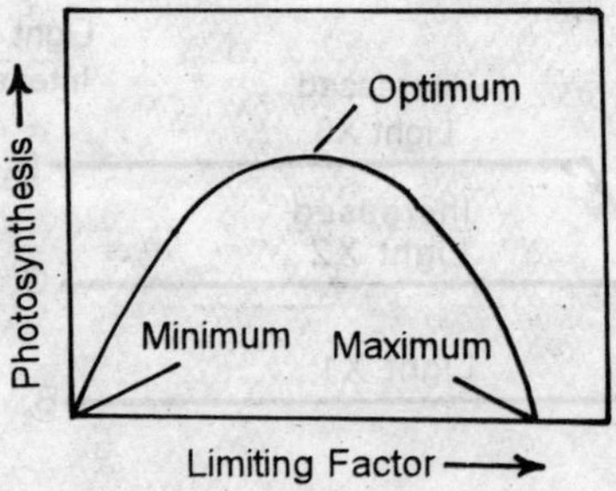

Fig. 16.13 Sach's concept of 3 cardinal points

The problem of influence of various factors and their optimum intensity was sought to be solved by Liebig(1843) who proposed the Law of the minimum which states that when a physiological process is conditioned by several factors, the slowest factor controls the rate of the process. Blackmann(1905) slightly elaborated this and proposed the Law of limiting factors which states "*when a physiological process is conditioned as to its rapididy by a number of factors, the rate of the process is governed by the pace of the slowest factor.*"

According to this principle, the magnitude of photosynthesis is controlled by only one of the factors at a time. This does not mean that other factors are not necessary. When they are available in abundance, the one that is available in the least governs the rate of the process. The rate of photosynthesis is proportional to the quantity of this factor. But this is not fixed; the rate of photosynthesis alters when the availability of another factor becomes least. Then this factor (which has become least) becomes limiting.

The factors that influence the rate of photosynthesis may be classified into External and Internal. The external factors are - Light, CO_2, Temperature, Water and Oxygen. The internal factors are - Chlorophyll content, Protoplasmic factors, accumulation of end products and availability of nutrients. We will discuss each one of these factors in some detail.

LIGHT

The solar radiation is the source of light, although photosynthesis can be carried out in the presence of artificial light also. The radiation from the Sun is available in the form of electromagnetic radiation much of which is shielded by the earth's atmosphere. Although some bacteria can carry on photosynthesis under infra red, much of photosynthesis is carried out under the visible light. Broadly speaking, three aspects of light affect photosynthesis. These are - light intensity, light quality and light duration.

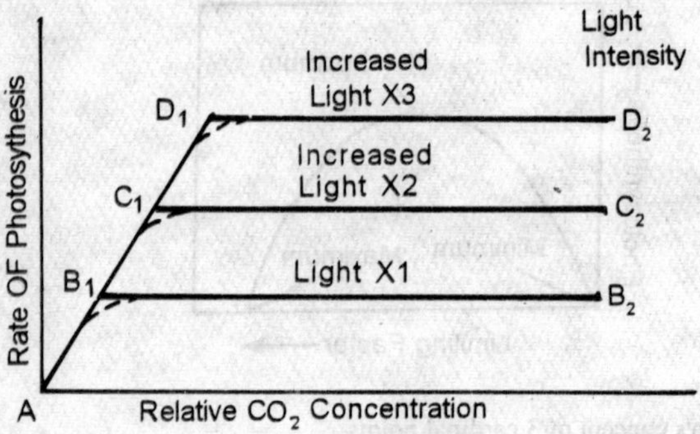

Fig. 16.14 Graph representing Law of limiting factors

Light Intensity: The morphology, architecture and internal structure of the leaf is best suited to utilize maximum amount of incident light. However leaves can utilize only a fraction of the light that falls on them. Of the total quantity of incident light, a part is reflected, a part is transmitted and only a little is absorbed. According to Seybold (1932) about 80% of the visible light falling on the leaves is absorbed. However most of the absorbed light is used up in raising the temperature of the leaf and is lost as heat. Only about 2% of the light perhaps is made use of in photosynthesis.

According to Wolkoff(1866), the rate of photosynthesis is driectly proportional to the intensity of light. But this is true only when other factors are not limiting. It has been observed that a maximum rate of photosynthesis can be achieved by leaves under a light intensity, less than the maximum.

Light quality: Normally the rays of white light between the range of 390nm and 670nm are useful in photosynthesis. Studies of Hoover have indicated maximum absorption for Chlorophyll in the range of blue (440nm) and Red(655nm). Photosynthetic bacteria however can carry on the process even at 900nm also. Most of the light in the green region is reflected indicating least absorption.

Some physioligists believe that the rate of photosynthesis actually depends on the energy absorbed irrespective of the wavelength absorbed. But quality of light is also important as seen from the fact, that the rate of photosynthesis is not high in the undergrowth of a thick forest as they get mostly the green light filtered through the leaves of tall trees.

Duration of Light: When light intensity and other factors are favourable, the rate of photosynthesis is directly proportional to the duration of light. It has been found out that the quantum yield is higher when plants are exposed to continuous daylight of 10-12 hours. Findings of Mitchell (1936), Boehring (1949) and others have indicated that plants continue photosynthesising for long periods without any ill effect on them. Devlin and Baker (1971) however have suggested that continuous light is damaging to photosynthesis. The rate of photosynthesis has been noted to be high under diffused light for longer duration, than under high intensity of light for shorter duration.

CARBON DI OXIDE

The quantity of CO_2 in the air is extremely small. It is about 3 parts in 10,000 or 0.3% by volume. This amount is constant and its concentration affects the rate of photosynthesis markedly.

The sources of CO_2 are plant and animal respiration, bacteria in the soil, water and ocean, decay of organic substances and combustion of various types of fuels. It is believed that the "golden age" for green plants as for as photosynthesis is concerned was the carboniferous age (300 million years ago). This is due to the fact that the humid and warm atmosphere of the carbonifeours age contained far more CO_2 than it is today and this supported a higher rate of photosynthesis. Much of the fossil fuel we have today represents the stored photosynthetic products of the carboniferous period.

Godlewski(1873) found that there is considerable increase in the rate of photosynthesis with the increased supply of CO_2. But this is subject to a certain limit. A higher CO_2 concentration may actually retard the process possibly by affecting other processes (Steeman-Nielson 1955).

Optimum level of CO_2 for maximum photosynthesis varies with plants. In *Triticum sativum*, maximum photosynthesis is achieved at 0.15% of CO_2. In water plants even 1.1% of CO_2 is conducive for a higher rate of photosynthesis.

TEMPERATURE

Like any other physiological process photosynthesis too is affected by fluctutations in temperature. It has a range when the process is active, a minimum as well as a maximum. This range however is not fixed. It varies with plants. While it is true that the light phase of photosynthesis is unaffected by temperature, a high temperature however not only retards photosynthesis, but might also even prove harmful to the plant itself. It is generally believed that photosynthesis is possible at a temperture range tolerated by protein compounds (O^0 C- 60^0 C). Between 6molecules C and 35°C the rate of photosynthesis increases with increase in temperature. At this range it follows Van Hoff's rule that for every rise of 10°C, the rate of photosynthesis doulbes (Q10=2). Many of the desert plants can photosynthesize comfortably at 55°C. In majority of the plants an optimum of 30-35°C is required for the maximum size of photosynthesis. Extremes of temperature cause injury or even permanent damage to the plants. A very cold temperature inactivates photosynthesis by blocking the activity of enzymes. Besides, in cold temperature ice formation takes place resulting in the change of colloidal nature of cytoplasm. Ice formation also disturbs the permeability of the membranes.

WATER

It is difficult to analyze the effect of water scarcity on photosynthesis largely because, plants absorb a large quantity of water of which only a minute quantity is utilized for photosynthesis. Further, scarcity of water makes its deleterious effect felt on the general survival of the plant before it influences the rate of photosynthesis. Thus due to water scarcity various other forces make the functioning of the photosynthesis impossible, before the effect (of water scarcity) on photosynthesis could be analysed.

Schneider and Childers (1941) noted almost 50% reduction in photosynthetic efficiency in dry soil conditions which favour high transpiration. Elimination of water results in general dehydration, closure of stomata, inactivation of enzymes etc. Beside, dehydration may cause irrepairable damage to the micromolecular strucrture of the chloroplast membranes.

OXYGEN

It is one of the byproducts of photosynthesis. While many plant physiologists have opined that CO_2 is necessary for photosynthesis, there does not seem to be any experimental evidence in support of this assumption. The accumulation of oxygen infact retards the rate of photosynthesis. Findings of Warburg(1920) have proved the inhibitory effect of O2 on photosynthesis. Photosynthetic inhibition by O_2 is called Warburg effect. Working on *Chlorella* plants, Warburg (1920) demonstrated such an effect. In Warburg effect, both O_2evolution and CO_2 assimilation will be inhibited in the presence of atmospheric O_2.Gibbs (1970) has demonstrated that high concentration of oxygen brings about competetive inhibition of Ribulose phosphate carboxylase. High concentration of oxygen also inhibits photosynthesis by encouraging photorespiration.

Seventeen

Respiration

The most basic requirement of all living organisms is energy. Without energy no life activity is possible. The solar energy that is trapped by green plants will be stored in cells in the form of carbohydrates and other nutrients. These are oxidized and the energy so released will be utilized to carry on the vital processes of living beings.

The process of energy release by the (oxidation of) substrate material is called respiration. The term respiration which has been in usage since a long time is derived from the latin word respirare = breathe. Breathing as a phenomenon of life has been noticed in animals long before its relation to respiration and its universal occurrence in all forms of life.

Breathing and respiration: Breathing and respiration are often used as synonymous. But it is not correct. Breathing as a separate process is noticed only in few organisms (animals), while respiration occurs in all organisms. Further, breathing is a physical process, while respiration is a chemical process taking place in all the cells. Breathing may be regarded as the last step found only in aerobic respiration, where atmosphere oxygen participates. There can be respiration without breathing, but the reverse is not possible.

Respiration and combustion: Respiration is often compared to the process of combustion, because in both the instances the substrate material gets burnt (oxidized) up. But the process of respiration differs from mere combustion in many respects. In combustion the burning up or oxidation of the substrate is sudden while in respiration it is stepwise. In combustion, the burning up or oxidation of large amount of energy is wasted as heat, while in respiration it is channelised and stored up for further usage. Hence respiration may be termed as an orderly combustion found in living organisms.

Cellular respiration: In a multicellular organism where there are thousands of cells, how does energy reach each cell? Each cell obtains its own energy by the oxidation of substrate material present within. Hence respiration takes place in all the cells wherever they may be present in animals and plants. This is called cellular respiration.

Cellular respiration is a complex process which includes various aspects. These are the following.

a) Absorption of oxygen (not in all)

b) Degradation of substrates such as carbohydrates to CO_2 and H_2O (oxidation)

c) Energy is released in the process of oxidation; some amount will also be lost in the form of heat.

d) Formation of many intermediate products which play different roles in metabolism.

e) There will be some loss of weight due to the process of oxidation.

From the point of view of plant metabolism, cellular respiration is of paramount significance as it is the basic biological process that provides energy for all the vital activities. In short, the following aspects may be mentioned as indicating the significance of respiration.

(i) Energy release in respiration is used for various metabolic processes. In fact most of the physiological activities are driven by respiratory energy.

(ii) Many intermediate compounds are formed which play a key role in the intermediate metabolism.

(iii) Insoluble food is converted into soluble form.

(iv) Potential energy is converted into kinetic energy.

(v) CO_2 which is released during respiration is an important step in maintaining the carbon balance of nature.

The overall formula for respiration is as follows

$$C_6H_{12}O_6 + H_2O + 6O_2 \longrightarrow 6CO_2 + 12H_2O + 686 \text{ K cal energy.}$$

Respiratory substrate: The energy rich material (food substrate) that is used for the purpose of oxidation is called the respiratory substrate. In other words a respiratory substrate is defined as "any organic plant constituent oxidized partially or completely with the concomitant release of energy together with CO_2 and H_2O".

Plants use a variety of organic constituents as their respiratory substrates. These may be a variety of carbohydrates (hexose sugars, disaccharides, poysaccharides), organic acids, fats etc. Under certain circumstances like starvation, even protiens may be used as respiratory substrates.

MITOCHONDRIA (Gr.mito = thread, chondrion = granule)

Popularly called 'power house' of cells, these granular or filamentous organelles are present in all eukaryotes. They are characterised by their special structure, and unique physiological function concering intracellular energy transformation.

Historically, mitochondria were first observed by Altmann (1894) who called them bioblasts. The present name was coined by Benda (1897) who developed the crystal violet staining. Bensely and Hoerr (1934) first isolated the liver cell mitochondria. The discovery that mitochondria are seats of cellular respiration was made by Hogebomm et al in 1948.

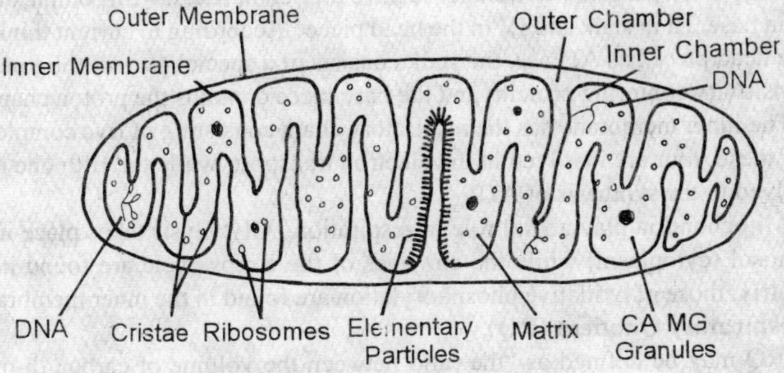

Fig. 17.1 Diagrammatic representation of a section through aa mitochondrion (ETPs are not shown on all cristae)

Ultra structure of mitochondria: The mitochondrion is enveloped by two membranes viz. inner membrane and outer membrane. The outer membrane is 6nm thick; internal to this is the inner membrane. Seperating these two is a space (outer chamber) filled with a watery fluid. This space is 40-70 A° thick. The inner membrane is folded into finger like projections called cristae. This membrane encloses the inner chambers or matrix, usually filled by dense proteinaceous material. The matrix may be homogenous or may have a granular substance. The cristae of the inner membrane incompletely divide the matrix into chambers. The side of the inner membrane facing matrix is called the M side, while the other side is called the C side. The cristae have an inner space called the inter cristae space which is in continuity with the intermembrane space.

Within the mitochondrial matrix are small ribosomes and the circular DNA. The ribosomes are of 70s type and similar to those of bacteria. These can synthesise proteins (Tyler 1973). The DNA called mDNA may be circular as in most of the animal cells or linear as in plant cells. Rabinowitch (1968) has demonstrated that the mitochondrial DNA has more G-C content than the nuclear DNA.

The mitochondinal DNA can replicate itself as it has DNA polymerase. It is due to this, that mitochondria can replicate themselves.

The inner membrane: The inner membrane has mumerous nail like or tennis racquet like projections protruding into the matrix. Each particle is about 70-100 A° in diametre. The particles are placed apart a distance of 100 A° and each is attached to the inner membrane by a short stalk of 35-50 A° long. The number of these particles per mitochondrion ranges from 104 - 105. These are called elementary particles, F1 particles or subunits. It was formerly beleived that these particles contain all the enzymes necessary for electron transport and oxidative phosphorylation. Hence they are named electron transport particles (ETP). Each particle was thought to have four complexes, with complex I and II in base, III in stalk and IV in the head piece. According to current thinking, the head piece has ATPase, the stalk consists of a special protein (oligomycin sensitivity confering protein) and the base piece contains the proton channel.

The inner membrane has the respiratory chain consisting of five complexes. Of these, four are involved in the electron transport, while the fifth one is involved in the synthesis of ATP.

Mitochondria play a vital role in respiration. Glycolysis takes place in the cytosol (cytoplasm) while the enzymes of the Krebs cycle are found in the matrix, those of oxidative phosphorylation are found in the inner membrane.

Respiratory Quotient (RQ)

RQ may be defined as "the ratio between the volume of carbon di-oxide given out and oxygen taken in simultaneously by a given weight of the tissue in a given period of time at standard temperature and pressure"

RQ may be calculated as follows

$$RQ = \frac{\text{Volume of } CO_2 \text{ evolved}}{\text{Volume of } O_2 \text{ consumed}}$$

The value of RQ depends upon the nature of the respiratory substrate, amount of oxygen present (in the substrate), and the extent to which the substrate gets broken. Normally the value of RQ should be unity (1), but many deviations are noticed due to the alterations in the levels of oxidation and reduction. The value of RQ also depends upon whether all the oxygen consumed is used up only for respiration or some of it is utilized for other purposes also. It is for these reasons that RQ may be unity, less than unity or even more than unity. RQ value also gives an idea of the respiratory substrate that is being used.

RQ for Carbohydrates: Carbohydrates are the principal respiratory substances in a large majority of organisms. These include a variety of monosaccharides (glucose, fructose etc), disaccharides (sucrose) and polysaccharides (starch, inulin etc). When simple sugars are available they are directly, channelised into the respiration stream, but when poly and disaccharides are involved, they are first hydrolysed into monosaccharides.

When hexose sugars are the respiratory substrate, volume of CO_2 evolved equals volume of O_2 used and hence the value will be unity. This will be illustrated by the following equation.

$$C_6H_{12}O_6 \longrightarrow 6O_2 + 6H_2O$$

$$RQ = \frac{CO_2}{O_2} = \frac{6CO_2}{6O_2} = 1 \text{ or unity.}$$

TYPES OF RESPIRATION

The involvement of atmospheric oxygen during respiration constitutes the basis for its categorization into aerobic and anaerobic types.

Aerobic respiration: The respiratory substrate is completely broken down into CO_2 and H_2O with the participation of atmospheric O_2.

$$C_6H_{12}O_6 + 6O_2 \longrightarrow 6CO_2 + 6H_2O + 686 \text{ k cal.}$$

Anaerobic respiration: Here there is no participation of atmospheric oxygen. Food substances are incompletely oxidized. Hence all the carbon in the substrate will not be released in the form of CO_2. The end products are not CO_2, but intermediate products like ethy alcohol, acids etc.

Aerobic	Anaerobic
1. O_2 necessary	O_2 not necessary
2. Common to most of the organisms	Restricted to certain organisms, and certain tissues.
3. Occurs always	Occurs temporarily in certain tissues,besides microrganisms
4. Oxidation complete	Oxidation incomplete
5. Large amount of energy is liberated	Less energy is liberated.
6. End products are CO_2 and water	End products are ethyl alcohol, organic acids etc and CO_2
7. The entire carbon content of the substrate comes out in the form of CO_2 (inorganic)	Due to incomplete oxidation of the substrate, carbon will still be lfet in the organic compounds (ethyl alchol)
8. Process non toxic to plants	Toxic to plants
9. Respiratory enzymes present in mitochondria and cytoplasm	Enzymes present only cytoplasm
10. Output is 38 molecules of ATP	Output is only 2 molecules of ATP

MECHANISM OF RESPIRATION

As has already been pointed out, during respiration the substrate material is broken down with the release of energy, CO_2 and H_2O.

In the mechanism, we will be studying the various stages of the breakdown of the substrate material, enzymes involved etc., and account for the release of CO_2 and energy released in the form of molecules of ATP. In the foregoing paragraphs various steps involved in the aerobic breakdown of carbohydrates are given.

Broadly categorizing, the following steps are identifiable in the oxidative breakdown of carbohydrates.

1.Hydrolysis. 2.Glycolysis 3. Pyruvic acid oxidation, 4.Krebs cycle or Citric acid cycle and 5. Electron transport system or oxidative phosphorylation.

The mechanism of anaerobic respiration consists of the following steps.

1.glycolysis and 2.anaerobic oxidation of pyruvic acid.

From the above account it can be understood that the step glycolysis is common for both aerobic and anaerobic respiration.

HYDROLYSIS

When starch is the reserve food material, it should first be hydrolysed into simple sugars before it can enter into the respiratory cycle. The process of breakdown of starch into simple sugars with the addition of water is called hydrolysis.

Glycolysis. Glycolysis is a term used to describe the sequential series of reactions present in a wide variety of tissues that start with a hexose sugar (glucose commonly) and end with pyruvic acid. The usual reaction steps are known as Embden - Meyerhof-Parnas (E.M.P) pathway after names of the three German scientists who traced them.

Carbohydrates are not utilized is respiration as such. As a result of interconversions and phosphorylation the carbohydrates first produce various types of hexose phosphates. The phosphorylation of hexose sugars is catalyzed by enzymes of transphosphorylase group with the help of ATP- the universal biological currency of energy transaction in a living cell.

1. Glucose is the respiratory substrate and is converted to Glucose-6-phosphate by reaction with ATP. This reaction is brought about by the enzyme known as hexokinase.

2. Glucose-6-phosphate changes, in the next step, to its isomer fructose-6-phosphate under the influence of the enzyme phosphohexose isomerase.

3. Further phosphorylation of 'Fructose-6-phosphate with the help of ATP results in the formation of Fructose-1-6-diphosphate. This reaction is catalysed by the enzyme phosphohexokinase which requires Mg++ as cofactor.

4. Under the influence of desmolytic enzyme aldolase, Fructose 1,6 diphosphate splits up to produce two molecules of triosephosphates viz. 3-phosphoglyceraldehyde (3-PGAL) and dihidroxy acetone phosphate. These two molecules are chemical isomers and are interconvertible by the enzyme triosephosphate isomerase.

5. In the next step 3-PGAL is converted into 1,3-phosphoglyceric acid. The reaction involves the incorporation of inorganic phosphate and the oxidation of aldehyde into acid by the enzyme phosphoglyceraldehyde dehydrogenase. The co-enzyme NAD is reduced to $NADH_2$.

6. In the next reaction, 1-3, diphosphoglyceric acid reacts with ADP and loses a phosphate group forming 3-PGA. ADP is converted to ATP. This reaction is catalysed by the enzyme phosphoglyceric transphosphorylase.

7. In the next step of glycolysis, 3-PGA is converted to 2-PGA by the enzyme phosphoglucomutase. This is accomplished by an intramolecular shift in the position of phosphate group.

8. The last compound loses a molecule of water and changes to Phosphoenol pyruvic acid. This reaction is catalysed by the enzyme enolase.

9. The final reaction of glycolysis consists in the dephosphorylation of P.E.P. by the enzyme phosphopyruvic transphosphorylase in the presence of the co-enzyme and ADP resulting in the formation of pyruvic acid and ATP. This reaction requires Mg++ and K+ as cofactors.

It may be emphasised that the above reactions of glycolysis do not involve atmospheric oxygen. This explains how the first phase of respiration can be the same in both anaerobic and aerobic respiration.

Pyruvic acid is a key compound in respiration. Its further breakdown proceeds in two different ways depending on the availability of oxygen.

Anaerobic brakedown of pyruvic acid. In the absence of oxygen, and under certain other conditions anaerobic oxidation of pyruvic acid usually occurs but the course of reaction differs in different tissues and organisms. In general, the products of anaerobic respiration of pyruvic acid are incompletely oxidized compounds such as alcohols and organic acids.

Pyrvuic acid is broken down in presence of enzyme carboxylase, the products being acetaldehyde and CO_2.

$$CH_3\text{-}CO\text{-}COOH \xrightarrow{\text{decarboxylase}} CH_3CHO + CO_2$$

The acetaldehyde formed, undergoes reduction in the presence of reduced coenzyme NADH-H to form alcohol and oxidised co-enzyme NAD.

$$CH_3\text{-}CHO \xrightarrow{\text{NAD.H.H}} CH_3\text{-}CH_2\text{-}OH + NAD$$

In the muscle tissue of animals and in some bacteria anaerobic respiration results in the conversion of pyruvic acid into lactic acid.

Aerobic oxidation of pyruvic acid. Under aerobic conditions pyruvic acid is ultimately oxidised to carbon dioxide and water by the free oxygen of air. In this process pyruvic acid undergoes a series of step-wise degradation involving decarboxylation and dehydrogenation processes. These reactions occur in a cycle known as citric acid cycle or Krebs cycle or Tricarboxylic acid cycle since it results in the formation of a number of tricarboxylic acids.

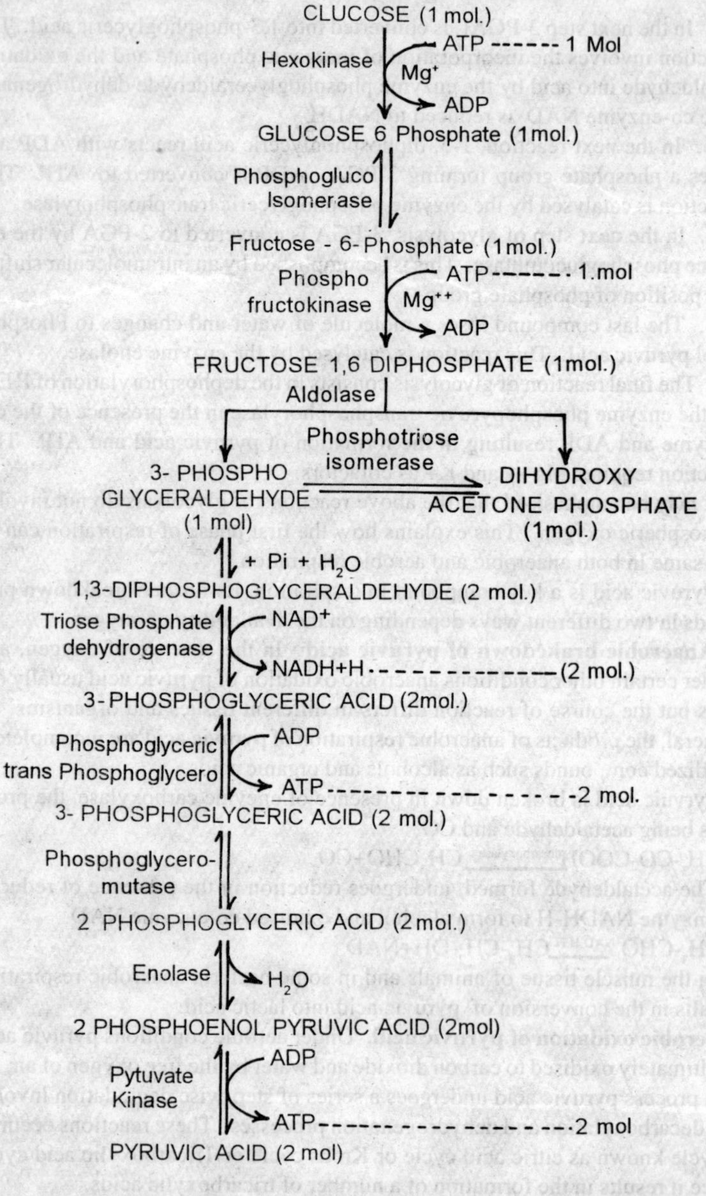

Fig. 17.2 Glycolysis- Various reactions leading to the formation of Pyruvic acid

Pyruvic acid by itself is not very active and hence it is first converted into a more active and energy rich compound called acetyl CoA. The formation of the latter is a very complicated process though it constitutes the major and only link between glycolysis and Krebs cycle.

Formation of acetyl CoA. Its formation occurs in the cytoplasm external to the mitochondria. The oxidative decarboxylation of pyruvic acid to form acetyl CoA is a very complex reaction, requiring the presence of at least 5 essential co-factors and a complex enzyme viz. The CoA Mg ions, Thiamine pyrophosphate (TPP) NAD+, NADP, and lipoic acid, The reactions taking place in the oxidative decarboxylation of pyruvic acid are as follows.

Pyruvate + TPP — TPP complex + CO_2

TPP complex + Lipoic acid — Acetyl-lipoic acid complex + TPP
(Oxidised form)

Acetyl-lipoic acid + CoA — Acetyl CoA + Lipoic acid
(Reduced)

Lipoic acid + NAD — Lipoic acid + $NADH_2$
(Oxidised form)

Krebs cycle. The Krebs cycle begins with the reaction of Acetyl CoA with oxaloacetic acid a four carbon dicarboxylic acid, under the influence of condensing enzyme forms a six carbon tricarboxylic acid called citric acid. A molecule of water is used up in the reaction and the CoA is liberated.

Citric acid first undergoes dehydration producing cis aconitic acid which in turn produces iso-citric acid by the reverse process of hydration accompanied by rearrangement of atom groups. The enzyme aconitase which catalyses this reaction requires co-factor Fe++.

Isocitric acid undergoes dehydrogenation (oxidation) in the presence of enzyme isocitric dehydrogenase giving oxalosuccinic acid. The coenzyme of this enzyme in most cases is not the usual NAD but NADP which is reduced to NADP.H.

The next reaction is the decarboxylation of oxalosuccinic acid by the same enzyme as in the last reaction and results in the production of a ketoglutaric acid and liberation of CO_2. Mn^{++} is an essential cofactor for the dehydrogenase.

The next reaction is the oxidative decarboxylation of a ketoglutaric acid by the specific enzyme a ketoglutaric dehydrogenase. Essential coenzymes are CoA and NAD and Mg^{++}. This reaction takes place in two steps. In the first step a ketoglutaric acid reacts with CoA and NAD and undergoes decarboxylation and dehydrogenation producing succinyl CoA which undergoes a series of complex reactions in the presence of inorganic phosphate and ultimately gives succinic acid releasing CoA. One molecule of ATP is generated in this process.

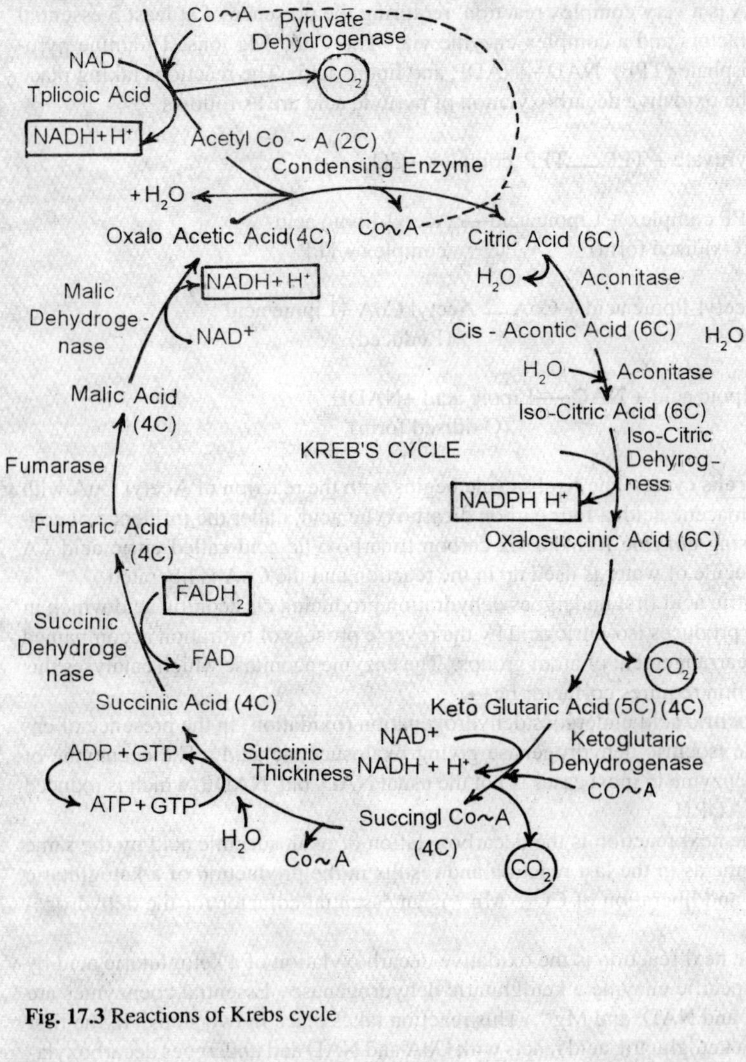

Fig. 17.3 Reactions of Krebs cycle

Next, succinic acid is oxidised (dehydrogenated) to fumaric acid by the enzyme succinic dehydrogenase. The coenzyme concerned is flavin (FAD) and not NAD.

Fumaric acid is hydrated by the hydrase enzyme fumarase and produces malic acid.

Finally malic acid undergoes oxidation to form oxaloacetic acid. Thus a complete cycle is formed. In this cyclic sequence of reaction oxaloacetic acid merely acts as a catalyst.

The over all reaction is as fcllows

$$CH_3\text{-CO-COOH-}+21/2\ O_2 \longrightarrow 3\ CO_2 + 2H_2O + \text{Energy}$$

If we add to the left side of the above equation $1/2O_2$ required to oxidise $NADH_2$ produced in glycolysis and to right side add H_2O on right side by doubling this we will obtain the overall equation for the complete oxidation of a hexose sugar with CO_2 on the left side and $6CO_2$ and $6H_2O$ on right side for each molecule of sugar oxidised as follows.

$$C_6H_{12}O_6 \xrightarrow{\text{Glycolysis}} 2CH_3\text{-CO-COOH}+2H_2O + \text{Energy}$$

$$2CH_3\text{-CO-COOH}+5O_2 \xrightarrow[\text{Cycle}]{\text{Kreb's}} 6CO_2+4H_2O + \text{Energy}$$

$$C_6H_{12}O_6 + 6O_2 \longrightarrow 6CO_2+6H_2O+\text{Energy}$$

Terminal Oxidation. The last step in the aerobic respiration of plant cells is the oxidation of reduced co-enzymes by molecular oxygen. The oxidation of one of the reduced co-enzymes in Krebs cycle viz $NADH_2$ takes place by the stepwise transfer of this electron to a series of electron acceptors forming ATP along the way. Finally, the electrons along with protons from Hydrogen ions are transferred to molecular oxygen, forming water. The metabolic pathway along which the electrons of reduced co-enzyme are channelled is known as the electron transport system or the respiratory chain.

For explaining the mechanism of terminal oxidation we will use the oxidation of malic acid to oxaloacetic acid as an example. In this reaction two electrons from malic acid are transferred to NAD which acts as a hydrogen acceptor and it is reduced to $NADH_2$. The latter reacts with second electron acceptor FAD. As a result of this, electrons are transferred from NADH to FAD resulting in the oxidation of former and reduction of latter.

In this process electrons step down by one step and a part of their energy is released. This energy is stored in ATP.

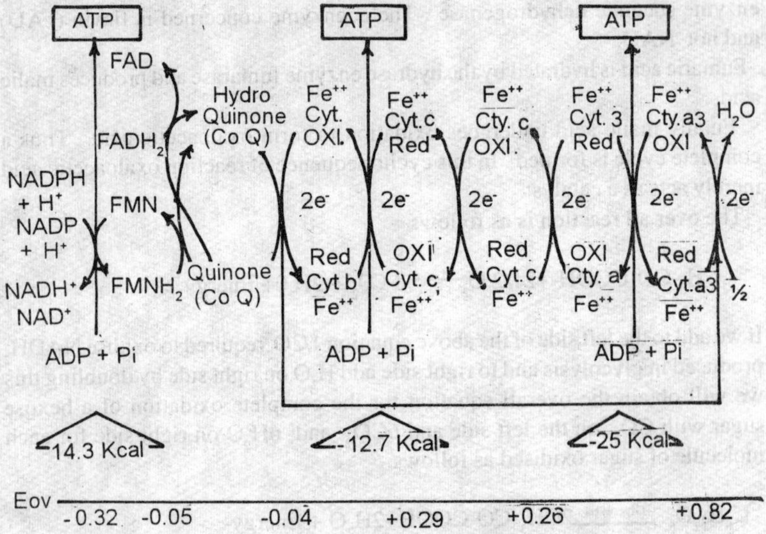

Fig. 17.4 Schematic representation of electron transport system in oxidative phosphorylation

The electrons are then transferred stepwise to a group of coloured compounds called cytochromes. There are 4 types viz. cyt b, cyt c, cyt a and cyt a3.

$FADH_2$ is stable in air but is readily oxidised to FAD by cyt b whose iron gets reduced to ferrous state. Then electrons are transferred through cyt c, to cyt a. Finally cyt a3 called cytochrome oxidase accepts the electrons. In the meantime, the energy stored in electrons are released into ATP. Cytochrome oxidase is most important because this atom is capable of oxidation of molecular oxygen. 95% of oxygen utilized by cells is via cytochrome oxidase.

In the mechanisms outlined above the oxidation of $NADH_2$ by the molecular oxygen is linked with the phosphorylation formation of ATP. This is called oxidative phosphorylation because oxygen is consumed in this process. All these processes occur in the mitochondria which is the power house and store house of energy.

Balance sheet of energy in Respiration (account of ATP molecules synthesized). The following is a total account of ATP molecules synthesized during complete break down of glucose to CO_2 and H_2O.

Glycolysis

Reaction sequence	ATP Used	ATP synthesized
1. Glucose to glucose 6 phosphate	1	-
2. Fructose 6 phosphate to Fructose-1,6-diphosphate	1	
3. 1.3 diaphosphoglyceric acid to 3 phosphogyceric acid (1X2)		2
4. Phosphoenol pyruvic acid to pyruvic acid (1X2)		2
5. Dehydrogenation of glyceraldehyde-3 phosphate to 1.3 diphosphoglyceric acid (through ETS with participation of NAD) 3x2		6
TOTAL	2	10

Net gain of ATP during Glycolysis = 8 molecules.
(Note: reactions 1 and 2 are for entire glucose molecule, while 3,4 and 5 are only for a triose hence they are doubled)

Pyruvic acid oxidation

Reaction sequence	ATP used	ATP Synthesized
Conversion of pyruvic acid to Acetyl CoA; dehydrogenation through ETS with initial acceptor being NAD (3X2)		6

Krebs Cycle
(There is no ATP loss in Krebs Cycle)

	ATP SYNTHESIZED
1. Isocitric acid to oxalosuccinic acid dehydrogenation through ETS with NAD participation (3 x 2)	6
2. Conversion of a ketoglutaric acid to succinyl CoA, ETS with NAD as the initial acceptor (3 x 2)	6
3. Conversion of succinly CoA to succinic acid; substrate level phosphorylation. ——— $6H_20+6CO_2$ a ratio of No ETS (1x2)	2
4. Conversion of succinic acid to Fumaric acod. ETS beginning	

with FAD and not NAD. Hence only two sites of ATP synthesis (2X2)	4
5. Conversion of malic acid to Fumaric acid. ETS beginning with NAD (3X2)	6
TOTAL	24

Total ATP used up during entire aerobic respiration 2.

Total ATP Produced during glycolysis, pyruvic acid oxidation and Krebs cycle.
10+6+24=40

Net gain of ATP molecule for aerobic oxidation of one molecule of glucose.
40-2=38

PASTEUR'S EFFECT: The main difference between aerobic and anaerobic respiration lies in the complete decarboxylation in the former and partial decarboxylation in the latter. The following formulae show that-

Anaerobic respiration

$C_6H_{12}O_6 \longrightarrow 2C_2H_5OH+2CO_2$

Aerobic respiration

$C_6H_{12}O_6+6O_2 \longrightarrow 6H_2O+6CO_2$

a ratio of 1:3 is maintained in the release of carbon dioxide between anaerobic and aerobic respiration. The noted scientist Pasteur however noticed that supply of O_2 to an anaerobic system does not increase the output of CO_2. This means that in anaerobic respiration the ration 1:3 is not maintained always. The amount of CO_2 released many a time in anaerobic respiration is far less when compared with the equations given above. This due to the presence of excess O_2 the system. This decreased rate of glucose breakdown in anaerobic systems caused by the presence of O_2 is called Pasteur's effect. In other words inhibition of sugar breakdown occurs due to the presence of oxygen under aerobic conditions (in anaerobic systems).

According to Dixon (1937), Pasteur's effect prevents excessive conversion of carbohydrates into products of fermentation and thereby maintains the level of carbohydrates. Pasteur's effect may be due to any one or all of the following reasons.

a) Prevention of funtioning of glycolytic enzymes.

b) Increased rate of glycolysis with decreased oxygen tension.

c) Release of excess CO_2 from degradation of compounds other than the respiratory substrate.

d) Formation of partial glycolytic products.

FACTORS AFFECTING THE RATE OF RESPIRATION

Like any physiological process respiration is also influenced by a number of external and intermal factors. These are

Internal factors- protoplasmic factors and concentration of respiratory substrate

External factors- Temperature, light, O_2 concentration, CO_2 concentration, water, injury, chemical, mechanical effects etc.

INTERNAL FACTORS:

Protoplasmic factors: The rate of respiration depends on the metabolic state of protoplasm. Young and meristematic cells which are actively dividing have dense non vacuolated cytoplasm. Besides, the presence of respiratory enzymes is another protoplasmic factor influencing the rate of respiration. In general the rate of respiration is more in undifferentiated cells than in mature cells.

Concentration of respiratory substrate: The availability of the respiratory substrate is very important for the proper rate of respiration. Poor supply of respiratory substrate retards the rate of respiration. When other conditions are favourable, the rate of respiration increases with the supply of respirable substance. Rate of respiration increases in a leaf after it is transferred from darkness to light (due to adequate supply of substrate as a result of photosynthesis). Etiolated leaves have a low rate of respiration.

EXTERNAL FACTORS

Temperature: The influence of temperature on respiration is the same as on other metabolic processes. Plants generally are capable of respiring in the temperature range between 0-45°C. Within this range any increase or decrease of temperature would have a similar effect on respiration.

In several experiments it has been noticed that the increase in the rate of respiration follows the Vant Hoff's law that the rate of reaction is doubled for every 10°C increase of temperature, if no other factor is limiting. In general an optimum temperature of 30°C is most suitable for respiration. Beyond this range there is no further increase but the rate flattens untill 45°C when it starts declining.

Light: It is generally believed that the influence of light on respiration is indirect even though physiologists like Emerson and Lewis (1943) have reported, that illuminated ***Chlorella*** cells have a higher rate of respiration than those kept in darkness. A similar effect on barley seedlings has been observed by Johnson (1944). But there is no stoppage of respiration indirectly by governing the supply of respirable materials via photosynthesis. Another influence of light, is that by increasing temperature, it increases the rate of respiration.

Oxygen concentration: Oxygen is a must for aerobic respiration, though it is not necessary for anaerobic respiration. Since the concentration of O_2 in the atmosphere is constant, slight variations in O_2 concentration do not seem to have a lasting influence on respiration. But if the concentration of O_2 goes below 2.0% it will retard the rate of respiration as there is rise in CO_2 level due to fermentation.

Carbon dioxide concentration: The atmosphere has a carbon dioxide concentration of 0.03% and it is almost constant. Hence generally its influence on respiratory rate is negligible. However any increase in the carbon dioxide concentration has a retarding influence on respiration. Soil has a higher concentration of CO_2 than atmosphere due to the activity of microbes. Hence seeds sown in such a soil generally fail to germinate. Ploughing the soil increases the aeration and enhances the rate of germination.

Decrease in respiration due to increased CO_2 concentration may be used to preserve vegetables and fruits (in CO_2 rich atmosphere) by keeping the rate of respiration at minimum. This is not however practised as a higher CO_2 concentration may prove to be poisonous. Higher CO_2 concentration also affects all other vital activities which require biological energy.

Water: Hydration of the medium is a basic requirement for all the metabolic activities and respiration is no exception. The rate of respiration is minimum in dry seeds, and as soon as they come in contact with water the rate of respiration registers a marked increase. Respiratory enzymes become active only in a hydrated medium. Water maintains proper turgidity of the cells and provides a suitable environment for respiratory reactions.

Injury: The works of Hopkins (1927) and others have demonstrated, that injury enhances the rate of respiration. It has been observed that sugar content of cells near the wounded region increases preparatory rate of respiration. In the wound healing process new cells have to be added, with the result there is increased metabolic activity, thereby accounting for increased respiratory rate. Potato tubers cut into pieces show enhanced rate of respiration due to the conversion of starch into sugars.

FERMENTATION: Fermentation is essentially anaerobic respiration found in microorganisms like certain Fungi and bacteria. In fermentation, the substrate for oxidation is extracellular (not present within the cell) since the organisms are basically saprophytic.

Fermentation as a process to produce alcohol was known to human beings since time immemorial, even though its scientific understanding dates back to only the middle of the 19th century. Experiments conducted by Louis Pasteur established that fermentation is carried out with the participation of living cells.

Buchner in 1897 found out that the extract of yeast cells (which does not contain living cells) can bring about fermentation of sugar solution. Subsequently it was discovered that the enzyme Zymase present in the yeast extract is capable of fermenting activity (break down of carbohydrates together with the formation of alcohol and CO_2).

The enzyme Zymase will be active only in the presence of phosphate. Sucrase and maltase are the other carbohydrate fermenting enzymes secreted by yeasts.

In fermentation also, as in anaerobic respiration atmospheric oxygen does not participate. Oxidation is accomplished by the intermolecular atomic shifts, with the result the product will have less energy than the original substrate. Much of the energy released during fermentation is wasted as heat, while the remainder is used for the conduct of metabolic activities.

The end product of fermentation need not always be alcohol, it may be various types of organic acids such as butyric acid, lactic acid etc. Depending on the end product, fermentation may be termed as alcoholic fermentation, lactic acid fermentation etc. The following are the formulae for these fermentations.

i) $C_6H_{12}O_6$— $2C_2H_5OH+2CO_2$ (ethyl alcohol)

ii) $C_6H_{12}O_6$—— $C_4H_8O_2+2H_2+2CO_2$ (Butyric acid)

iii) $C_6H_{12}O_6$ — $23H_6O_3$ (Lactic acid)

iv) $C_2H_5OH+O_2$—— $CH_3COOH+H_2O$+ energy (acetic acid)

In the formula (iv) given above, for the formation of acetic acid from ethyl alcohol, O_2 is necessary. Hence acetic acid fermentation is different from other types of fermentation.

Fermentation is carried out by microorganisms, a point which has been emphasized earlier. The following is a list of organisms which carry out the various types of fermentation.

Organism	**Type of fermentation (based on end product)**
1. *Bacillus butyricus* and *Clostridium butyricum*	Butyric acid
2. *Bacterium lactic acidi*	Lactic acid
3. *Acetobacter aceti*	Acetic acid.

Eighteen

General Account of Growth

One of the unique p0roperties of living organisms is growth. It is this that is responsible to turn a microscopic, tiny zygote into a giant sized *Sequoia* tree. There is a fundamental difference between plants and animals in the manner of growth. While animals have a fixed growth i.e., they grow up to a certain stage in the life and from then on just maintain themselves, plants continue to grow indefntely.

What is Growth? Growth is a complex phenomenon and represents the culmination or end result of metabolic pathways bringing about an overall irreversible change. While it is difficult to give a precise definition for growth to encompass all its nuiances, the following are some of the attempts.

(i) "Permanent, irreversible change in the size of a cell, organ or whole organism usually accompanied by an increase in dry weight".

(ii) "Finished product of metabolic loom" (Blackmann).

(iii) "A permanent increase in weight which is followed by a permanent change in form" (Miller).

(iv) "Increase in size involving development and increase in volume which is irreversible" (Stiles).

(v) "Increase in volume intimately combined with change in form (Sachs).

(iv) "Irreversible increase in volume"(Thimann).

Growth or development is not merely an increase in volume, size or weight. As we realize these are only quantitative parameters. Growth however has a qualitative dimension also. Qualitative dimension is in fact the most significiant aspect, for without it growth results in an unorganised mass of cells. Morphogenesis and differentiation are the most important aspects of growth which start operating soon after the first cell division of the zygote. Morphogenesis deals with the development of typical organs, while differentiation operates at the cellular level and produces various tupes of cells from the standpoint of structure and function.

Growth regions in plants: In the case of majority of plants growth is always at the apices like the root and the shoot spex. Apical meristems present at these regions are responsible for the growth in length and differentiation of various types of cells and tissues. This type of growth is called apical growth (Primary growth) and is characteristic of a large majority of plants. In some of the plants belonging to the family *Poaceae* (Bamboos, grasses etc) the primary growth may be intercalary being brought about by intercalary meristems present at the nodal regions. Intercalary mesistems are actually regarded as parts of apical meristem separated by permanent tissues during ontogeny.

The plants (Dicotyledons) have another kind of growth called 'secondary growth' brought about by meristematic tissues arising late in the course of development. These are called lateral meristems. A typical example of a lateral meristem is *Cambium.*

Cambium brings about secondary growth in thickness, while primary growth is linear bringing about growth in length. New cells and tissues produced due to the activity of meristems develop into various organs depending on nutritional balance, hormonal activity and environmental and genetic factors. The interaction of the external and internal factors with the genetic components of the plant will result in proper and integrated growth.

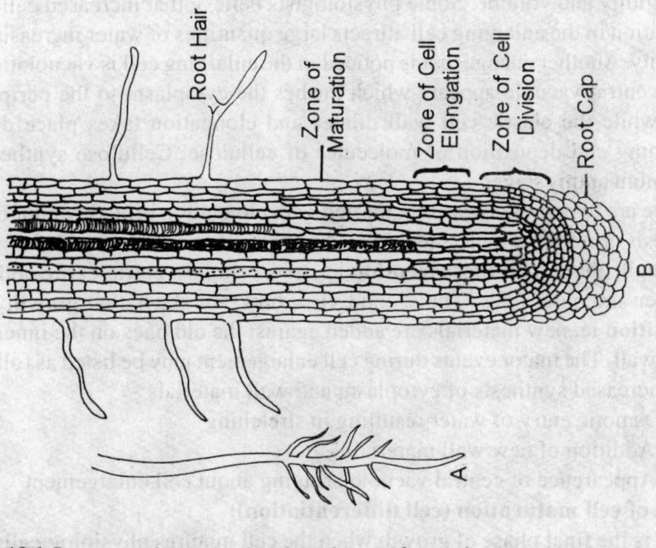

Fig. 18.1 Structure of root tip showing phases of growth

Phases of growth: Growth comprises of several complex biochemical and morphological steps. A meristematic cell has to pass throught all these stages before it attains its mature, shape, size and structure. Broadly speaking, these various stages of growth have been categorized into three phases. These three phases of growth are :-

(i) Phase of cell formation or cell division

(ii) Phase of cell elongation or enlargement

(iii) Phase of cell differentiation or maturation

Phase of cell formation (cell division): This is the first phase of growth. New cells are formed by the division of meristematic cells. The meristematic cells are thin walled, isodiametric and have dense non vacuolated granular cytoplasm together with a prominent nucleus. Cell division for growth is always mitotic hence it is also called vegetative cell division(meiosis is called reproductive cell division). Synthetic activity will be maximum before mitotic cell division, as extra materials have to be produced to be distributed between the two daughter cells. The daughter cells immediately after their formation will be smaller than the parent cell and they have to pass through the next two stages before they become an integral part in the structure and function of the plant system.

Phase of cell elongation (cell enlargement): This is the second phase of growth when cells attain their maximum size. The enlarging cell has a large quantity of solute particles and as a result water enters into the cell increasing its turgidity and volume. Some physiologists believe that increased cellulose deposition in the enlarging cell attracts large quantities of water increasing its turgidity. Another change that is noticed in the enlarging cell is vacuolation. A large central vacuole appears which pushes the cytoplasm to the periphery. Meanwhile the elastic cell wall dilates and elongation takes place due to stretching and deposition of molecules of cellulose. Cellulose synthesis is maximum at this stage.

There are two views regarding the cell wall elongation brought about by the synthesis of cellulose molecules. According to the first view cell wall elongation is brought about by **intussusception** ie., new cellulose particles are added in between the older ones. The second view however states that growth is by **apposition ie.,**new materials are added against the old ones on the inner side of the wall. The major events during cell enlargement may be listed as follows.

(i) Increased synthesis of cytoplasm and wall materials

(ii) Osmotic entry of water resulting in stretching

(iii) Addition of new wall materials

(iv) Appearence of central vacuole bringing about cell enlargement

Phase of cell maturation (cell differentiation):

This is the final phase of growth when the cell matures physiologically and morphologically. It assumes its final shape, size, and structure. Depending on the requirement, location, genetic factors etc differentiation sets in resulting in the formation of specilization in the mature tissue. Thus a meristematic cell matures into a parenchyma, sclerenchyma, xylem, phloem etc. Secondary cell walls are laid down during cell maturation.

Structure of the root tip and zones of growth:

A long section of the root tip shows several regions or zones which can be identified with the three phases of growth mentioned above. The terminal most part of the root has the root cap which protects the growing region.A little behind this region is the meristematic zone consisting of actively dividing cells.This zone can be identified with the phase of cell formation or cell division. A little behind (above) this region is the zone of cell elongation corresponding to the phase of cell enlargement. Above the zone of cell elongation is the zone of cell maturation, when the cells attain maturity after undergoing differentiation. This zone corresponds to the phase of cell maturation. Behind the zone of cell maturation is the root hair zone.

Grand period of growth:

The three phases of growth mentioned above viz., cell formation, cell elongation and cell maturation are characterized by the three aspects of growth namely (1) an initial period of growth (growth initiation),(2) a period of maximum growth (cell elongation) and a period of steady growth maintenance (cell maturation). These three (3) periods of growth together constitute what is called the grand period of growth. This grand period of growth is of universal occurence. Whether it is the growth of individual cell or a tissue or an organ or in an entire individual, grand period of growth can be observed.

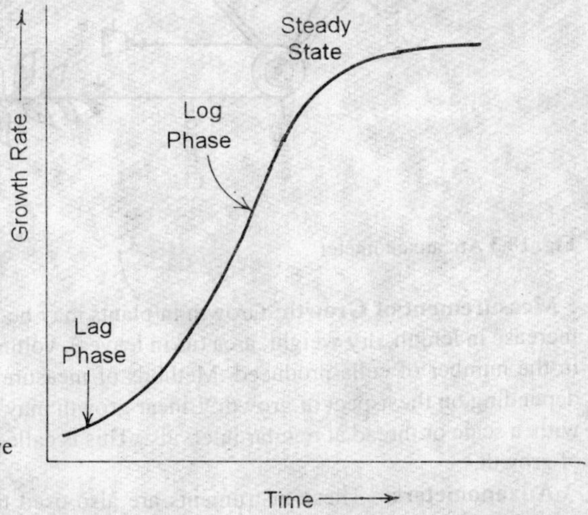

Fig. 18.2 Growth curve

Growth curve:

Studies conducted on growth in unicellular organisms, or organs (in multicellular organisms) like leaf, stem root etc have revealed the existence of a characteristic pattern. This pattern of growth rate resulting in increase of height, weight, volume etc have a perfect correlation when considered in the context of time. The results obtained by the study of growth pattern when plotted against time graphically would yield what is known as the growth curve. Studies conducted by plant physiologists have shown that the growth curve is always sigmoid or 'S' shaped in all the living organisms without any exception. The growth is slow initially and then is followed by a rapid growth rate and finally settles down to a slow rate or steady state in all the organisms. These three phases of growth can be marked on the growth curve as lag phase, log pahse and the steady state phase. The lag phase is characterised by a slow rate of growth (cell division) followed by the log pahse when there is geometric increase of growth (cell elongation) to be followed by a steady state phase (cell maturation) when the curve flattens out.

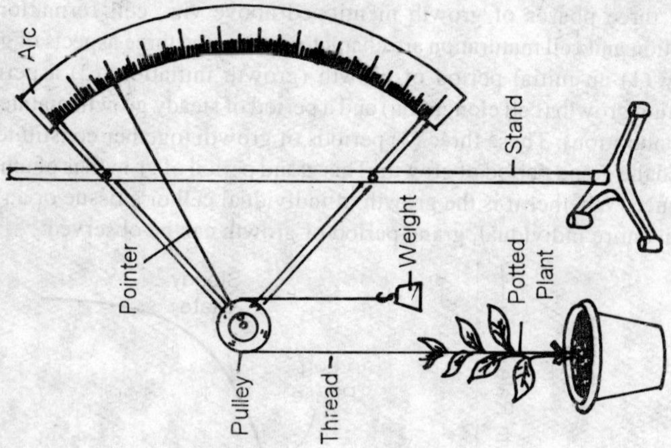

Fig. 18.3 Arc auxanometer

Measurement of Growth: Growth in plants may be measured in terms of increase in length, dry weight, area (as in leaves), volume (fruits) or increase in the number of cells produced. Methods of measurement of growth vary depending on the aspect of growth. Linear growth may directly be measured with a scale or thread at regular intervals. This is called direct measurement of growth.

Auxanometers: These instruments are also used to measure the linear growth. There are two types of auxanometers- Arc auxanometer and Pfeffer's auxanometer.

Arc auxanometer: The instrument consists of a tripod stand and a pully connected to a pointer which moves on an arc scale. A thread is made to pass over the pulley tightly. One end of the pulley is tied to the tip of a young growing plant kept in a pot. The other end of the thread is tied to a weight to keep the thread stretched (the thread should not slacken). As the stem tip grows, it will push the thread up and the tight thread will turn the pulley. The pulley inturn will move the pointer over the graduated arc scale. The extent of movement from the initial to the final in a given period of time can be measured. The distance moved by the pointer does not give the actual linear growth, because the arc is divided into 90^0 and not linear measurements.

Further, the measurement on the graduated arc is magnified because of the length of the pointer. The actual growth is calculated as follows.

Initial position of the pointer - A
Final position of the pointer-B
Actual distance travelled by the pointer during a given period of time B-A

$$\text{Actual growth in length} = \frac{\text{Distance travelled by the pointer X Radius of the pulley}}{\text{Length of the pointer as measured from the centre of the pulley}}$$

If the pulley has a 4" radius and the pointer is 20" long from the centre growth will be usually magnified ten times.

Ninteen

Factors affecting growth

Growth is influenced by a number of external and internal factors. These are environmental, physiological, nutritional and genetic. Growth in fact is the sum total of the interplay of all the metabolic activities. Some of these factors are-

Environmental factors: Light, Temperature, Water supply and oxygen supply are the chief environmental factors that influence growth.

Light: Light by itself has no influence on growth. In fact high intensities of light may even retard the growth of a plant. Plants grown in complete darkness however show etiolation i.e., yellowing of the leaves due to poor development of chlorophyll. Absence of light, leads to elongation of internodes, tenderness in stem and poor development of mechanical tissues. On the other hand, plants grown in the presence of light are strong statured. Light also helps in morphogenesis. Duration of light affects the flowering process and this is called photoperiodism

Temperature: Growth is the final morphological expression of various metabolic activities taking place in the plant. Temperature influences all these metabolic activities and hence influences growth also. Generally a temperature range that is suitable for other metabolic activities is ideal for growth also. Though growth occurs in the range of 4°C - 45°C cptimum range is 28°C - 35°C. Very low temperature is injurious to the plant (cold injury), so also very high temperature. Temperature also influences the process of flowering (vernalization).

Water: Adequate supply of water is necessary for all physiological activities including growth. Plants under water stress cannot grow properly.

Oxygen supply: Proper supply of oxygen will ensure the process of growth by providing adequate energy to the dividing cells by way of respiration. It is possible that 0_2 may have some influence on the G_1 stage of cell division.

Nutrition:Proper availability of nutrition, both organic and inorganic is necessary for growth. Plants under starvation have retarded growth. Supply of essential elements is necessary for the functioning of many enzymes which influence growth.

C/N Ratio: Growth pattern is controlled by the ratio of carbohydrates to nitrogenous compounds. A higher nitrogenous material in relation to carbohydrates is known to produce more of protoplasm, less mechanical tissues and strong vegetative growth. A higher amount of carbohydrates in relation to nitrogenous matter on the otherhand favors the production of wall material, mechanical tissues etc.

Genetic factors: Genetic characters like the types of genes that an individual plant inherits has an enormous influence on growth. Proper growth takes place due to the interplay of genetic and environmental factors.

Chemical factors: Studies on growth have revealed that certain chemicals synthesized by the plants themselves have a profound influence on growth. These are organic compounds synthesized in minute quantities but are very potent growth regulators. These are called Growth hormones (Phytohormones, growth promoters).

GROWTH HORMONES

Among the many factors influencing growth, the most important ones are phytohormones. The term hormone is of Greek origin - *harmao* meaning to urge or to stimulate. Thimann (1948) suggested the usage of the term phytohormones. Since phytohormones primarily promote growth they are also called growth promoter's growth hormones etc.

Growth hormones are defined as "organic compounds synthesized in minute quantities in one part of the plantbody and are transported to another part where they are active". Philips (1971) defined hormones as-"substances which are synthesized in particular cells and which are transferred to other cells where in extremely small quantities they influence the developmental process."

Growth hormones are classified as follows-Naturally occurring growth hormones and synthetic (hand made) growth regulators. While naturally occurring hormones always promote growth, some of the synthetic growth regulators may even retard growth. The following is a list of natural as well as synthetic growth hormones/regulators.

1. Auxins (IAA, IBA etc)
2. Gibberellins (GA)
3. Cytokinins (Kinetin, Zeatin etc)
4. Ethylene
5. Dormins (ABA)
6. Flowering hormone (Florigen etc)
7. Phenolic substances
8. Synthetic growth regulators

AUXINS

Charles Darwin (1880) the great evolutinoist had hypothesized the presence of growth substances in plants. In his book 'The power of movement in plants' Darwin has described the experiments concerning the growth movements in canary grass (*Phalaris canariensis*). He had noticed the unilateral bending of the shoot tip (Coleoptile) towards the source of light. Other significant observations made by Darwin on the phototropic responses in grasses are

(i) Seedlings grown in dark do not bend

(ii) Curvature occurs only when the tip is illuminated

(iii) Decapitating the shoot or covering the shoot tip with a light insulated material does not cause curvature.

According to Darwin, when seedings are exposed to a lateral source of light, some inducer is transmitted from the upper part to the lower part causing the curvature.

The following are some of the names concerned with the discovery of aunixns. Boysen Jensen (1910, 1911, 1913): Conducted significant experiments concerning the phototropic responses in *Avena* (oats) coleoptile. He found that decapitated tips do not exhibit curvature, but replacement of the tip on the cut end of the shoot reintroduces curvature under the influence of lateral light. In another experiment he cut off the coleoptile tip of the seedling are replaced it with a thin plate of gelatin between the tip and the cut stem and observed that the coleoptile tip could still bend towards the source of unilateral light. Although he did not give any explanation, it was clear from his experiment, that a material substance which can pass through gelatin acted as the stimulus for curvature.

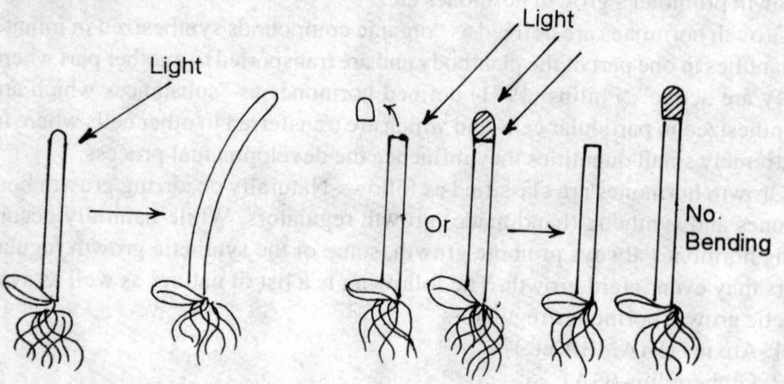

Fig. 19.1 Diagram showing Darwin's experiment with phototropism

Paal (1918) demonstrated that replacement of the coleoptile tip on the cut end of the shoot, unevenly would cause curvature towards the otherside eventhough there is no light (experiment being conducted in darkness). This indicates the presence of a diffusible substance at the tip.

F.W. Went (1917) finally provided the proof for the presence of a curvature inducing substance. He cut off the coleoptile tips and placed them on Agar blocks for sometime. When these agar blocks were kept asymmetrically on the cut ends of coleoptile, there was curvature even in darkness. This experiment clearly showed that some substance from coleoptile tip diffused into agar blocks. Went named this substance auxin and opined that no growth can take place without these auxins.

The technique of asymmetrically keeping the agar blocks on *Avena* coleoptile tips, to measure the auxin activity is the best bioassy for auxins and is popularly called the "Went Avena coleoptile test".

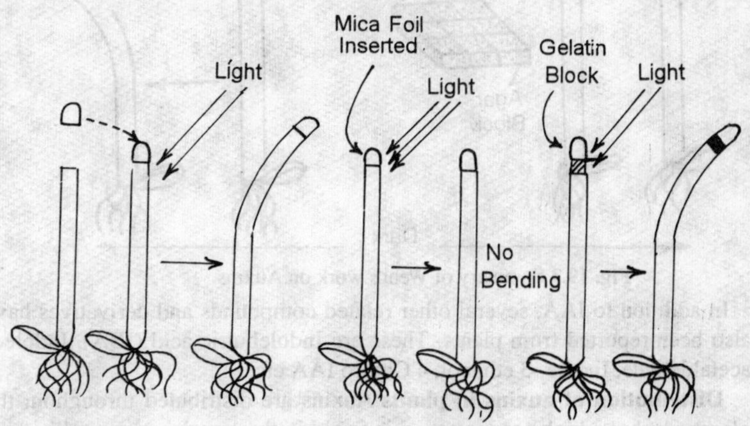

Fig. 19.2 Diagrammatic representation of Boysen-Jensen's work

Went (1928) also compared the quantity of the auxin in the illuminated and the non illuminated regions in the AVena coleoptile tip and found out that the shaded area has more (57%) auxin content than the illuminated area (28%). Auxins always move towards the base from the tip even if the orientation of the plant is reversed.

Chemical nature of auxins

Isolation and chemical characterisization of auxins was done by Kogl (1931) and Thimann (1935). In their studies on human urine, they isolated two compounds and named them auxin a and auxin b. Studies conducted on the culture media extracts of the fungus *Rhizpous* (Thimann 1935) helped in the isolation of a substance viz Indole acetic acid (IAA).

Kogl and Hagen Smit (1931), isolated about 40 mg of auxin a from nearly 150 iitres of human urine. Later auxin b was also isolated from corn germ oil. Heteroauxin, another type of auxin was also isolated from human urine subsequently.

Chemically, auxin a is auxentriolic acid ($C_{18}H_{32}O_5$) while auxin be is auxeniolonic acid ($\mathbf{C_{18}H_{30}O_4}$). In plants however auxin a and b are hardly known. Heteroauxin which is chemically IAA ($\mathbf{C_{10}H_9O_2N}$), is the principal naturally occurring hormone in higher plants.

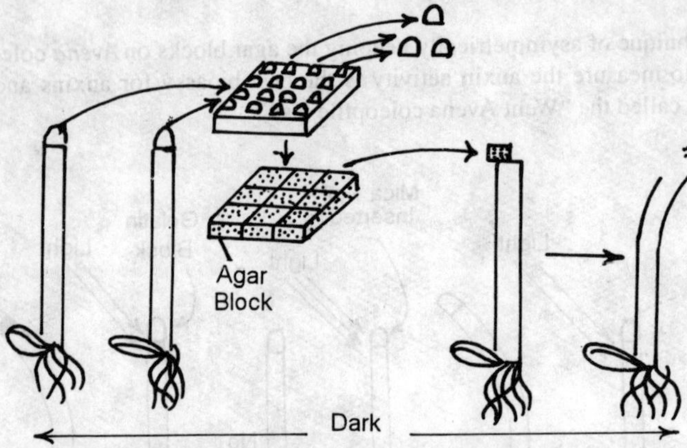

Fig. 19.3 Summary of Went's work on Auxins

In addition to IAA, several other related compounds and derivatives have also been reported from plants. These are Indolebutric acid (IBA), Indole 3 acetaldehyde, Indole -3 ethanol, 4 Chloro IAA etc.

Distribution of auxins in plants:Auxins are distributed throughout the plantbody, but a high concentration is found in the growing regions like root and shoot apices. Meristematic tissues have a high concentration of auxins.

Mechanism of auxin action: The mechanism of auxin action has not been fully understood. The following are some of the hypotheses-

(i) IAA increases the plasticity of cell walls bringing about elongation.

(ii) IAA induces DNA to synthesize mRNA which codes enzymes for cell elongation.

(iii) IAA increased the rate of oxidative phosphorylation and thereby increases energy supply. This enhances the rate of growth.

(iv) IAA induces the production of etheylene which in turn promotes growth.

(v) According to some of the recent finds in, the target cells (auxins do not act on all cells. The cells on which auxins act are called target cells) have a specific protein called receptor protein which selectively binds auxins to a given cell. Auxin receptor proteins have been detected in many target cells (Dodds and Hall, 1980, Rubery 1981).

A theory has recently been proposed to explain the influence of auxins on cell wall. According to this theory, called the Acid-growth hypothesis (Tiaz 1984) auxins influence target cells to secrete H+ into the surrounding primary walls. This lowers their pH and loosens the wall material allowing them to elongate.

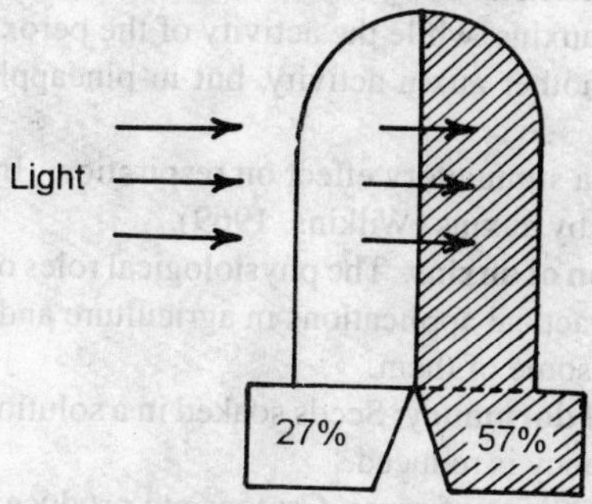

Fig. 19.4 Diagram showing differential concentration of Auxins in the light stroke shaded area of Avena coleoptile tip

Physiological Effects (Role) of Auxins

1. **Apical dominance**: In a great majority of flowering plants, if the apical bud is intact, it suppresses the development of lateral buds. When the apical bud is removed, lateral buds become active. The phenomenon of suppression of lateral buds. by apical bud is called apical dominance. It has been found out, that apical dominance is not due to the mere physical presence of the apical bud, but due to the production and transportation of auxins in/from them. It has been reported that during their polar migration auxins promote the production of axillary buds.

2. **Cell division and cell elongation**: Auxins stimulate cell division in cambium and also in the primary meristems. Cell elongation brought about by additional production of wall materials (see mechanism of auxin action) is an important physiological role played by auxins. Another important characteristic of auxin is its differential action on shoot tip and root tip. While a higher auxin concentration promotes cell division at shoot tip, it actually inhibits cell division at root tip.

3. **Root formation**: Auxins induce root formation at low concentrations. This is made use of by horticulturists to induce root formations in stem cuttings. The concentration is very critical because a higher concentration inhibits root growth.

4. **Prevention of abscission layer**: Auxin prevents leaf fall by suppressing the formation of abscission layer.

5. **Parthenocarpy**: Parthenocarpy or seedless formation of fruit is induced by auxins. Spraying of auxins on flowers induces parthenocarpic development of fruits in tomato, apples, cucumbers etc.

6. **Xylem differentiation**: Axins can induce the differentiation of xylem vascular differentiation in callus, produced during grafting.

7. **Miscellaneous activities**: Enzymes concerned with aspartic acid metabolism are activated by auxins, while the activity of the peroxidases is lowered. Floral inhibition is another auxin activity, but in pineapple auxin promotes flowering.

Auxins also have a stimulatory effect on respiration. Increased synthesis of m RNA is induced by auxins (Wilkins, 1969).

Practical application of auxins: The physiological roles of auxins explained above have several practical applications in agriculture and horticulture.

The following are some of them.

1. **Germination and dormancy**: Seeds soaked in a solution of auxins germinate quickly as dormancy is reduced.

2. **Rooting**: Stem cuttings of roses, Crotons etc produce adventitious roots profusely if soaked in auxin solution of appropriate concentration.

3. **Parthenocarpy**: Treatment of flowers with IBA has induced parthenocarpy is *Solanum melongena.* Improved fruit setting is noticed in several plants by the use of IAA. Treatment with auxins like IAA, IBA, 2-4D is known to prevent premature drop of fruits.

4. **Weed Control**: This is one of the major applications of auxins in agriculture. Many of the synthetic auxins (artifically produced)like 2-4D (2-4 dichlorophenoxy acetic acid) are selective weed killers and are employed in fields. 2-4D kills broad leaved plants, while it is relatively ineffective against narrow leaved ones. Hence 2-4D can be used to kill dicotyledonous weeds in a field with monocotyledonous crops. It has been demonstrated that 2-4D increases carbohydrate metabolism (by increased rate of respiration) so that plants die due to exhaustion of nutrients. Another advantage of 2-4D is, it is less toxic to animals and human beings.

The following is a summarised list of the role of auxins in agriculture and horticulture.

(i) Vegetative propagation of cuttings, grafts etc
(ii) prevention of fruit fall
(iii) Induction of parthenocarpy; Increased fruit set
(iv) Maintaining dormancy in buds
(v) Breaking of dormancy in seeds
(vi) Control of flowering
(vii) Prevention of defoliation
(viii) Selective weed killing

GIBBERELLINS

Gibberellins are another kind of growth hormones discovered first in Japan. Some of the rice seedlings in the field of a Japanese farmer (Konishi 1898) were growing unusually tall due to the infection from Deuteromycetaceous fungus - *Gibberella fujikorii* (= *Fusarium moniliforme*). The seedlings generally collapsed due to the unusual elongation of the slender stem. The disease was called the *bakanae disease* (foolish seedling disease). Kurosawa (1926) who was working in Formosa observed that the extract of the fungus (sterile filterate) could cause the same disease in rice seedlings. Later Yabuta (1935) and Yabuta and Sumiki (1938) isolated two gibberellins from the fungus viz., Gibberellin A and B. Subsequently gibberellins were isolated from a number of higher plants. West and Phinney (1956) isolated gibberellins first from the immature seeds of *Phaseolus cocineus.* Curtis and Cross (1954) of ICI (Imperial chemical laboratories) in U.K. isolated gibberellins and called them gibberellic acid (GA). Subsequent work on gibberellins revealed that there were at least six of them and they were named GA_1, GA_2, GA_3, GA_4, GA_7 and GA_9. As of today about 60 different types of Gibberellins have been isolated from a variety of plants and plant organs.

Chemistry of Gibberellins: All the known gibberellins possess a gibbane skeleton. Chemically Gibbane skeleton is made up of cyclic diterpenes. The molecular formula of various gibberellins is as follows-

(i) $C_{19}H_{24}O_6$ (GA_1)
(ii) $C_{19}H_{26}O_6$ (GA_2)
(iii) $C_{19}H_{22}O_6$ (GA_3)- Gibberellic acid
(IV) $C_{19}H_{24}O_5$ (GA_4)

Structurally gibberellins contain a r lactone ring on a cyclohexane ring.The various types of gibberellins differ in their attachment of groups like OH, COOH etc.

Types of gibberellins: As has already been pointed out, about 60 chemical types of gibberellins have been isolated and they are named GA_1, GA_2 ...GA_{60} etc., Of these, *Gibberella fujikorii* accounts for about 20 and about 51 types are reported from higher plants. Gibberellic acid (GA_3) has been reported from variety of plants and plant parts such as ears of *Hordeum vulgare* and *Festuca pratensis*; seeds of *Echinocystis macrocarpa*; roots of *Eichhornia crassieps*; petals of *Cassia fistula*; endosperm of *Cucurbita maxima*, young seed and seedlings of *Pisum sativum*, *Ricinus communis* and *Zea mays* etc.

Mechanism of Gibberellin action: Several possible modes of action have been suggested. Some of these are listed below

(i) Gibberellins induce auxin synthesis and consequent ly enhanced growth.

(ii) Gibberellins act at the gene level, possibly activating certain genes by bringing about derepression. This leads to enhanced mRNA synthesis and eventually enzymes. Broughton (1968), Johri and Varner (1968); Grishaber and Fellenberg (1972) have provided evidences for GA induced mRNA synthesis in dwarf pea plants. Another evidence which favours this, is the inhibition of mRNA synthesis retarding the influence of GA. Mozer (1980) has reported that in barley, GA increases the transcription of gene that codes for amylase.

(iii) Gibberellins promote growth by increasing the rate of hydrolysis of starch, fructans and other polysaccharids into simple sugars.

(iv) Wall plasticity is increased by the action of gibberellins.

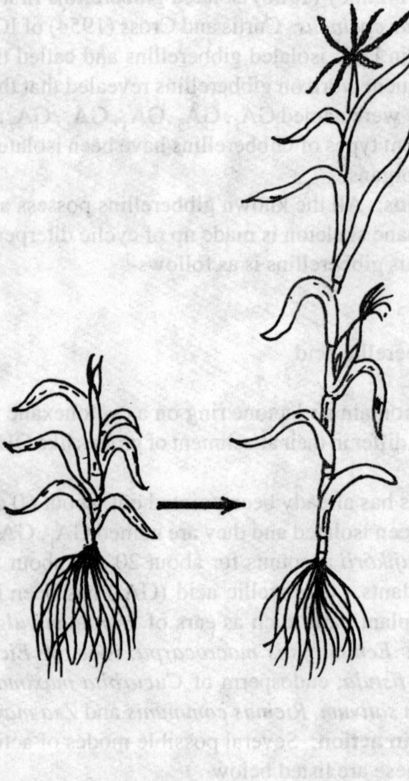

Fig. 19.5 Foliar spary of Gibberelins causing intermodel alongation

Physiological Effects (Role) of Gibberellins

1. **Dormancy and seed germination**: Gibberellins are known to break dormancy in seeds and thus hasten the process of germination (Ross and Bradbeer, 1968) Dormancy in potato tubers is also broken by GA.

2. **Root Growth**: Genrally GA inhibits the growth of roots (Stow and Yamaki, 1957). It has been reported that in lettuce and cucumber, radicle gets elongated due to GA action.

3. **Stem elongation (Internodal elongation):** By far the most pronounced physiological effect of GA is in stem elongation particularly at the internodes. Seedlings of tomato, lettuce etc sprayed with GA show elongation of stem. This property of GA has been made use of in treating genetic dwarfism in plants. Moose, Osgood and Ginoza (1980) have reported that rosette sugarbeet treated with GA3 reached normal height with improved sugar content. Similar have been the results of experimentation with dwarf varieties of pea and corn.

4. **Apical dominance**: The role of gibberellins in apical bud dominance is not certain. While the reports of Scott et al (1967) suggest that decapitatated shoots supplied with GA continued growth, the work of Hillman (1970 suggests an opposite effect. But in some instances atleast gibberellins might promote apical growth by blocking the action of dormins (endogenous growth retardants that promote dormancy).

5. **Bolting**: Development of elongated shoot (bolting) in dwarf plants is encouraged by the use of GA3.

6. **Role in flowering**: GA treatment to plants can substitute long day exposure necessary for flowering. GA can also replace the effect of vernalization.

7. **Cell division**: GA is known to activate cell division in vascular cambium. The work of Lin and Loy (1976) has demonstrated that GA promotes cell division by stimulating cells in the G1 phase to enter the S phase and shortening the S phase itself.

8. **Male Sterlity**: GA3 is known to induce male sterility in plants by suppressing the growth of androecium. Spraying of gibberellins in grapes supreses the development of stamens.

9. **Parthenocrapy**: GA promotes the synthesis of the enzyme amylase in the aleurone layer of endosperm in germinating barly grains. This has been experimentally proved by the *in vitro* treatment of aleurone layer by GA.

10.**Synthesis of amylase**: GA promotes the synthesis of the enzyme in the aleurone layer of endosperm in germinating barley grains. This has been experimentally proved by the in vitro treatment of aleurone layer by GA.

Agricultural and Commercial applications of Gibberellins

(i) Enhancement of seed germination by breaking dormancy
(ii) Reversal of genetic dwarfism
(iii) Expansion of leaf areas in commercial crops
(iv) Intoducing male sterility and parthenocarpy
(v) Elongation of peduncle length in thompson seedless grapes. Application

of GA3 is known to increase the size of berries directly as well as indirectly. GA increases the length of the pendEuncle in between the flowers. As a result fruits will be less crowded and they grow larger due to available space. Less crowding of the fruits will also allow aeration and prevent decay during transportation. This practice on a very large scale in the Central and Imperial valleys of California (USA).

(vi) Gibberellins are used in breweries to enhance 'malting' due to their starch digesting activity.

ETHYLENE

It has been known since a long time that certain gases stimulate fruit ripening eventhough its scientific study dates back to the beginning of the 20th century. Ancient Chinese knew that fruits stored in a room where incense is burnt ripen quickly (Salisbury and Ross, 1986). Denny (1921) reported that ethylene might help in fruit ripening. A conclusive evidence however was ultimately provided by GAne (1934) who reported the role of ethylene synthesized by plants themselves contributing to the ripening of the fruits.

Ethylene may be regarded as a volatile hormone as it is a gas under ordinary temperature. Maximum rate of ethylene production has been reported in many fruits just before senscence. Ethylene production has also been seen in etiolated pea seedlings. In light however ethylene production drops down.

Biosynthesis of ethylene: Available evidences seem to point out that methionine, one of the amino aacids may act as the precursor of ethylene (Morris and Mapson, 1964). Methionine is first converted to S - adenosylmethionine (SAM), the reaction requiring the participation of ATP. This is next converted to ACC (1-aminocyclopropane -1-carboxylic acid) under the influence of the enzyme ACC synthase (Adams and Yang, 1979).

Philips (1971) has suggested two alternate pathways of ethylene synthesis with methionine as the precursor.

(i) Methionine —FMN / Light—> Menthional —Peroxidase / H_2O_2—> Ethylene

(ii) Methionine —Trans-—> 4 Methyl thio butyric acid + H_2O_2 —Peroxidase—> Ethylene

Physiological Roles of Ethylene

1. **Abscission**: Ethylene promotes the changes that occur before the abscission of leaves, flowers and fruits. It promotes cellulase activity.

2. **Ripening and climateric rise**: Climateric rise is a term employed to denote the increased rate of respiration in mature and senescing tissues such as ripe fruits. Originally it was believed that ethylene is a product and not cause of senescence. Recent studies however have shown that ethylene production begins even before the climacteric rise.

3. **Activation of enzymes**: Ethylene activates the action of certain enzymes like malic enzyme and pyruvate decarboxylase. Chlorophyllase, an enzyme that degrades chlorophyll is also activated by ethylene.

4. **Elongation of roots and stems**: Ethylene inhibits root and stem elongation in dicots.

5. **Effects on flowering**: Although, ethylene generally inhibits flowering in most plants, in some of the bromeliads like-pineapple, ethylene is known to initiate the flowering. An ethylene releasing substance has been commercially produced (ethrel) to promote flowering in pineapple. Ethylene also produces the degreening effect in citrus and banana fruits.

ABSCISSIC ACID

Abscissic acid is a naturally occuring growth (inhabitor) hormone playing an important role in abscission and dormancy. Evidence for the occurrence of a natural growth inhibitor first came from the work of Hamberg (1949) who showed that dormant tubers of potato contain several inhibitors; while the levels of inhibitors was considerably less in sporuting tubers. Originally these inhibitors were identified as b inhibitor complex. Subsequently it was demonstrated by Millborrow (1967) that the main inhabitor in the inhibitor complex was Abscissic acid (ABA). It was Addicott et al (1963) who first isolated and characterized ABA from maturing cotton fruits. They identified two compounds Abscissin I and Abscissin II (ABA).

Chemical structure and biosynthesis of ABA

ABA is a 15 carbon sesquiterpenoid. It can exist as + or - enantiomers. Naturally occuring ABA is always in the cis form and more active than the *trans* form.

Biosynthesis of ABA is believed to follow the mevalonic acid pathway common to many other isoprenoids. There is evidence to indicate, that photodegradation (sometimes enzymatic also) of violaxanthin (in chloroplast) might give rise to ABA . Violaxanthin first forms Xanthoxin which gets converted into ABA.

ABA can be inactivated by the esterification of the carboxyl group by glucose to form an ABA - glucose ester.

ABA is translocated both through phloem as well as through xylem. Parenchyma cells surrounding the vasculature also help in ABA translocation, as the migration is not polar.

Mechanism of ABA action: ABA has three routes of influence in plant metabolism.

(i) Inhibition of RNA synthesis (transcription)

(ii) Inhibition of protein synthesis (translation)

(iii) Influence on root plasma membrane by making them more positively charged.

Physiological Effects (Role) of ABA

1. **Bud dormancy**: Experimental results of Wareing et al (1965) have clearly revealed the inhibitor role of ABA in leaves and buds. Application of ABA to non dormant buds causes dormancy in *Betula*, Acer etc.

2. **Senescence**: ABA encourages leaf senescence in *vitro*. Whether endogenous AGA performs the same function *in vivo* is not yet proved.

3. **Abscission**: Observations of Frederic Addicott (1964) have revealed that ABA is responsible for abscission in cotton fruits.

4. **Flowering**: ABA induced flowering during long days has been noticed in several short day plants such as *Ribes, Phorbitis, Fragaria* etc. But this influence is not universal as reported by its (ABA) failure to initiate flowering in a typical short day plant such as Xanthium.

5. **Stomatal functioning**: Membrane functioning is altered by influencing the rapid loss of K+ ions from guard cells.

6. **Release of Ethylene**: ABA influences the release of ethylene and causes senescence.

7. **Inhibition of other hormones**: ABA counteracts the influence of GA and possibly IAA also. Hydrolase production influenced by GA is inhibited by ABA.

Twenty

Introduction to plant ecology

Plants growing on the surface of the earth exhibit different characters in different environmental conditions. This is because the environment plays a profound role in the growth and development of plants. Plant and animal life is affected by various factors of the environment and they in turn modify the environment in various ways. An intimate relationship exists between plants (animals also) and their environment. In short, plants and environment are interdependent.

Environment determines the density of plants and whether they can exist or not. The plants growing in any particular area, fully adapt themselves to that habitat. In other words they are in perfect harmony with the environment.

What is Ecology?

The science of Ecology also called environmental biology is extremely diversified and has transcended the limits of biology. In the broadest sense it deals with organisms to relation to their environment.

The credit of coining the word Ecology goes to Reiter (1868), who actually called it Oekologie, by combinig two Greek words *Oikos* (house or dwelling place) and *logos* (study), to denote the relationships, between organisms and their environment. Thus ecology may be said to be the study of organisms 'at home'

Definition of Ecology

Various authors have defined ecology in various ways. Some of these are given below.

Haeckel (1869)- Ecology is the science treating of reciprocal relations of organisms and the external world.

Warming (1905)-Study of organisms in relation to their environment.

Clements (1916)- Ecology is the science of community.

Elton (1927)- Ecology is the scientific natural history dealing with the sociology and economics of organisms.

Woodbury (1955)- Ecology is a science which investigates organisms in relation to their environments.

Odum (1963)- Ecology is the study of structure and function of nature.

A simplest definition of ecology would be to define it as the study of interactions of organisms with their environment.

Scope of Ecology: Of all the branches of science that come very close to human life is ecology, because human beings also interact with the environment and are influenced by it. Present day human problems are directly or indirectly related to ecology as their solution requires an understanding of ecology. Ecology has been contributing very much these days to socio economic, political and other policies of the world. The concern to conserve the environment has awakened many people to rise against the destruction of natural wealth. In this context one may mention the controversy regarding the silent valley of Kerala and Bedthi area of Karnataka. In both these areas hydroelectric projects were planned, but the environmentalists have been fighting against it as vast areas of natural wealth will be destroyed.

The changes in climate, weather, monsoon are all related to changes in the environment. A thorough knowledge of ecology is very much essential to monitor all these. Ecology thus plays a very vital role in human welfare in having its influence in agriculture, management of grass lands, forestry, biological surveys, pest control, fishery biology, pollution control, soil conservation, wild life management etc. In short, one can say that ecology influences every field of human endeavour.

Approaches to Ecology:

Ecological studies are based on three principal aspects viz., Taxonomic affinities., Habitat and Levels of organisation. Accordingly, ecology may be sub divided into the following.

(1)Based on taxonomic affinities, ecology may be divided into Plant ecology and Animal ecology. Modern studies however do not distinguish between these two as both plants and animals are also interdependent.

(2)Based on Habitat, ecology may be classified into fresh water ecology, marine ecology, forest ecology etc.

(3)Based on levels of organisation, ecology may be sub divided into Autoecology and Synecology.

Autoecology is the study of an individual organism and its relationship with the environment. It may be called ecology of individuals. In autoecology, individuals are the units of study. Individuals are studied for details of their distribution, morphology, pattern of life cycle and the environmental factors which influence these.

Synecology is the study of groups of organisms together with reference to their environment. Here, groups of organisms rather than the individual, constitute the units of study. Since organisms live in nature as groups and not as isolated individuals, synecology is believed to reflect the natural conditions better than autoecology. Synecology also takes into consideration the interactions between organisms.

Synecology may be further sub-divided into, population ecology, community ecology, ecosystem ecology etc.

Units of Ecological study: Ecological studies are made at different levels. The main aim of study at these different levels is based on different ecological units like individual ,groups of individuals etc., The ecological units are as follows.

Species: This is the lowest stable unit in taxonomic hierarchy. This forms the main aim of study when an individual species has to be studied in relation with others. Ecological study where an individual species is the centre of attention is known as autecology.

Population: A population may be defined as a group or collection of individuals belonging to a single species. The units of study are populations. Study of interaction between individuals in terms of growth, reproduction, dispersal etc., constitutes population ecology.

Community: A communituy may be defined as populations of different species bound together in a common habitat. The members of a community may be plants as well as animals. Only the living components of a community however are studied here mainly to analyse the nature of interdependence.

Ecosystem: An ecosystem may be defined as the total environment including both biotic and abiotic components. In any environment a mere study of interactions between living organisms will not give a total picture of the complexities because in the growth, development and other parameters of living beings the abiotic components play great role. The role of the abiotic components such as climate, soil , temperature etc., have to be analysed before one can understand the total parameters of living beings. It is this realisation of ecologists that has lead to the concept of ecosystem. The concept of ecosystem was first proposed by Tansley (1935) who defined it as "the system resulting from the integration of all living and non living factors of environment". Thus according to Tansley not only the living communities but also the non living components constitute an integral part of ecosystem. Odum (1971) defines ecosystem as the basic functional unit of organisms and their environment interacting with each other and within their own components.

Concept of Biosphere:

The term biosphere appeared in scientific literature in 1875 in monographic work devoted to the geological structure of the alps. The author of the monograph Edward Sueb (1831 - 1914) wrote "one thing seems to be foreign on this large celestial body consisting of spheres, namely - organic life.....on the surface of constiuents it is possible to single out a self contained biosphere.......

Sews coined the term concurrently with two other similar terms hydrosphere and lithosphere. Russian mineralogist Vladimir Ivanovich Vernadsky has elaborated the concept of biosphere since 1911.

Biosphere may be defined as that part of the earth which consists of living matter. In his various works Vernadsky gave several definitions of the biosphere always emphazizing its two distinctive features. The first feature of the biosphere is the envelope of life, i.e. the area of existence of living matter...."

The second feature " The biosphere can be regarded as the area of the earth's crust occupied by transformers which convert cosmic radiations into effective terrestrial energy: Electric, chemical, mechanical, thermal etc.

Extent of biosphere: Life is present all over the earth, except in very dry deserts, frozen ice peaks and toxic salt basins. Biosphere extends down in the oceans up to 9000 meters, below the surface of the soil over 300 meters and up in the air upto 1500 meters. Compared to the volume of earth, life is confined to thin layer of earth's surface.

Divisions of biosphere: Biosphere is further divided into (a) lithosphere - land (b) hydrosphere - water and (c) atmosphere - air.

Lithosphere: The outer layer of the earth crust called soil forms the lithosphere. It supports life and contains eight important elements like O_2, Si, Al, Fe, Ca, Na, K, Mg. In addition traces of other elements are also present.

Hydrosphere: It is the water, covering about 70% of the earth's surface. The greater part of hydrosphere is ocean, rest is formed by river, lakes and ponds. Ocean water is saline and contains about 3.5 Kg. of salts in every 100 kg. ocean water. Large number of organisms are found in hydrosphere compared to lithosphere. It is believed that life first originated in the hydrosphere and then entered into the lithosphere.

Atmosphere: It is a gaseous envelope called air surrounding the earth. It also permeates into earth crust and water. Atmosphere is denser near the earth and lighter at higher atitudes. Similarly is the atmospheric pressure. Dry atmosphere has N 78% O_2 20% Argon 0.93% CO_2 .033% and trace components abut 0.033%. Ozone layer is present at a height of about 32 - 48 Km. It acts as a screen and checks the entry of ultraviolet rays. Atmosphere does not have permanent fauna or flora in it like other two components biosphere.

Features of biosphere: The important features of biosphere are (1) nutrient cycling and (2) energy dissipation.

Components: The biosphere consists of two main components namely non living environment and living organisms.

Twenty one

Ecosystems

There is an intimate relationship between the different components of the environment. The relationship is so close and so intimate, that any disturbance to any component has its effect on the entire environment. In the beginnings of ecology, interrelationships and interdependencies between the organisms was stressed as forming the foundation of ecological sciences. But as knowledge progressed regarding the various components of the environment, it became clear that the non living components of the environment have a pivotal role in governing the interrelationships between organisms. Hence any proper appreciation and understanding to the interrelationships between organisms must take into account, the prevailing abiotic components of the environment also. It is this awareness among ecologists that led to the concept of ecosystems. In the beginning of ecological study, an organism and its interrelationship was the centre of ecological study. As this did not provide any meaningful understanding, population and later community were the units of study. Modern ecologists always regard an ecosystem as a proper unit of study for the proper comprehension of all forces (biological as well as abiological) in nature. The concept of ecosystem was first proposed by A.G.Tansley (1935) who defined it as 'The system resulting from the integration of all living and non living factors of the environment;. Thus according to Tansley, not only the living communities, but also the non living components constitute an integral part of the ecosystem. But the concept of ecosystem, (though not in the ful sense as we understand it today) seems to have been there in human mind as revealed by the use of terms like biocoenosis (Karl Mobius, 1877) microcosm (Forbes 1877) etc.

Some definitions of ecosystem:

Woodbury (1954): "Ecosystem is a complex in which habitat, plants and animals are considered as one interacting unit, the material and energy of one passing in and out of others".

Odum (1971) "The ecosystem or the ecological system is the basic functional unit of organisms and their environments interacting with each other and within their own components".

The size of an ecosystem: It varies with several factors such as area and type of plants, animals or microorganisms involved. An ecosystem may be as small as a drop of pond water - (a micro ecosystem) or perhaps an entire ocean. For the purpose of study, a unit area having certain characters may constitute an ecosystem. For example a pond, a lake, a river, a grassland, an ocean etc., But considering the total influence of all the organisms on the total environment of the earth and also having regard to the fact that the various ecosystems mentioned above (pond, lake etc) are interconnected with one another, the entire global environment may be regarded as constituting a single ecosystem.

An ecosystem may be said to be self sufficient or closed, when the components can exist, without influencing or without being influenced by other ecosystems. But in nature such an ecosystem can rarely occur. The adjacent ecosystems are always mutually influenced.

Components of an ecosystem : From the discussion about the ecosystem given above, it must be clear that there must be two components in an ecosystems viz. living and non living. From the structural point of view, Odum (1971) recognises four components in an ecosystem . These are:-

1. Abiotic components
2. Biotic components
   a) Producers
   b) Consumers and
   c) Decomposers

From a functional point of view an ecosystem has two components viz:-

1. Autotrophic component and
2. Heterotrophic component

The autotrophic component of the ecosystem traps the energy and prepares complex food materials. The heterotrophic component obtains energy from the autotrophic component and redistributes it in nature after utilization for its own maintenance.

We will now briefly analyse each one of these components.

***Abiotic Components*:** These are, the radiant energy of Sun, inorganic elements and compounds like atmospheric gases (CO_2, O_2, N_2 etc), soil etc. Source of energy of an ecosystem . Hence every ecosystem for its survival must have a community to trap the solar energy.

Biotic Components:

a) **Producers:** These are the most important components of an ecosystems. Producers of an ecosystem may be defined as those individuals which depend directly on the abiotic component for their survival and production of nutrients. Green plants are the chief producers of a negotiable energy source in the nature which could be utilized by all other living organisms. By their photosynthetic activity, the green plants trap the solar energy and convert it into chemical energy which is available in their body tissues.

Apart from the green plants, photosynthetic and chemosynthetic bacteria are the two types of autotrophic organisms in the ecosystem .

b) **Consumers:** The living components of the ecosystem which depend on producers for their nutrition are called consumers. All the animals and certain plants are included in this category. There are different types of consumers. These are:

a) Primary consumers
b) Secondary consumers
c) Teritary consumers, and
d) parasites and saprophytes

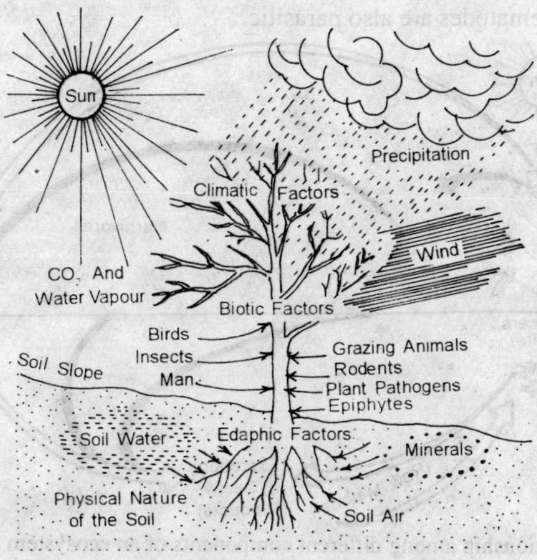

Fig. 21.1 The various components of an ecosystem

Primary consumers: All the herbivorous animals like rodents, cow, elephants, deer, goats etc, which directly consume the plants are called primary consumers. Among the aquatic animals, certain kinds of fish, crustaceans, molluscs etc which survive on phytoplankton are also primary consumers. According to Elton (1939), herbivorous animals are key industry animals, because all other animal's life is dependent on these primary consumers.

Secondary consumers: Carnivorous and omnivorous animals belong to this category. These animals predate on herbivorous animals. Omnivorous animals eat herbivorous animals as well as plants. Sparrow, Crow, Fox, Wolves, Cat, Dogs, Snakes etc belong to this category.

Tertiary consumers: They are strictly carnivorous animals that prey upon carnivores, herbivores, and omnivorous organisms. Lions, Tigers, Vultures etc are regarded at tertiary consumers.

Decomposers: These are also called transformers as they transform organic compounds into inorganic or simple compounds. Saprophytic fungi and bacteria belong to this category. They act upon the dead bodies of plants and animals and decompose them to their elemental stage. These in turn can be used by producers for their existence and photosynthetic activity. The decomposers occupy a pivotal place in the ecosystem as they indirectly support the producers.

Parasites: Plants and animals that infect other living components of the ecosystem and survive on them are regarded as parasites. Various types of fungi, bacteria and a few flowering plants are parasitic. Several protozoans, insects and nematodes are also parasitic.

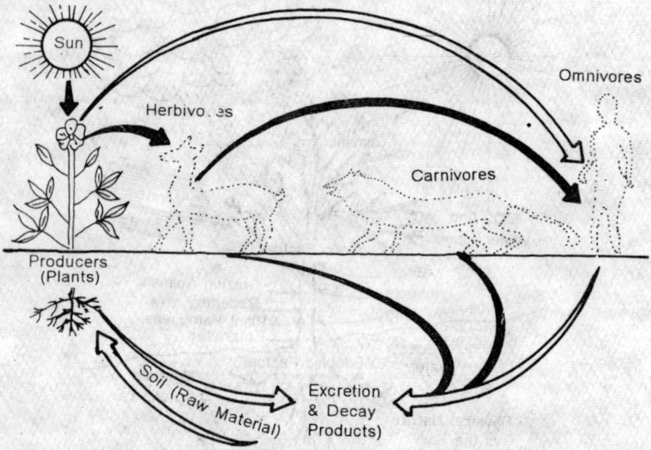

Fig. 21.2 Relationship among different components of an ecosystem

Development and evolution of ecosystem : Ecosystem development is otherwise known as ecological succession or plant succession. There are three important parameters in the development of an ecosystem . According to Odum (1971), these are:

1. It is an orderly process of community development that involves changes in species structure and community process.

2. It results from modifications of the physical environment by the community, that is succession is community controlled even though the physical environment sets the pattern.

3. It culminates in a stabilized ecosystem in which maximum biomass and symbiotic functioning between organisms are maintained per unit of available energy flow.

The complete sequence of communities, beginning with the pioneers and culminating with the stabilized ecosystem are called *seres*, while each com munity is called *seral stage*. Succession takes place because the seral communities change the environment to such an extent, that they no longer can live there. They render the conditions suitable for other populations until an equilibrium is achieved between the biotic and abiotic components of the ecosystem .

Evolution of ecosystem : As in the case of development of ecosystem their evolution also is governed by two forces viz., allogenic and autogenic forces. Allogenic forces are described as outside forces not produced by the biotic component. Geological, land and climatic changes are regarded as allogenic forces. Autogenic are inside forces resulting from the activities of the biotic components of the ecosystem.

The first ecosystem perhaps originated about three billion years ago. These were anaerobic heterophic microbes that sustained on the organic matter produced by abiotic processes. The origin of autotrophic microbes such as algae converted the reducing atmosphere into a oxygenic one, by the constant release of oxygen by their photosynthetic processes. From then on, through long geologic ages, ecosystems have become more complex and diverse, not only in their bodily structure but also in their functions. Within these communities evolution occurred primarily at the species level due to the process of natural selection.

TYPES OF ECOSYSTEM : The whole earth may be regarded as a giant ecosystem the biosphere. For the convenience of study however the ecosystem may be classified as follows:

I. Natural ecosystems: These work under natural conditions, if there is no interference by man. This is further classified into

a) Aquatic: ecosystem developing in water, Aquatic ecosystem depending upon the body of water in which is found may be

(i) *Fresh water* - in ponds, lakes rivers etc. Fresh water ecosystem is said to be lotic when found in moving water and is known as lentic when found in standing water.

(ii) *Marine* - in salt water such as oceans, estuaries etc.

(b) Terrestrial: These develop on land, Depending on the type of land the terrestrial ecosystem is classified into

(i) Forest ecosystem

(ii) grassland ecosystem and

(iii) desert ecosystem .

II. Artificial ecosystem : These are man made. They are created by man for his utilisation. The dominant factor governing these ecosystems is the human being. Croplands such as field of maize, sugarcane rice etc., belong to this category. Constructions of dams, lakes etc also may be called artificial ecosystems.

In this book, pond marine, grassland and forest ecosystem are described in some detail.

Pond ecosystem :

This is an example of a lentic fresh water ecosystem. The pond is a self sufficient ecosystem in which the interactions of biotic and abiotic components maintain a perfect equilibrium. The ecosystem may be divided into biotic and abiotic components.

Biotic components :

1. Producers: Algae such as blue greens and chlorophycean members, many higher plants and many photosynthetic bacteria are the producers. These trap the solar energy and prepare complex organic compounds. Producers are of two kinds. Macrophytes and Phytoplanktons. Many large hydrophytes such as *Trapa, Typha Elecoharis, Vallisnaria, Nymphaea, Chara, Nitella, Potomogeton* and a whole range of other green aquatic plants belong to the category of macrophytes. Free floating algae such as *Chlamydomonas, Scendesmus, Pediastrum, Chlorella, Volvox*, Diatoms, blue green algae are called phytoplanktons.

2. Consumers: These are heterotrophic organisms which depend on the producers for their nutrition. Consumers may be primary, secondary and tertiary. Primary consumers are fish, insect larvae, beetles etc. Many of the zooplanktons such as rotifers (*Brachionus*), protozoans (*Cyclops*) and crustaceans (*Coleps*) are also primary consumers.

Secondary consumers are the carnivores which feed on primary consumers. Insects and certain categories of fish belong to this category. Tertiary consumers are those that feed on secondary consumers. Certain large fishes belong to this group. In the pond ecosystem fishes occupy all the three trophic levels (primary, secondary and tertiary).

3. Decomposers: These include microorganisms such as saprophytic fungi, Actinomycetes and bacteria. They help in decomposing actively many inorganic compounds such as minerals are released into the pond to be utilized by producers. By various methods such as Martin's medium, Thornton's medium and Jenson's medium, fungi, bacteria and actinomycetes may be isolated and analysed. Some of the fungi present in a pond ecosystem are *Saprolegnia, Rhiocarpus, Aspergillus, Cladosporium, Pencillium, Pythium, Fusarium, Cephalosporium, Curvularia, Alternaria, Circinella* etc.

Abotic components: Abiotic components of a pond are water itself, Temperature of water, pH value of water, the dissolved organic and inorganic compounds, gases such as O_2, CO_2 ,N_2 etc. Many mineral elements are also present in dissolved state in water. The depth of the water and the fact whether light can penetrate and reach the bottom, decides to a great extent the development of submerged hydrophytes. Various methods are adopted to determine the pH, and the dissolved nutrients in water.

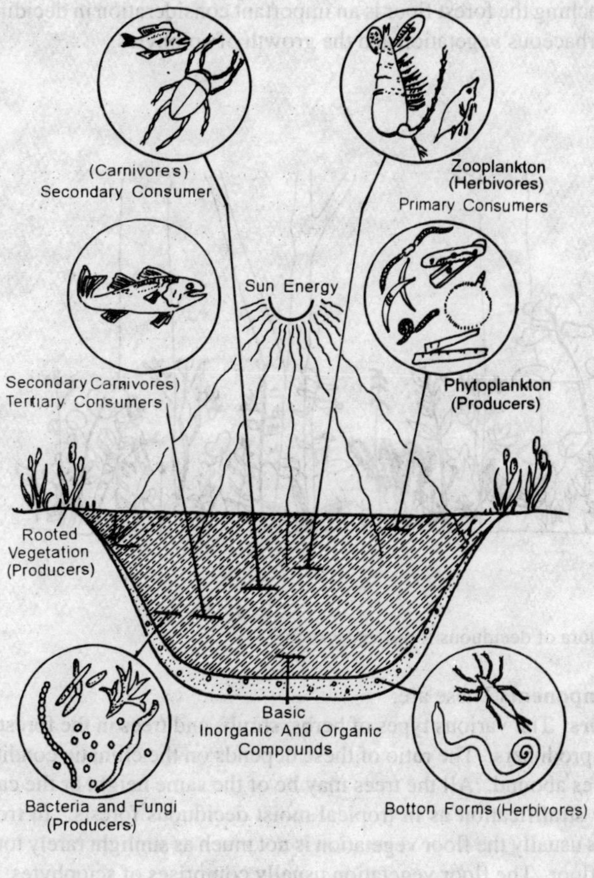

Fig. 21.3 Pond ecosystem

Tropical deciduous Forest ecosystem :

This is one of the most important of the terrestrial ecosystems. Forest occupy roughly 40% of the land. In India however forest cover is woefully less and it occupies about 10-11% of the land surface. The various components of the forest ecosystem are:

Abiotic component: The organic and inorganic contents of the soil and atmosphere constitute the abiotic components. Mineral elements, organic debris in the form of humus are the components, of the soil. Atmospheric humidity, precipitation, light are the other abiotic components. The amount of light reaching the forest floor is an important consideration in deciding the type of herbaceous vegetation and the growth of epiphytes.

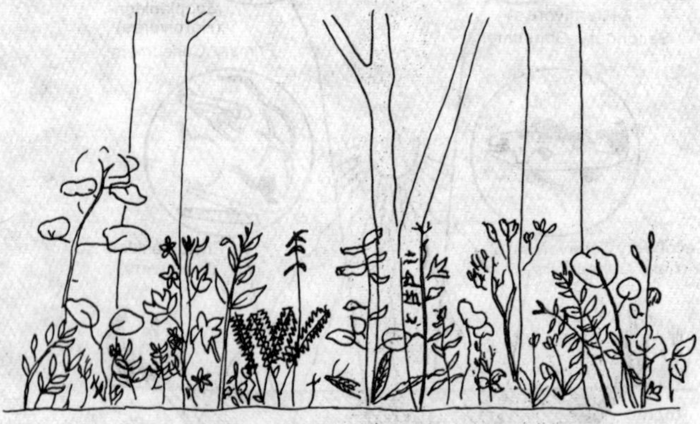

Fig. 21.4 Flora of deciduous forest ecosystem

Biotic components These are:

Producers: The various types of herbs, shrubs and trees in the forest constitute the producers. The ratio of these depends on the climatic conditions. Mostly trees abound. All the trees may be of the same height or the canopy may show stratification as in tropical moist deciduous forests. In tropical rain forests usually the floor vegetation is not much as sunlight rarely touches the forest floor. The floor vegetation usually comprises of sciophytes. Herbaceous plants requiring sunlight grow as epiphytes on the branches of trees. Other nonarborescent heliophytes become climbers or lianes. The important trees in a forest ecosystem are - *Tectona grandis, Butea frondosa, Shorea robusta, Lagerostromia parviflora etc in deciduous forests Terminalia tomentosa, Diospyros melanoxylon, Acacia arabica, Chloroxylon Spp, Bauhinia etc* in dry deciduous forest *Quercus, Acer, Betula, Thuja, Piecea* etc in temperate deciduous forests and*Abies, Picea Pinus, Cedrus, Juniperus, Rhododendron*, etc in temperate coniferous forests. The forests are mainly classified into various categories based on the climate and species components in the vegetation cover. These are desert ecosystems,, Grassland, Rain forest, Tropical moist deciduous, Tropical dry deciduous, Alpine forests, Alpine scrubs etc.,

Consumers: As usual these are of three categories- primary, secondary and tertiary. Herbivorus animals feeding on the leaves, tender shoots and fruits of producers are the primary consumers. These are ants, flies, beetles, locusts, leafhoppers, bugs, spiders etc., among small animals, Squirrels, Flying foxes, Mongooses etc are also primary consumers. Secondary consumers are carnivorous like snakes, birds, lizards, foxes etc. Birds may be primary consumers also. Tertiary consumers include top carnivores like lion, tiger etc. feeding on primary and secondary consumers.

Decomposers: Decomposing microorganisms like fungi and bacteria abound in the forest soil, particularly where leaf litter is plenty. *Aspergillus, Coprinus, Polyporus, Ganoderma, Fusarium, Alternaria, Geastrum* etc., among *fungi, Bacillus, Clostridium, Pseudomonas* etc., among bacteria and antinomycetes like *Streptomyces*, are the chief decomposers in a forest ecosystem. Tropical and subtropical forests have a rapid rate of decomposition than the temperate forests.

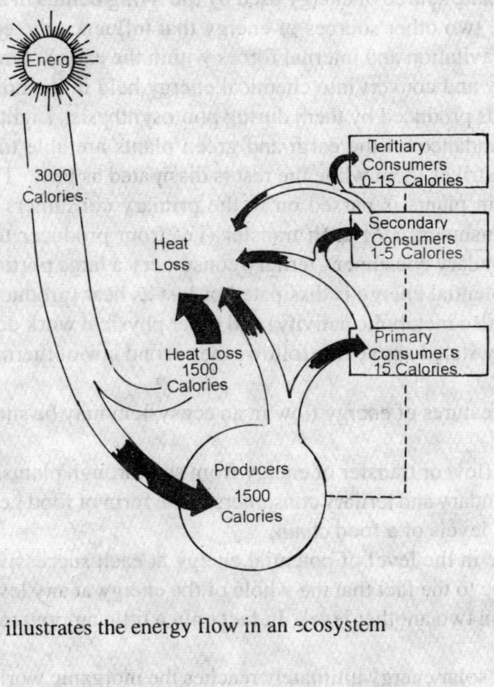

Fig. 21.5 It illustrates the energy flow in an ecosystem

Energy flow in an ecosystem :

Energy may be defined as the capacity to do work. There are two kinds of energy Potential and Kinetic. Potential energy is energy at rest, while Kinetic energy is energy at work or motion. Work that - results as a result of expenditure of energy can either store energy or rearrange matter without storing energy. The storage and expenditure of energy is described by two laws of thermodymanics. The first law called the law of conservation of energy states that ' Energy can neither be created nor destroyed'. It may change forms, assume shapes of various matters but there will not be any gain or loss. For example when wood is burned energy is released in the form of heat and released to surrounding. The heat so released will not be able to do any further work and it is called entropy. Dissipation of energy in a disorderly manner when it is incapable of any further work is called entropy.

The second law of thermodynamics states that whenever energy is transformed from one step to another step entropy increases and potential energy decreases. For instance when coal is burned to heat water, not all the heat goes into boiling water, some dissiptates into surroundings. The same thing happens to energy in an ecosystem.

Sun is the ultimate source of energy used by the living beings in an ecosystem. (There are two other sources of energy that influence the ecosystem. These are the gravitation and internal forces within the earth). Green plants trap solar energy and convert into chemical energy held in the bonds of organic compounds produced by them during photosynthesis. Light energy is available in abundance on the earth and green plants are able to use only about - 3% that strikes them while the rest is dissipated as heat. The energy that - is stored in plants is passed on to the primary consumers and from producers to consumers. At each transfer (i.e. from producer to primary consumer - secondary consumer - teritary consumer) a large portion - about 80 to 90% of potential energy is dissipated or lost as heat (produced during respiration and also metabolic activity) and other physical work done by the body. As already stated above this follows the second law of thermodynamics.

The essential features of energy flow in an ecosystem may be summarized as follows

(i) A constant flow or transfer of energy from sun through plants, primary, consumers, secondary and tertiary consumers in the form of food i.e., through different trophic levels of a food chain.

(ii) A decrease in the level of potential energy at each successive trophic level. This is due to the fact that the whole of the energy at any level cannot - be transferred in two another level. In fact only a little amount reaches the next level.

(iii) The entire solar energy ultimately reaches the inorganic world as heat.

Twentytwo
Tropic organisation

Trophic organisation refers to the nutritional relationship between the different components of the ecosystem . Basically all living components of the ecosystem require energy which is available to them in the form of food. In any ecosystem there will be producers who receive energy from the abiotic components and pass it on to biotic components in a stepwise fashion. In this chapter we shall study the following aspects of trophic relationships in the ecosystem :-

a) Food Chain - food web and

b) Ecological pyramids

Food chain:

In an ecosystem energy flows from one trophic level to another. A trophic level represents a group of organisms, which are either predators or they get consumed. All organisms in an ecosystem are linked to one another based on their nutritional needs. This relation between the individuals constitutes a linear chain called the food chain. Ultimately a food chain always begins with the producers. The various components of the food chain can be defined as a group of organisms in which there is transfer of food energy through a series of repeated eating and being eaten

Two types of food chains are recognized in nature. These are Grazing food chain and Detritous food chain.

Grazing food chain: The grazing food chain starts from green plants passes on to herbivorous primary consumers and ends with carnivorous animals. Thus the nutrition produced by plants utilizing energy
of solar radiation are used in the following three ways

(i) used for respiratory activity of the plant

(ii) get decayed

(iii) may be consumed by herbivores

In a grazing food chain there are basically four trophic levles - Auto trophic, primary consumers, secondary consumers and tertiary consumers. Grazing food chain is the most important food chain in nature.

Detritous food chain: Detritous or decomposing food chain begins with dead organic matter goes to micro organisms and then passes on to organisms that feed on detrivores (organisms that - eat detritus) and their predators. Ecosystems of this type are less dependent on direct solar energy. On the other hand they depend on the supply of organic matter produced by another ecosystem . Detritus food chains are seen in those areas where there is plenty of organic matter in the soil as in temperate forests.

In a detritous food chain organic matter to start the chain comes from outside. The main source of organic matter is the leaf fall from the trees producing lot of litter. The fallen leaf fragments are acted upon by saprophytic fungi and bacteria and are eaten by a small group of animals. These animals include crabs, copepods, insect larvae, grass shrimps, nematodes etc. All these animals are called detritus consumers or detrivores. These animals are in-turn eaten by small fish. These small fish form food for large fish, fish eating birds constituting the large carnivores.

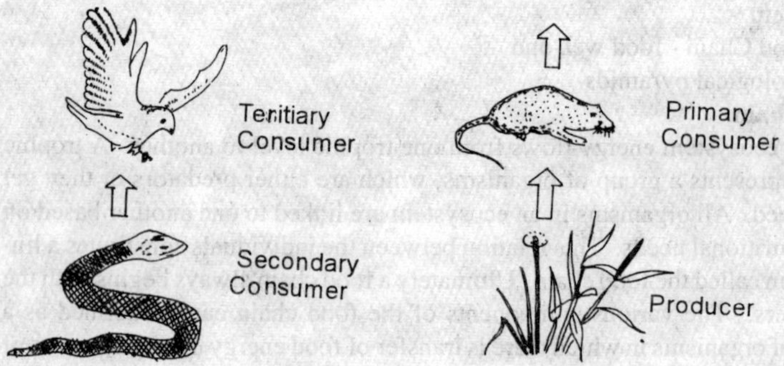

Fig. 22.1 A simple food chain in an aquatic medium (marine)

The above description clearly shows that fallen leaves generally regarded as waste contribute to a large food chain.

There are some similarities and differences between the two chains. The two chains begin different - The grazing chain begins with the energy entry from an inorganic source; the detritus chain receives energy from organic source (i.e. decaying leaves). But both the chains end up similarly i.e. carnivorous eating lower level animals. The two food chains i.e. grazing and detritus are ultimately linked to the same energy source i.e. solar radiation.

Food Webs:

A food chain represents a simple, linear one way relationship between one trophic level and the other. But in nature trophic relationships do not operate is such a simple manner, because if for some reason the link is broken between one trophic level and the other in a food chain it disturbs the whole ecosystem

In order to assure the stability of relation between different trophic levels, several links are established both in a linear and horizontal fashion and different food chains are interlinked.

A food Web may be defined as a net work representing the multiple inter links between various trophic levels is an ecosystem .

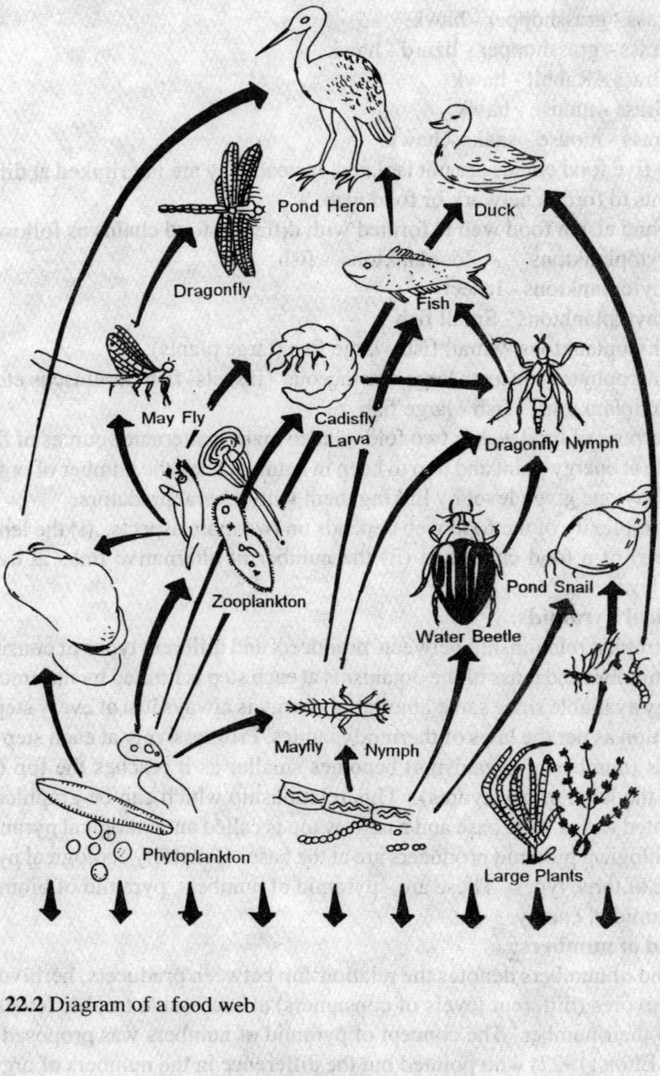

Fig. 22.2 Diagram of a food web

The trophic relationship in an ecosystem is more a web (network) than a chain. For instance in a grazing food chain the grass is consumed by rabbit as well as mouse. The mouse is turn may be eaten either by hawk or by snake which in turn is eaten by hawk. In other words these are always alternate sources of energy at different trophic levels.

In a grass land ecosystem there are as many as five linear food chains which are inter linked. These are -

(i) Grass - grasshopper - hawk.

(ii) Grass - grasshopper - lizard - hawk

(iii) Grass - Rabbit - hawk

(iv) Grass - mouse - hawk

(v) Grass - mouse - snake - hawk.

These five food chains are not isolated, instead they are interlinked at different points to form a network or food web.

In a pond also a food web is formed with different food chains as follows:-

(i) Phytoplanktons - Zooplanktons - fish

(ii) Phytoplanktons - Insects

(iii) Phytoplanktons - Small fish

(iv) Phytoplanktons - small fish - large fish (large plants)

(v) Macrophytes - Insects large crustaceous - insects- fish molluscus etc.

(vi) Zooplanktons - fish - large fish

The purpose of food web is two fold - one to make a alternate sources of food available at energy point and two to keep in natural check the number of organisms at any one given level by linking them with several predators.

The complexity of the food web depends on two main aspects. (i) the length or (levels) of a food chain and (ii) the number of alternative links at every trophic level.

Ecological Pyramids:

In the trophic relationship between producers and different types of consumers, the number and mass of the organisms at each step is limited by the amount of energy available since same amount of energy is always lost at every step of the reaction as per the laws of thermodynamics. Progressively at each step the bio mass (number of organisms) becomes smaller as it reaches the top (i.e. reaches the level of carnivores). This relationship which can be graphically represented with a wide base and a narrow top is called an ecological pyramid. In an ecological pyramid producers are at the base. Generally ecological pyramids are of three types. These are - pyramid of numbers, pyramid of biomass and pyramid of energy.

Pyramid of numbers:

Pyramid of numbers denotes the relationship between producers, herbivores and carnivores (different levels of consumers) at successive trophic levels in terms of their number. The concept of pyramid of numbers was proposed by Charles Elton (1927) who pointed out the difference in the numbers of organisms at different levels. The producers (autotrophic plants) will occupy the base and they are the largest in number. They support the herbivores whose number should be less than their own (plants) number. Still higher in the pyramid secondary level consumers are lesser in number than herbivores. Finally only a very small number or carnivores occupy the top of the pyramid.

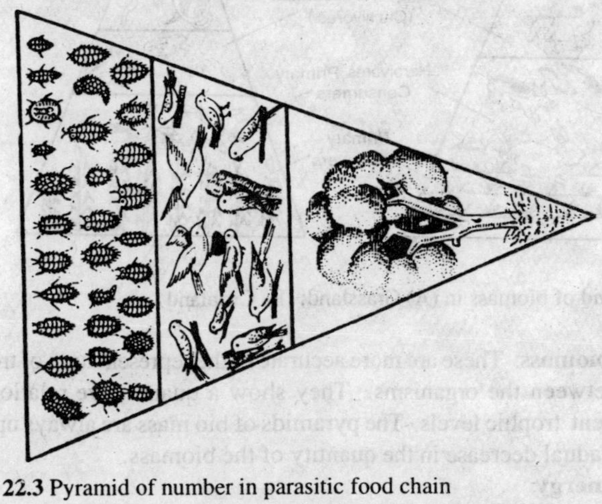

Fig. 22.3 Pyramid of number in parasitic food chain

A situation similar to the one mentioned above exists in a pond also. Here successively starting from the phytoplanktons - zooplanktons - Crabs, larvae - small fish - large fish, the number gets reduced.

In the above two situations the pyramid is said to be upright. If we apply the concept of pyramid of numbers to a forest ecosystem the situation will be slightly different. Here for instance the producers will be large sized trees which are generally less in number from the base of the pyramid. Fruit eating birds, Insects etc. which depend on the tree are more in number, but then the carnivores which depend on the birds and insects are smaller in number and the pyramid once again becomes upright. In parasitic food chain however the pyramids are always inverted. This is because a single plant may support many herbivores and each herbivores in turn supports many parasites.

A specific comment seems to be necessary in the case of inverted pyramids. Pyramids of numbers do not - reflect the actual functional relationship between different trophic levels. For instance in a parasitic chain even there is only a single plant at the base and a number of parasites at the top in reality the total biomass of a single plant is more than the accumulated bio mass of all the parasites.

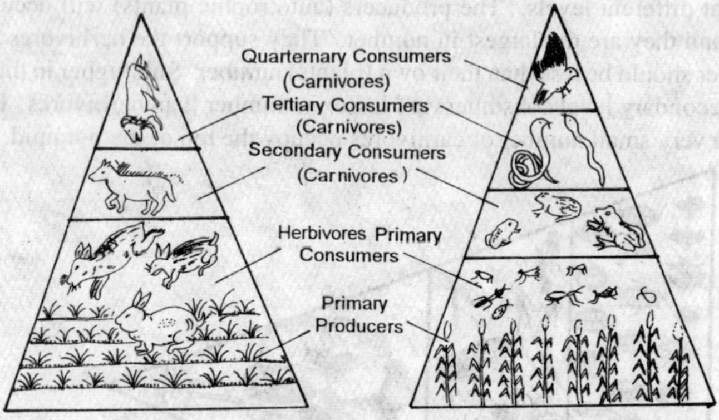

Fig. 22.4 Pyramid of biomass in (A) Grassland; (B) Cropland

Pyramid of biomass: These are more accurate in the representation of trophic relationship between the organisms. They show a quantitative relationship between different trophic levels. The pyramids of bio mass are always upright as there is a gradual decrease in the quantity of the biomass.

Pyramid of Energy:

Among the three types of ecological pyramids the pyramid of energy perhaps gives the best picture of the overall functioning of the ecosystem . At the entry point i.e. producers, energy is always more and at every level quantum of energy transferred gets reduced and it is reflected in terms of biomass. In contrast to the other two pyramids, this pyramid (of energy) represents the rate of flow of energy (food) through the food chain. The shape of the pyramid of energy is always upright as the energy gets tapered as it passes on from one level to another.

BIOGEO CHEMICAL CYCLES

The cycling of essential elements: The cycling or circulation of essential elements of life from the environment to the organisms and back to organisms in a characteristic path in the biosphere is called the Boigeochemical cycle.

Several inorganic salts are essential for life (protoplasm). The chemical elements including all essential elements of the protoplasm tend to circulate in the biosphere in definite circular paths. These circular paths are called inorganic-organic cycles or biogeochemcial cycles or nutrient cycling. Every type of nutrient cycling has two phases viz., (a) abiotic phase- in the environment and (b) biotic or organic phase-in the organisms through food chains.

There are two types of biogeochemical cycles with reference to abiotic phase.

They are. (a) atmosphere reservoir cycle or gaseous cycle and (b) sediment reservoir cycle or sedimentary cycle. The atmosphere reservoir cycle has a dominant atmospheric phase. In this cycle, the nutrient cycles through atmosphere, organisms and soil (sediment) Eg. Nitrogen cycle, carbon cycle etc. Sediment reservoir has a dominant sedimentary phase and atmospheric phase may be absent. The elements circulate from sediment to the organisms and back. Eg. Phosphorus and sulphur cycle. In sedimentary cycles the main reservoir is sediment or soil, whereas in gaseous cycle, it is air. Both cycles involve abiotic and biotic agents and are tied to water cycle.

Gaseous cycles are of following types 1. Oxygen cycle, 2. Carbon cycle, 3. Nitrogen cycle.

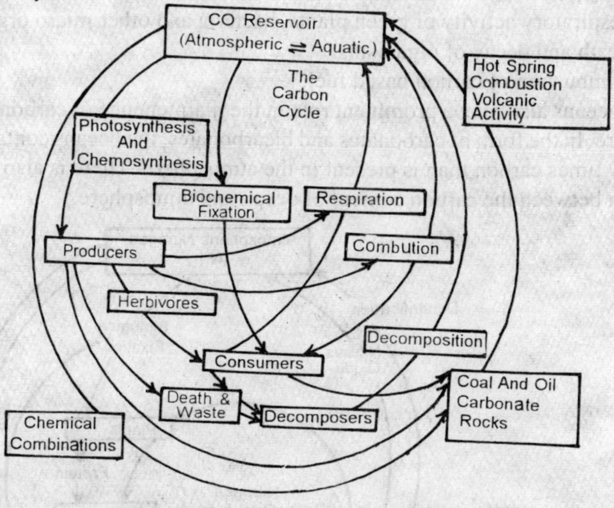

Fig. 22.5 Carbon cycle

Oxygen cycle

Oxygen is one of the major components of the life or protoplasm. It occurs in combination in H_2O, carbohydrates lipids etc. The major supply of O_2 to life is from atmosphere. The atmosphere gets oxygen from two sources viz., (a) photo dissociation of water vapour in which hydrogen escapes into the outer space (b) photosynthesis it is active only after the origin of life on the earth. The cycling of O_2 is very complex. It circulates as a constituent of CO_2 and H_2O. It also combines with many elements to form their oxides. In photosynthesis, water is split to force oxygen. This oxygen is reconstituted during respiration. Some part of the oxygen in higher atmosphere is converted into ozone.

Carbon cycle:

Carbon is one of the most basic components in nature. It is also present in living organisms. The source of Carbon are two:- a) gaseous form (CO_2) in air and b) dissolved form in water. The gaseous carbon is fixed in living organisms (Primary Producers-green plants) during photosynthesis. From the producers, carbon is shifted to primary consumers (herbivores), secondary consumers (carnivores) and to decomposers. The decomposers return the carbon to the abiotic environment.

CO_2 is present in air in minute quantities (0.03 to 0.04%). In spite of its absorption by green plants, the level is maintained because of its regular return to the environment by decomposers and other agents. In nature there is a dynamic balance between carbon consumers (green plants) and carbon producers decomposing bacteria, fungi etc. Carbon is returned to the environment by the following ways.

1) Respiratory activity of green plants, bacteria and other micro organisms.
2) Death and decay of organisms.
3) Combustion of carbon based fuel.

The oceans also play a prominent role in the maintenence of carbon balance in nature. In the form of carbonates and bicarbonates, the oceans contain thirty to forty times carbon than is present in the atmosphere. There is also an equilibrium between the carbon content of ocean and atmosphere.

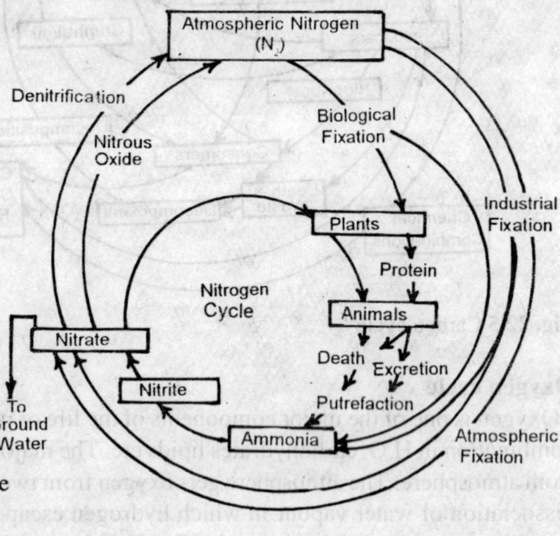

Fig. 22.6 Nitrogen cycle

Nitrogen cycle

Nitrogen is an essential constituent of the organic molecules and also a major component of atmosphere. Though nitrogen is very abundant in its gaseous state, it cannot be used in that stage. Nitrogen cycle follows a regular path.

Atmospheric nitrogen is fixed in two ways. One is by very high energy fixation by lightening, cosmic radiation etc. In this, nitrogen combines with oxygen and hydrogen of water in the presence of energy to form nitrates and ammonia. Second is biological fixation; in this bacteria (*Azatobacter* and

Clostridium) and blue green alga (*Nostoc*) can fix atmospheric nitrogen and convert it into organic substances. The bacteria may be symbiotic (*Rhizobium*) or free living.

Plants absorb ammonia and nitrates from the soil or may get them from symbiotic bacteria (of leguminous plants) and incorporate them in their tissue. These are incorporated as amino acids (proteins). This is called nitrate assimilation. Animals feed on plants and they assimilate the plant amino acids. When animals excrete ammonia, urea and uric acid: the decay bacteria convert these into ammonia. They also decompose the nitrogenous waste compounds of dead plants and animals into NH_2, CO_2 & H_2O.

Ammonia formed by the decay bacteria (putrefaction) is oxidised by nitrite bacteria (*Nitrosomonas*) into nitrites. This is further oxidised by nitrate bacteria (*Nitorbacter*) into nitrates. The process is called nitrification and the bacteria involved are called nitrifying bacteria. Nitrates are added to the soil and thus cycle is completed. Some of the NH_3 is converted into nitrogen by certain bacteria. This process is called dentrification.

SEDIMENTARY CYCLE

Sedimentary cycle or mineral cycle passes mainly through soil or sediment. This forms the main reserve. This cycle has two phases; the salt solution phase and rock phase. Following are the important mineral cycles (a) phosphorus cycle (b) sulphur cycle.

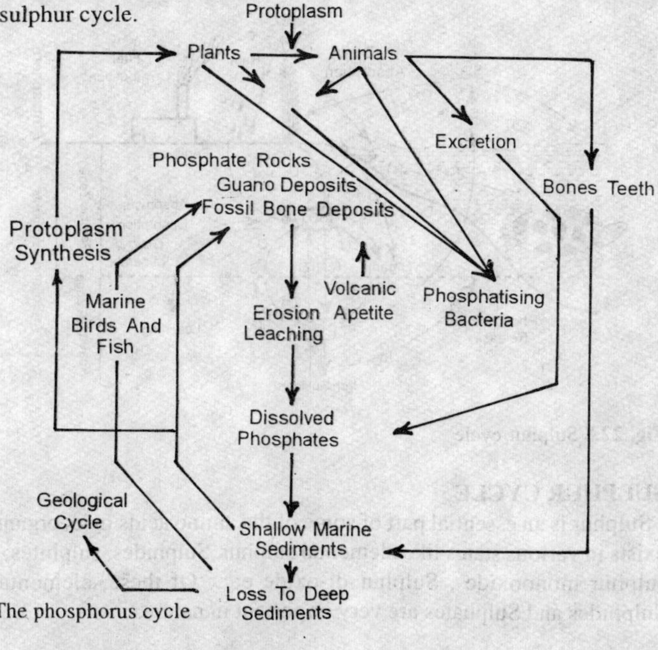

Fig. 22.7 The phosphorus cycle

PHOSPHORUS CYCLE:

Phosphorus cycle has no atmospheric phase. It occurs naturally as phosphate in the soil. It is released by erosion or weathering and gets dissolved in water. Plants absorb phosphates through roots and incorporate them in their tissue. This phosphate passes through the food chain.

The ultimate source of phosphate in the eco system is from crystal and rocks. As these get eroded phosphate will be available to living organisms in form of phosphate ions. Autotrophic plants absorb phosphate through the root systems and incorporate it in the living tissue. From these plants the Phosphate then passes along to the grazing food chain in the same way as nitrogen. The excess phosphate is excreted along with the feces. An extreme example of fecal phosphate is the guano deposits built up by birds on the desert west cost of South America. Sometimes Phosphates are also released as particulate matter from forest and grassland fires.

In the detritus food chain large organic molecules containing phosphate are degraded and the phosphate is liberated as inorganic ions of phosphate. In this form it is readily available to autotrophs or sometimes it may be incorporated as a sediment particle either in the terrestrial ecosystem or in an aquatic ecosystem . The sedimentary phase of phosphorous cycle is generally slow in its movement than the organic phase.

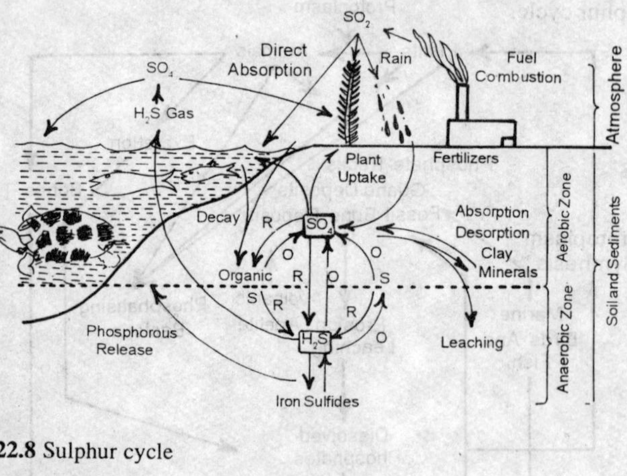

Fig. 22.8 Sulphur cycle

SULPHUR CYCLE

Sulphur is an essential part of some of the amino acids in its organic form. It exists in various states like elemental sulphur, Sulphides, sulphites, sulphates, sulphur monooxide , Sulphur di-oxide etc. Of these, elemental sulphur, Sulphides and Sulphates are very important in nature.

The sulphur cycle exists in both sedimentary and gaseous phases. The sedimentary phase is long termed and in this the sulphur is tied up in organic and inorganic deposits. When these deposits undergo weathring and decomposition sulphur is released and is carried into terrestrial and aquatic ecosystems in the form of salt solution. The gaseous phase of sulphur cycle is less pronounced and it allows global circulation.

Sulphur enters the atmosphere from various sources such as combustion of fossil fuels, volcanic eruption, surface of the oceans and gases released by decomposition. Initially sulphur enters the atmosphere in the form of Hydrogen sulphide which gets quickly oxidised into sulphur di-oxide. This sulphur di-oxide is carried back to earth along with rain water as weak sulphuric acid (H_2SO_4). In any case sulphur is absorbed by plant roots mostly in the form of sulphates. Within the plant sulphur gets incorporated into amino acids and then to proteins. From the plants sulphur gets into the consumers the excess being excreted.

Death of the Organisms releases the sulphur in the living material back to the soil and also to aquatic ecosystems like ponds lakes etc., Here the organic material is decomposed by bacteria wherein the sulphydryl group is broken and is released from the amino acids. Aerobic bacteria oxidize Hydrogen sulphide is to sulphate.

$$H_2S + 2O_2 \longrightarrow SO_4 + 2H$$

The Sulphate produced by the above reaction can be re-absorbed by the autotrophic plants.

In anaerobic environments such as the bottom of lakes, ponds, etc., sulphide cannot be oxidized by the above method as the process requires oxygen. But if infra red radiation is present, photosynthetic bacteria can oxidize sulphide eithere to elemental sulphur or sulphate.

The series of reactions which take place within the organic phase provide a ready mechanism for regulating the availability of sulphur to autotrophic plants.

From the organic phase sulphur is removed in the form of elemental sulphur which is insoluble and accumulates sediments. In these sediments if iron is also present it can combine with sulphide to form iron sulphides.

Twenty three

Environmental problems

There exists a balance between organisms and their environment in the biosphere. Man due to his activity has disturbed this balance. Human activity includes destruction of wild life and forests, construction of dams and buildings, pollution and overpopulation. This has caused crisis in physical, biological and cultural environments of man himself and threatened his own existance.

Physical environment: In physical environment there is a crisis of resources. The crisis faced is as follows.

(a) Building activity like construction of cities, roads and canals have reduced the available space for the plant growth (b) Burning of coal, petrol and wood has reduced atmospheric oxygen. Continued oxygen deficiency can harm the animal and human population. (components) Burning of fuel adds to CO_2 concentration. This not only affects health but increases earth's heat resulting, - in melting of polar ice, rise of sea level, and inundation of coasts. (d) Pollution of air, land and water with industrial and domestic waste has become a global problem. It affects human plant and animal life. (e) Both renewable and non renewable resources are affected. Renewable sources like land, water, air, vegetation and wild animals are facing destruction due to pollution. Non renewable resources like minerals, oil etc are decreasing due to excessive use. (f) Cutting of forests reduces rainfall and may lead to periodic famines.

Biological environment

Over population and under nourishment are the twin crisis faced in the biological environment.

Overpopulation: Biosphere is facing the problem of overpopulation. Human population now doubles every 35 years. It is now about 4 billion and it is estimated that within 2010 AD it will touch 8 billions. Overpopulation is the root cause of all other ecological crisis. Food production cannot keep pace with the human production. Hence in future famines will be more common. Over population not only results in food problem but also, energy crisis, space problem etc,.

Undernourishment: It is linked with the over population. Two thirds of population of under developed countries are facing the problem of undrnourishement. No one can forget the catastrophe which struck in Africa (Ethopia). Thousands of people suffered from the malnutrition, Same catostrophe may strike India also if the crisis continues. Under nourishment is the result of less per capita income which is due to over population. The lack of funds for farm implements results in the less farm production and undernourishment.

Cultural environment: It has produced a crisis of science and technology. Every scientific and technological discovery benefits human life. But also introduces risks and disadvantages. For example invention of automobiles helps mankind but causes pollution, similarly the atomic power. We should find out ways to reduce or eliminate risks before using scientific discoveries. Man is facing the environmental crisis due to indescriminate use of scientific advances and misuse of nature.

Population Explosion

Human population is increasing at an alarming rate. A brief survey of world population shows that human population is ever increasing.

Year	Population
1 A.D.	0.25 billions
1000 A.D.	0.50 billions
1750 A.D.	1.00 billions
1965 A.D.	3.30 billions
2000 A.D.	7.00 billions estimated.

Above table indicates that after 1750 A.D. Population increase is taking less time to become double. Recent periods have seen the population explosion.

Factors for Growth of Population

In the early periods the human population was checked by environmental hazards and diseases. In later periods invention of tools for hunting, agriculture revolution, industrial revolution, modern medicine form the primary factors for growth of population. Advancement of scientific technology, in 1950's resulted in better conditions, more food, and less death rate, caused increase in population.

Population Growth in different countries:

Growth of human population varies in different countries. Nations are divided into 2 categories.

Developed countries: These are industrialised or advanced countries and included U.S.A, Canada, Japan and European countries.

These countries have low population rate, more per capita income and better life facilities. The population growth is almost nil since 1950 but per capita income has increased almost- doubled or tripled.

Under developed countries: There are under developed less advanced countries. They include India, African countries, Egypt, Latin America etc., They have high rate of population growth, and worst life facilities.

Developed countries constitute about 1/3 of the world population but still control the rest of the world. Problems of developing countries is linked with their high rate of population growth. All their national income is spent on food and shelter and less funds are available for other developmental activities.

Developing countries, instead of looking for aid, must check their population growth, then only they will have funds for farm machinary, industry, education, medical care and other facilities.

Consequences of Population growth

Increase in population creates many consequences or problems, which affects basic necessities of life like space,food etc.

Space: Increase in population requires more accomodation. This results in conversion of agricultural land into urban land for habitation and clearing of forests. This has other effects like soil erosion, less rainfall etc.,

Food: Families with more children and less income, similarly & thickly populated nations with less per captia income face the problem of food. Their main problem is food and hence cannot give interest in any other directions. People suffering from malnutrition develop into less fit members of the society.

Employment: Population increase leads to large scale under employment. All schemes of Government are made useless by the over population.

Medical Aid: Larger families cannot afford better medicai aid because of their less income. similarly it is very difficult to provide better medical facilities to ever increasing population in any country.

Sanitation: Increase in population or urbanisation has caused the problem of sanitary disposal and resulted in pollution.

Population Control:

Population can be controlled by two ways.

1.**Planned Control of Population**

The practical method of control of population is by reducing the birth rate. This can be achieved by (a) Education (b) Family planning (components) and Enunciation of laws.

(a) *Education*: People of all ages, both in urban and rural areas must be educated about the hazards of population explosion and advantages of small family. Mass media and educational institutions can play vital role in this regard.

(b) *Family Planning*: Most of the Governments in the developing countries have taken up family planning programmme on a war footing. But results are not encouraging. This is because of superstition, illeteracy and communal feeling. Even in this 20th century, many educated persons feel that it is against their religion to follow family planning. India with all this has achieved a limited success in achieving 38.4% increase against 41.7% during the year 1976.

(c) *Law Enactment*: Unless it is forced by the government no programme can get success. Countries like Egypt have achieved success in this regard. Government must pass laws against having more children. If people with less children are provided with tax rebate, and other facilities naturally they follow family planning methods.

2. **Catastrophic Control of Population**

When a population overgrows its carrying capacity of its environment, it rapidly declines. The carrying capacity is the ability of the environment to provide food shelter and other basic needs. The population declines because, the basic needs are not provided. This type of catastrophic control of human population is a fearsome likelihood. It is estimated that 15 billion is the carrying capacity of the earth. If this is reached human population will die out of starvation.

Population explosion is the result of ecological arises. If this is not checked by reduction, in birthrate the fate of human civilisation will be in danger.

POLLUTION

A clean and pure environment is a basic necessity for the happy and healthy life of man. The fact that the present day environment around the globe is rapidly becomıng unsuitable for human beings is to a very great extent due to man himself. Rapid progress in the quality of life and unending needs brought about by industrialization have let into the environment as waste products, effluentes etc., different kinds of solids, liquids and gases which have vastly changed the clean environment. Factories let out exhausts into the air, release obnoxious effluents into water. Transport vehicles of various kinds release poisonous gases into the atmosphere. Oil leaks into ocean waters have greatly endangered the aquatic life. All these are continuously polluting the environment. In short, one can say that in his quest for knowledge, discovery, progress etc, man himself is destroying the very environment on which he survives. This can be compared to the cutting of the stem of a tree while standing on a branch of that very same tree.

Pollution may be defined as the presence of new extraneous liquid, solid or gas in a quantity more than the normal so as to endanger the different forms of life including human beings.

Environmental pollution is a great hazard as it affects us directly as well as indirectly. Increase of poisonous gases in the atmosphere induces various kinds of respiratory and other diseases, discharge of effluents into water kills the acquatic life including fish and agricultural production gets decreased as plants are sensitive to atmospheric pollutants.

In this context it will be worthwhile to quote the report of Environmental Panel of USA. "Environmental pollution is the unfavorable alteration of our surroundings, wholly or largely as a by product of man's actions, through direct or indirect effects of changes in energy patterns, radiation levels, chemical and physical constitution and abundance of organisms. These changes may affect man directly or through his supplies of water and of agricultural and their biological products, his physical objects or his opportunities for recreation and appreciation of nature".

Causes of pollution: Human population explosion, unplanned urbanisation, industrialisation and deforestation are some of the causes of pollution. Increasing human population produce more waste and pollute the atmosphere. Urbanisation causes sanitary problems; the sanitary waste is let out which may mix with water and pollute it. Industrialisation including modern agriculture (insecticides and herbicides) adds continuously both gases, liquids and other materials to the atmosphere. They not only pollute the environment but also modify it. For example industrial waste released into a stream kills all the aquatic fauna and flora.

Creators of pollution: Materials which create pollution are called pollutants. Odum, from the point of view of ecosystem, classified these into 2 types (1) Nondegreadable pollutants and (2) Biodegradable pollutants.

The nondegradable pollutants are materials or poisons such as aluminum cans, mercurial salts, DDT and phenolic chemicals which do not degrade, or degrade very slowly in the natural environment. These substances accumulate in the atmosphere or combine with other compounds to form toxins.

Biodegradable pollutants such as domestic sewage and heat (human waste) rapidly undergoes natural process of decomposition or can be engineered (by muncipIe sewage treatment plant) to decompose; problem arises only when the input of these substances is more rapid than that of the natural degradation. This may be due to over population or urbanisation.

Pollutants are also classified according to the types of materials that create pollution. (1) Deposited matter e.g. soot, smoke, tar, dust, etc. (2) Gases e.g. Sulphur di-oxide, carbonmonoxide, hydrogen sulfide, flourine, chlorine etc. (3) Chemical compounds e.g. arisines, aldehydes etc., (4) Metals like mercury, iron, zinc, lead, etc. (5) Ecological (6) Sewage (7) Fertilisers (8) Radioactive substance (9) Noise and heat.

Types of pollution:

Generally pollutions are classified according to the environment and also on pollutants into following types:

1. Air pollution
2. Water pollution
3. Soil or land pollution
4. Radioactive pollution
5. Noise pollution.

Air pollution:

Increase in the concentration of gases (other than O_2) and particulate solid waste in the air is called air pollution. It is caused due to industrilisation, automobile exhaust and burning of coal, petroleum, etc. These cause the accumulation of sulphur dioxide (SO_2), smoke and suspended particles.

Causes: Pollutants are produced by four major sources (1) automobiles and transporters (2) electric power plants (using petrol or coal) (3) industries, and (4) heating plants.

Kinds : Air may contain pollutants of different sized particles and gases. Accordingly they are classified into many types.

(a) Coarser particles of over 200 m in diameter like carbon particles and dust.

(b) Fine particles of less than 100 m in diameter like metallic dust, carbon particles, sulphates etc.

(c) Sulphur compounds

(d) Halogens

(e) Radioactive substances

(f) Oxygen compounds

Common air pollutants are sulphur dioxide and combustion gases.

Sulphur dioxide: Is produced as a result of coal, petrol and oil combustion. Sulphur dioxide does not remain as gas for a long time. Soon it reacts with moisture to form sulphuric acid. This is the cause of acid rain-fall in many advanced countries.

Twenty four

Development and conservation of ecosystems

India is one of the developing nations. It is facing a problem of ecological crisis. It is striving hard to compete with advanced countries. When India got independence its main objective was to surge forward in the field of industry. Thus concentration was given to developing industries of all types. As a result of industries developed but food problem arose, as it happened in 1960. India in the beginning was importing foodgrains. So emphasis was shifted towards agriculture. Now India is self sufficient in the food production.

All this has been achieved at the cost of nature. Industrialization was the call of the nation during 1950's. This resulted in deforestation, depletion of forest products (like bamboo for paper industry) and pollution. Modern agriculture in India resulted in green revolution in Punjab, white revolution in Khira district of Gujarat (Amul) and cultivation of extra acres of land. This is due to building of big dams, use of pesticides like DDT, conversion of forests, pastures into agricultural lands. All this development in India has brought advancement and self sufficiency. But slowly we are facing problems. Rains are becoming less and less, due to deforestation. Famines and draughts are more common. More and more agricultural lands are becoming less productive, because of continuous usage and erosion.

Thus development has resulted in the ecological problem. If this is not solved, immediate problem would be less rains, no drinking water, no fire wood for cooking. Now people in Government are understanding their mistake. Appointment of a minister for pollution and environment in Central Governments indicates this. Though realisation is late but it is of great importance.

Science and technology should be used in such a way that the ecology of our country should be preserved. Development activities should not sacrifice natural environment. In this regard afforestation, pollution control, and checking the population rate, must be done without delay. There should be balance between preservation of ecology and development. This conflict is found not only in India but in all developing countries. A good example is that of Ethopia (Africa) where thousands of people faced death due to starvation. Those countries sacrificed nature of development. It should be a lesson for all the developing countries.

Soil Conservation Measures :

Soil conservation measures should aim at preventing or atleast minimising the soils loss. In order to do this proper land utilisation coupled with agricultural practices should be adopted.

Broadly categorizing there are two methods of soil conservation. These are biological and mechanical. The biological measures are again divided into Agronomic, Agrostilogical and Dry farming, we shall study these measures in some detail.

SOIL CONSERVATION MEASURES

Biological			Mechanical
Agronomic Practices	Agrostological	Dry forming	Afforestation
1. Contour farming 2. Tillage 3. Mulching 4. Crop rotation 5. Strip cropping protection a. Contour strip cropping b. Field strip cropping c. Wind strip cropping d. Buffer strip cropping	1. Ley farming 2. Grass cultivation	Channel terrace Broad based ridge terrace Narrow based ridge terrace Bench terrace	1. Basin listing 2. Sub soiling 3. Contour terracing 4. Contourtrenching 5. Terrace outlet 6. Gully control 7. Ponds 8. Stream bank

Agronomic practices: Normally, the land will possess a vegetational cover so as to prevent cerosion. The measures to be followed must be patterned along the nature's own methods of conservation. The following are some of the methods.

Contour farming: Crops are cultivated along the contour of the land. The plough marks will be on level and can hold the rain. Even in heavy rain, the run off is checked by the plants growing along the contour.

Tillage: contour tilling will prevent the excess run of water.

Mulching: Interculturing operations will kill weeds and soil mulches help the plants to be rooted firmly in the soil.

Crop rotations: Alternatively growing a cereal and a legume in the same field will not only increase the yield, but also increase the fertility of the soil. They also help in checking soil erosion.

Strip cropping: This is an agricultural practice of growing plants in suitable strips in the field. This is of the following types.

Contour strip cropping: This is cultivation of soil protecting crops in strips alternating with erosion permitting crops. The strips should be across the slope.

Field strip cropping: Plants are cultivated in parallel strips across the slopes.

Wind strip cropping: Crops are planted across the slopes to prevent soil loss. These may be legumes or grasses.

Agrostological measures: Cultivation of grass in a land which is heavily eroded is called an agrostological measure. This is of two types. In ley farming grass is cultivated in rotation with regular crops. This helps in soil protection as well as produce fodder to cattle. If a land is heavily eroded it is best to allow it to the growth of grasses for few years. This will help in the checking of erosion.

Dry farming method:

This may be practised where rainfall is low, indefinite and variable. In dry farming methods only crops are grown that can sustain even a very low rainfall. The most important aspects of dry farming is conservation of soil moisture and fertility.

Mechanical -Measures: The main aims of mechanical measures are to allow for the absorption of run off, dividing the slope into short ones and protection against run off. A few of the mechanical measures are discussed below:

Basin listing: Small basins are formed along the contour with an implement called basin lister. These will hold water for sometime.

Sub soiling: Soil is broken with a sub soiler into fine grains to increase their absorptive capacity.

Contour terracing: Along the contour, series of ridges or bunds of mud are formed to check the run off. This is of four types. In channel terrace a shallow channel is dug and the mud is deposited along the lower edge of the canal. In broad base ridge terrace a canal is formed on the contour by exavating the mud. The canal is wide. If it is narrow it is called narrow based ridge terrace. In bench terracing a series of platforms are formed along the contour across the general slope of the plant.

Contour trenching: Several 2 feet by one foot trenches are formed across the slopes at suitable intervals. Tree seedlings are to be planted above the trench.

Terrace outlet: Outlets are to be constructed for the safe disposal of run off water.

Gully control: Suitable water conservation measures are to be taken so as to prevent the formation of gullies.

Ponds: Construction of small ponds at suitable places to store water is a good practice.

Stream bank protection: Banks of channels or rivers usually cave in during floods. To prevent this, construction of stone or concrete protective walls should be undertaken. In addition to this, planting some useful tree species will also prevent stream bank erosion.

Afforestation:

Forests occupy a central position in nature. They restore ecological balance of all ecosystems, maintain biological diversity, help in water and soil conservation by acting as catchment, prevent floods and safeguard the livelihood of tribal people. In order to meet these needs and to have required percentage of forest cover a massive afforestation programme has to be launched.

What is afforestation?

In order to maintain proper ecological balance and sustenance certain percentage of the total land area of any country should have a forest cover. If this is reduced it results in both floods and famine. In afforestation programme forest trees are planted in suitable areas and also on waste land. Exotic and fast growing tree species which do not disturb the ecological balance are planted. *There are two major goals for forestry. These are*

(i) Supply of forest goods (timber etc.) to people and Industry by a well thoughout plan of production.

(ii) Long range goal of ecological safety through conservation of forest cover and its restoration.

Among the goals mentioned above the first one may be achieved by planned development of suitable forest cover and its controlled utilization. There are various kinds of afforestation methods to suit particular requirements.

1. *Limited production forests*: Areas which are less fertile, more than 1000 mtrs in altitude and having a hilly topography can be used for development of forests. A part of the annual production of the forests may be used in a limited way so as not to damage soil and trees. Even after utilization the health of the forest should not be affected.

2. *Production forests*: These are forests meant for high degree of production. Felling and logging should be done on scientific lines. In these areas efforts are made to generate forests and not plantations.

3. *Commercial forestry*: The basic aim is to supply goods and services and meet the needs of people, for firewood, fodder, food, fibre, timber etc. In our country where the forest cover is only about 14% another 70 million hectares of additional forest cover has to be created in order to meet the demands. This can be done by the following methods.

Intensive plantations: Planting of suitable trees on all the available land including road sides to remove pressure on natural forests. Social forestry, Agro forestry programmes are included in this category. This would meet all the needs of fuel, fodder, fertilizer, food, fibre, timber etc. and prevents denudation of valuable natural forests.

Social forestry: Social Forestry means development of forests on community land solely to be used by the community to meet their needs. There are two main objectives in social forestry -

1. Using the public and common land of the village to produce in a decentralised manner firewood, fodder and small timber to meet the local needs (particularly poor people) and also to conserve soil and water.

2. Reducing the pressure on natural forests. In the absence of alternate resources village people tend to exploit the natural forests. This is unavoidable as their needs and that of their live stock (grazing cattle) has to be met by the nearby forests. If an alternate forest is developed with their co-operation they would also be interested in protecting the natural forests.

Social forestry involves the following steps.

(i) Educating the villagers the necessity to protect and conserve the natural forests. In particular they need to be told the vital role forests play in the conversation of water and soil.

(ii) Villagers, school children and landless labourers of the village should be made to directly involve themselves in social forestry programmes.

(iii) A nursery should be developed near the village to grow useful, fast growing plants keeping in mind the need of fire wood, fodder, fruits, mannure etc.

(iv) Identification of suitable village land to develop social forestry.

(v) Proper utilization of the products of social forestry should be made available to all particularly the need.

In our country social forestry is practiced in a big way and about 160 million hectares of land is under social forestry programme.

Advantages of social forestry:

1. Educates the people about the need to protect and conserve natural forests.
2. Provides them alternate source of firewood and fodder so that they need not go to the natural forests.
3. Species selected for social forestry like Acacia, grow very fast and produce usable biomass within a short period of time.
4. Protects the natural forest from grazing animals.
5. Waste land in and around the village will be put to proper use.
6. Involvement of rural people in the social forestry programme would generate in them a natural interest towards the conservation of natural forests.
7. Provides land labourers an avenue of employment.

WILD LIFE CONSERVATION

The expression of wild life is a very broad one and includes all non cultivated and non domesticated living beings . With in the perview of wildlife both plants and animals are included. Even fish which are rare are also often included.

Conservation and management of wildlife is a special field of applied ecology that should rank very high in public interest. In general the following may be said to be the aims and objectives of wildlife management.

1.Protection of natural habitats of organisms through controled exploitation.

2.Maintainence of rare species in protected areas such as national parks, santuries etc.,

3.Establishment of specific biosphere reserves for endangered plants and animals.

4.Protection of wild life through legislation such as banning hunting etc.,

5.Imposing specific restrictions on export of endangered plants and animals or their products.

6.Educating the public about the need to protect and preserve the environment as a long range goal for the welfare of future generations.

The above aims and objectives may be realised by the following methods.

1. Establishment of national parks and sanctuaries including preservation of the original habitats of the organisms.

2. Establishment of zoos for captive breeding and maintainence of exotic animals which are not native to our country.

3. Enactment of suitable legislation to protect the wildlife.

Establishment of national parks and santuries: Conservation of wildlife is of two types - *in situ* conservation and *ex situ* conservation. In the former the rare species are conserved in their original habitat where as in the second method the endangered species are conserved in green houses or glass houses if they are plants or in zoos and such other protected areas if they happened to be animals.

In conservation of biological diversity National parks and Santuaries have played a great role. Basically there is no difference between a national park and sanctuary. Generally however a sanctuary is species oriented . For eg. sanctuary for the conservation of pitcher plant or conservation of great Indian bustard etc. In National parks however the basic aim is to preserve the entire habitat with a variety of wildlife such as tiger, lion etc.

In the year 1981 there were only Nineteen national parks and Two hundred and two sanctuaries in India extending over 74,763 square kilometers amounting to about 2.3% of the total Geographic area. In the year 1989 the number of national parks increased to 67 and that of sanctuaries to 394 covering an area of about 1,41,298 Sq. kms. amounting to about 4% of Geographic area. The following table gives a list of important national parks and sanctuaries of our country.

NAME OF THE STATE		NAME OF THE SANCTUARY/N.PARK
Andhrapradesh	:	Kawal, Pocharam, Pakhal, Neelapattu
Arunachalpradesh	:	Namidapha
Assam	:	Kaziranga, Manas
Bihar	:	Hazaribarh, Betla
Goa	:	Mollen
Gujarath	:	Gir, Velavadar, Wild ass, NalSarovar
Haryana	:	Sultanpur lake
H.P.	:	Sechu-tum-Nallah, Gobind sagar
J & K	:	Dachigam
Karnataka	:	Bandipur, Nagarhole, Ranganthitoo
Kerala	:	Periyar, Waynard, Neyyar
M.P.	:	Kanha, Shivpuri,Bandhavgarh
Maharastra	:	Tadoba, Pench, Nawegaon, Bori, Dhakna Kolkaz, Kamala.
Manipur	:	Keibul Lamjao
Meghalay	:	Balpakram
Mizoram	:	Dampa
Nagaland	:	Intangki
Orissa	:	Simlipal, Satkasia, Chilka lake
Punjab	:	Abohar
Rajasthan	:	Ranthambore, Ghana, Sariska
Sikkim	:	Kanchenjuga
Tamil Nadu	:	Guindy, Mundumalai, Annamalai, Vedanthangal, Vettangudi
U.P.	:	Corbett, Dudwa
W.Bengal	:	Sajnkhali, Mahanandi, Jaldapara, Deer parks, Sunderban.

Some of the special projects which have been instituted to protect certain selected endangered wild life are as follows:

1.**Project Tiger** : It has been estimated that in the begining of this century the population of Royal Bengal Tigers in our country was close to 40,000 and this came down to a mere 2,500 by the year 1982. Considering the alarming decline of the number of tigers the Indian board of wildlife initiated a project to save the tigers from extinction. This was named project tiger and several tiger reserves have been established. In 1973 ther were 9 tiger reserves in the country occupying a total area of 13,017 Sq,kms. Presently there are 17 tiger reserves which not only provide sanctuary to tigers but also protect several endangered species of plants and animals. Among the plants are some rare

Orchids and ferns and among the animals the Lion -tailed macaque, Nilgiri langur, Nilgiri Tahr, Wild elephant, gaur and leopard are included. The following table gives a state wise break up of reserves.

	State	Name of the reserve
1.	Karnataka	Bandipur
2.	U.P	Corbett
3.	M.P.	Kanha
4.	Assam	Manas
5.	Maharastra	Melghat
6.	Bihar	Palamau
7.	Rajasthan	Ranthambore
8.	Orissa	Simlipal
9.	W.Bengal	Sunderbans
10.	Kerala	Periyar
11.	Rajasthan	Sariska
12.	W.Bengal	Buxar
13.	M.P.	Indravati
14.	Andhra Pradesh	Nagarjuna Sagar
15.	Arunachal Pradesh	Namdapha
16.	U.P.	Dudwa
17.	Tamil Nadu	Kalkad Mundanthurai

Lion Project: The Gir forest in Gujarat is the only surviving habitat for the Asian Lion. To protect Asian Lion a sanctuary was established in the year 1972 by the G•government of Gujarat . The total area of Gir sancturary consists of about 1,412 Sq.Kms. and the natural habitat for the Lions was protected to include a core area of 140 Sq.Kms. As a result of this the Lion population has increased to about 200.

Crocodile conservation: There are three species of Crocodiles in India. (i) salt water or estuarine crocodile (*Crocodylus porpsus*), (ii) freshwater, swamp crocodile or mugger (*C.plaustris*) and (iii) gharial (*Gavialis gangeticus*). The population of crocodiles declined due to illegal poaching as the skin is in great demand. In order to preserve the crocodiles the sale of the skin is legally banned. Several sanctuaries have been established. These include Krishna Sanctuary in Andhrapradesh, and chambal sanctuary including areas from U.P. M.P. and Rajasthan.

Rhino Conservation: This scheme to protect the rare one horned Rhino of Assam was started in the year 1987.

Snow Leopard Project: This project has been started to conserve the habitat of snow leopards through out Himalayas.

Biosphere reserves (Habitat preservation): The biosphere reserve is a better method of preservation of wild flora and fauna than the national parks. This is an ecosystem oriented program and generally not opened to tourists. The biosphere reserve includes a core zone and a manipulation zone. The core zone represents a natural area where no exploitation of natural resources is allowed. The manipulation zone includes silvicultured forests and also trible settlements.

The first biosphere reserve to be established was the Niligiri biosphere reserve covering an area of 5,520 Sq.Kms. Other biosphere reserves are Nandadevi biosphere reserve, Great Nicobar reserves etc.,

The Niligiri biosphere reserve includes two national parks viz., the Bandipur national park and the Nagarhole national park.

Estabishment of Zoos: This is an age old method of Preserving and protecting wild life. In this, rare forms of wildlife are kept in cages or protected open areas. But according to environmentalists this is mainly for recreation of man rather than the natural protection of wildlife. Inspite of the limitations imposed, zoos the world over have done yeomen services in captive breeding of many endangered species. Preservation of wildlife by the establishment of the zoos can be regarded as second best when compared to national parks.

Legislation to protect Wildlife: The Indian board for wildlife was constituted in the Year 1952 and at its first meeting the board made a suggestion for a unified legislation to protect wildlife in India. The wildlife protection act was enacted in 1972 and it has been adopted by all the states.

There is a Central board for wildlife preservation headed by a Director To assist the Director, there are four regional directors one each at New Delhi, Bombay, Calcutta and Madras. They are responsible for the Implimentation of Convention on International Trade in Endangered species (CITES) . These directors also give suggestions to local offices in the protection of wildlife and regulating commercial activity in wild flora and fauna.

A number of legislations have been enacted to protect the wild life. These are

1. Madras Wild Elephant Preservation Act, 1873
2. All - India Elephant Preservation Act, 1879
3. The Wild Birds and Animals Protection Act, 1912
4. Bengal Rhinoceros Preservation Act, 1932
5. Assam Rhinoceros Preservation Act, 1954
6. Indian Board for Wildlife (IBWL) 1952
7. Establishment of National Parks, Sanctuaries, and Zoological Gardens.

PART TWO
ZOOLOGY

Twenty five

Histology

There are two kinds of organisms in the animal kingdom, these are : *acellular* or *unicellular* and *multicellular*-

Unicellular organisms are characterised by protozoans while all higher animals are called metazoans. The protozoans are also called acellular because the entire body is made up of a single cell and all the vital functions are performed by it. In metazoa however the body is made up of a large number of cells with division of labour where different cells perform different functions. When several cells perform a common function they are referred to as a tissue or a group of cells. Most of the metazoans have tissue organisation.

What is a tissue?

The body of metazoans is made up of numerous types of cells. The human body for example consists of nearly 200 types of speicalised cells. These cells of one or more types aggregate together in a characteristic manner and perform a specific function. Such groups of cells are known as tissues. A tissue may be defined as " group of cells having identical structure function and origin". Some times more than one type of cells may aggregate together to perform a common function and these are also called tissues. Tissues which have only one kind of a cell are called simple tissues while those which have more than one kind of cell are called complex tissues.

In metazoans tissues are organised together to form organs and organs in turn combine to form organ systems.

The branch of zoology that deals with the study of tissues is known as Histology. Study of histology requires the use of the microscope. The term tissue was first coined by French surgeon Maire Francois-Xavier Bichat (1771-1802). It was Mayer (1819) who coined the term histology.

The cells in a tissue are generally held together by a non-living substance secreted by the cells themselves. This substance may be fluid, jelly-like or tough. Usually the cells in a tissue are bathed with a ground fluid called tissue fluid. The tissue fluids are blood and lymph.

Classification of tissues:

The animal tissues are basically classified into four types these are:

I Epithelial (Epithelium)

II Connective

III Muscular

IV Nervous

I. EPITHELIAL TISSUE

An epithelial tissue may be defined as a tissue composed of one are more layers of cells covering the external or internal surface of the body or may be embryologicaly derived from such a surface. Ruyesh was the first person who coined the term epipthelium.

Structure: Epithelial tissue has cells of various shapes which are closely connected leaving no intercellular space; these form continous sheets. Usually the epithelial cells rest on a delicate non-cellular gelatinous basement membrane or basal lamina. The basement membrane is made up of protein firbres matted together in a polysaccharide ground substances. The basement membrane is secreted by underlying connective tissue. In order to maintain the integrety of the epithelial layer the plasma membranes of adjacent cells are held together by means of intercelluar such as desmosomes, inter digitions, tight-junctions etc. In some places the adjacent cells form gap junctions for intercellualr communications and exchange of chemcials.

The cells of the epithelium tissue are not provided with blood vessels but still they possess a remarkable capacity of repair and regeneration. After an injury the epithelial cells receive useful materials and excrete waste products to the blood flowing in the blood vessels of the connective tissue present near the basement membrane.

Origin : It is beleived that epithelium was the tissue to be evolved in the animal kingdom. All the three primary germinal layers viz., ectoderm, mesoderm and endoderm give rise to the epithelium. The epithelial layer that lines the inner surface of blood vessels, lymphatic vessles and heart are usually referred as endotheliem. Mesothelium is another term used to designate the layer that covers the surface of the membrane lining the Coelom.

The mucous membrane which lines the canals and cavities of the body communicating to the exterior are composed of a superficial layer of epithelium together with a layer of connective tissue. The alimentary canal, respiratory tract, and the urinogenital tract have the lining of mucous membrane.

General features of the epithelial tissue:

1. They cover the surfaces and line the cavities and as such they protect other tissues from mechanical injury, drying up, infection and harmful chemcials.

2. They form many useful exoskeleton structures like scales, hairs, feathers, claws, nails, etc.

3. They have secretory activity; many epithelia produce glands and secrete juices like mucous, gastric juice etc.

4. Epithelia that line the uriniferous tubules help in the elimination of the nitrogenous waste materials as urine.

5. Epithelia lining the intestine help in the absorption of digested food.

6. They also help in exchange of gases in the respiratory tract. They bring about exchange of carbon di-oxide and oxygen between blood and air.

7. Many epithelia form a part of the sense organs as in nasal chambers.

8. Epithelia also help in movement of fluids and mucous in the organs where they are present.

9. They also take an important part in the production. Germinal epithelia help in the production of eggs and sperms.

10. Epithelial cells with pigmentation are responsible for the darkness of the cavity of the eye ball.

Classification of Epithelial tissue

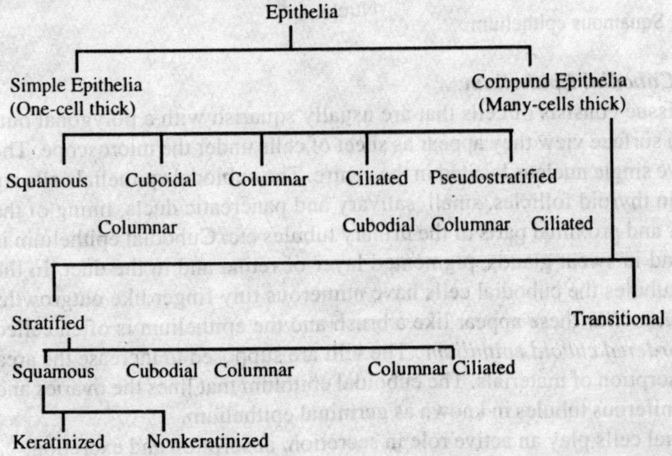

Basically the epithelial tissue may be classified into two main categories namely simple epithelium and compound or stratified epithelium.

SIMPLE EPITHELIUM : These are made up of single layer of cells resting on the basement membrane. Usually they cover moist surfaces where there is no wear and tear by friction. Based on their form and function five types of simple epithelia are recognised. These are- Simple Squamous Epithelium, Simple Columnar Epithelium, Simple Cuboidal epithelium, Simple Ciliated Epithelium and Simple Glandular Epithelium.

Simple Squamous Epithelium :

This is made up of flat disk like polygonal cells that are closely packed leaving no intercellular spaces. They appear like the arrangement of the tiles on the floor hence the epithilum is also known as the pavement epithilum. The cells of the tissue are connected by fine protoplasmic bridges and are slightly thicker at the centre. The nucleus is slightly flattened and is located at the centre of the cell.

Simple Squamous Epithelium is present in the inner lining of Bowman's capsule in the kidneys, alveoli of the lungs, blood vessels etc.

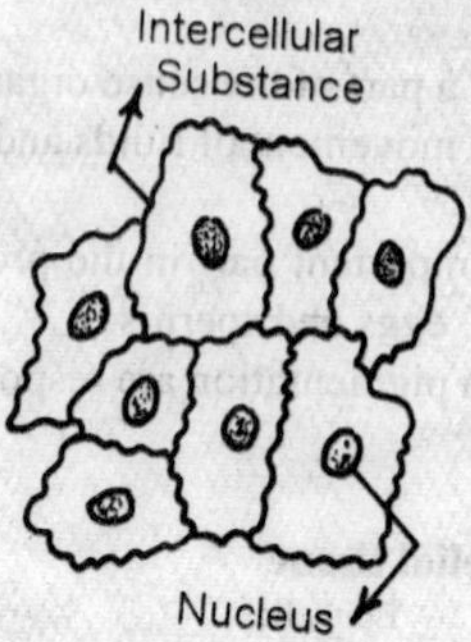

Fig. 25.1 Squamous epithelium

Simple Cubodial Epithelium:

This tissue consists of cells that are usually squarish with a polygonal outline. In a surface view they appear as sheet of cells under the microscope. The cells have single nucleus located in the centre. The cubiodal epithelial cells are located in thyroid follicles, small salivary and pancreatic ducts, lining of the inner ear and proximal parts of the urinary tubules etc. Cubodial epithelium is also found in sweat glands, pigmented layer of retina and in the duct. In the urinary tubules the cubodial cells have numerous tiny finger like outgrowths called *microvilli;* these appear like a brush and the epithelium is often called *brush bordered cuboid epithilium*. The villi are supposed to increase the area for reabsorption of materials. The cuboidal epithlum that lines the ovaries and the seminiferous tubules in known as germinal epithelium.

Cubodial cells play an active role in secretion, absorption and excretion.

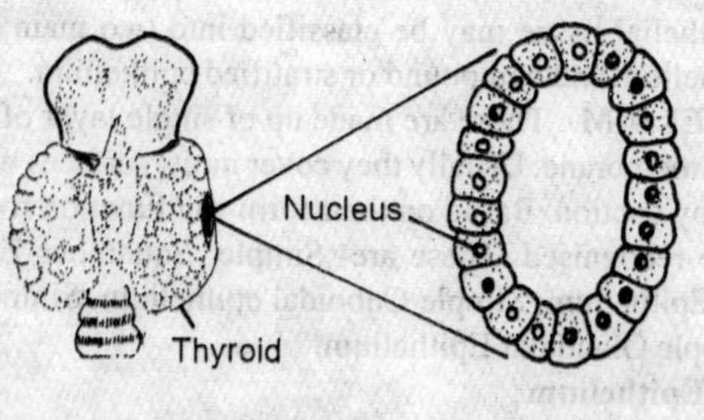

Fig. 25.2 Simple cuboidal epithelium

Simple Columunar Epithelium : As the very name indicates this epithelium consists of columnar cells that are longer (taller) than broad. The tissue is made up of a single layer of cells. The outlines of the cells appear polygonal in a surface view. Each cell has a flattened and an oval end. There is a single nucleus in each cell which is elongated and usually located either in the centre or a little away from the centre. Usually some mucous secreting cells called *Goblet cells* are found in between the columnar epithelial cells that line the stomach and intestine. The goblet cells have vacoule filled with mucous. The epithelial tissue containing the mucous secreting cells along with the underlying supporting connective tissue is usually refered to as the ***mucous membrane*** or mucosa.

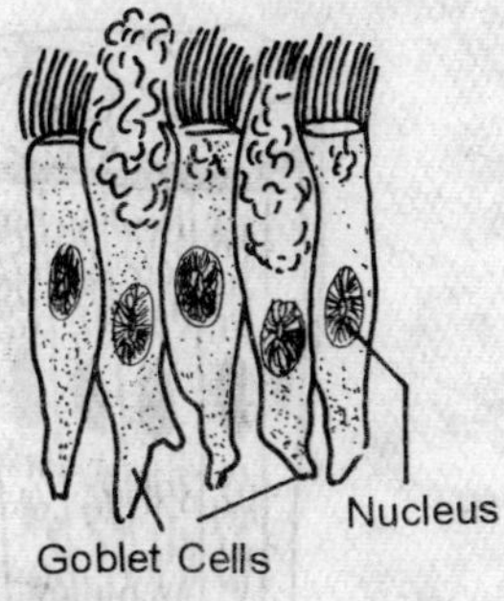

Fig. 25.3 Goblet cells and ciliated cells

Simple Columnar epithelial cells are found distributed in various organs like the lining of the gall bladder, bile ducts, gastric glands, intestinal galnds, pancreatic lobules etc.

In the intestine, the columnar epithelial cells that line the inner cavity are called microvilli and they are specialised for the absorption of water and digested food. The microvilli increase the surface area of absorption. These cells present in the small intestine constitute the brush bordered columnar epithelium.

Simple Ciliated Epithelium:

In this tissue the epithelial cells bear at their free ends thin elongated cytoplasmic processes called Cilia (Sing.cilium). Each cilium arises from a minute particle called basal granule or blepharoplast that lies internal to the cell membrane. In between the ciliated epithelial cells are present mucous secreting goblet cells. The mucous spreads over the epithelium as a thin coating. The cilia have a beating action (move like the lashes of a whip) as a result of which the mucous and other substances are transfered over the epithelium from one place to another . The cilia also help in shifting small solid particles entangled in the mucous.

The ciliated epithelium is classified into two types- *cuboidal ciliated epithelium* and *columnar ciliated epithelium.*

The Cuboidal ciliated epithelium consists of cuboidal cells with cilia on their free surface. These are found in the urinary tubules of the kidney. The cilia by their lashing action keep the urine moving. The columnar ciliated epithelium has columnar cells with cilia. They are found in the fallopian tubes (oviducts), termianl bronchioles and central canal of the spinal cord of the embryo.

The main function of ciliated epithelium is to help in the movement of substances that they come in contact with. All the cilia of the epithelium do not move together; instead they move in quick succession bringing about a wave like motion which helps in the continous movements of the substance.

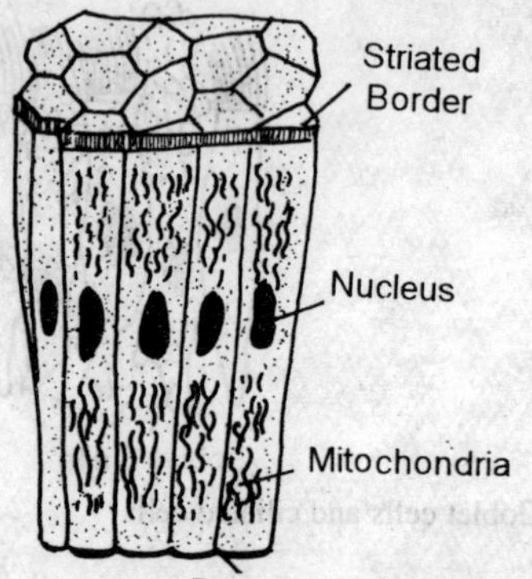

Fig. 25.5 Simple columnar epithelium

Glandular Epithelium: The Glandular epithelium consists of cells which are either cuboidal or columnar. These are specialised for the secretion of chemical substances like mucous or enzymes or hormones. The Epithelial cells fold in different ways to form the glands.

Glands are classified into two basic types. These are *unicellular glands* and *multicellular glands.* Unicellular glands are isolated cells and are distributed between the columnlar cells. These are also called the goblet cells and secrete mucous. Unicellular glands are found in the epithelium lining the intestine. Multicellular glands are made up of a number of cells. These are formed by the sinking of the glandular epithelial cells into the underlying connective tissue. Multicellular glands are further classified into two types viz., *tubular* and *saccular.*

Tubular: The tubular glands are tube like in appearence and are of two types - *Simple tubular glands* and *compound tubular glands.* The simple tubular glands may be straight coiled or branched. Their secretory parts open into a single duct. Eg. sweat glands in the skin and glandular epithelium in the intestine.

Compound tubular glands have numerous secretory parts, the ducts of which are fused to form ducts of higher order finally leading to a single duct. Examples of compound tubular glands are the milk glands in mammals.

Examples :

Simple straight tubular glands: eg. male frog's nuptial pads

Simple coiled tubular glands: eg. Sweat glands in the mammalian skin.

Branched tubular glands: eg. Sweat glands present in the armpits.
Compound tubular glands: eg. Milk glands of egg laying mammals.

Fig. 25.5 Simple ciliated epithelium

Saccular glands: (Alveolar glands) These glands are flask shaped in appearance and have a wide spherical secretory part called the acinus or alveolus at the inner end of a tubular duct. These glands may also be classified into simple and compound.

The simple alveolar glands may be branched or unbranched. They are found in the cutaneous glands of frog.

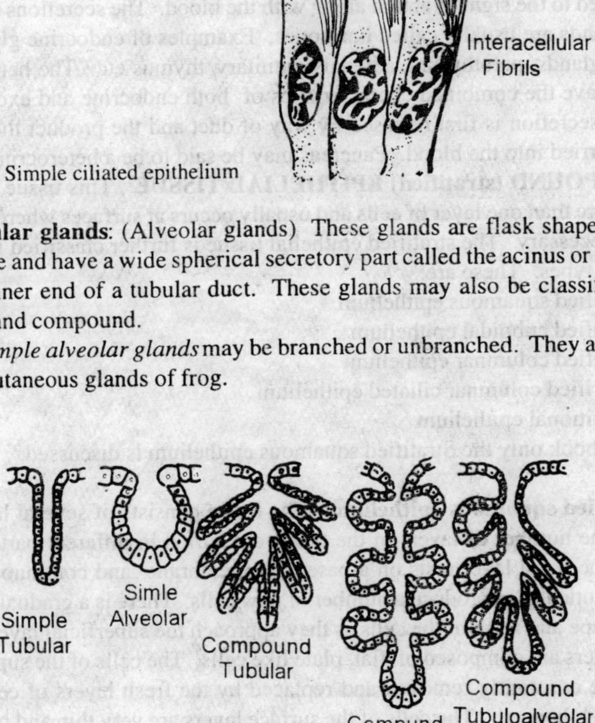

Fig. 25.6 Different types of multicellular glands

The compound alveolar glands are highly branched and consist of several lobules each having many acini. Each acinus opens into the lobule by a short ductlet which ultimately leads to the main duct. Examples of compound alveolar glands are the milk glands of the human beings.

There may be also a third category of glands called compound tubuloalveolar glands which bear both tubes and alveoli. Eg. Bruner's glands, Cowper's glands, Bartholin's glands, pancreas etc.

Glands may also be classified into three catagories based on the mode of releasing the secretion. These are exocrine, endocrine, and heterocrine. The exocrine glands send their secretions through the ducts to the sites of action. These include salivary glands, gastric glands, intestinal glands, tear glands, mammary glands etc., The endocrine glands are often called ductless glands because they discharge their secretions directly into the blood stream and they are carried to the sight of action along with the blood. The secretions of endocrine glands are usually called hormones. Examples of endocrine glands are thyroid glands, parathyroid, adrenals, pituitary, thymus etc. The heterocrine glands have the combined characteristics of both endocrine and exocrine in that the secretion is first released by way of duct and the product then is directly carried into the blood. Pancreas may be said to be a heterocrine gland.

COMPOUND (stratified) EPITHELIAL TISSUE: This tissue is made up of more than one layer of cells and usually occurs at surfaces where protection is necessary. The stratified epithelial tissue is further classified into five different types. These are:

1. Stratified squamous epithelium.
2. Stratified cuboidal epithelium
3. Stratified columnar epithelium
4. Stratrified columnar ciliated epithelium
5. Transitional epithelium

In this book only the Stratified squamous epithelium is discussed.

Stratified squamous epithelium: The tissue consists of several layers of cells. The number of layers in the tissue are varied in different parts of the body. The basal layer rests on a basement membrane and continuously divides mitotically to produce a number of new cells. There is a gradual change in the shape and form of the cells as they approach the superficial layers. The upper layers are composed of flat, plate like cells. The cells of the superficial layers are constantly removed and replaced by the fresh layers of cells produced from beneath. The cells of the surface layers are very thin and have flat nuclei.

The stratified squamous epithelium may be classified into two categories- *Keratinized* and *Non Keratinized.*

Keratinized stratified squamous epithelium: The cells of the outer layers have their cytoplasm replaced with a hard waterproof protein called keratin. The process of replacement of cytoplasm is called keratinisation or karnification. The keratinised cells are dead cells and are referred to as horny layers eg. the epidermis of the skin in land vertebrates or terrestrial vertebrates.

Non Keratinized Stratified Squamous epithelium: This tissue lines the buccal cavity, pharynx and oesophagus . Here the cells are living even at the surface because cytoplasm is not replaced by keratin. This layer however is not waterproof and affords some degree of protection.

Differences between Keratinised and Non Keratinised Squamous Epithelium.

Keratinised Epithelium	Nonkeratinised Epithelium
1. Constitutes skin epidermis in terrestrial vertebrates	Lines the inner cavity of pharynx, Oesophagus etc.
2. Surface cells are non-living ; no cytoplasm, replaced by keratin.	Surface cells are living and cytoplasmic
3. It is water proof .	Allows entry of water
4. Offers sufficient protection due	Can offer only moderate protection as it is soft

CONNECTIVE TISSUE

Connective tissue is of mesodermal origin and develops from mesenchyma. Structurally the connective tissue consists of various shapes of cells lying in a large amount of non living intercellular or extracellular material called the matrix. As a contrast to the epithelial tissue connective tissues have large intercellular spaces. The matrix or the intercellular material is secreted by the cells themselves. The connective tissue roughly constitutes 20% of the tissues in the body.

In the connective tissue the bulk of the tissue is made up of intercellular substance produced by the cells. This intercellular material consists of fibres and large amount of amorphous, transparent ground substance. Chemically the intercellular matrix is made up of acidic mucopolysaccharides bound with proteins to form hyaluranic acid. In the connective tissue several types of cells are found. They may be freely wandering in the matrix or fixed to a position.

Functions of connective tissue:

1. The main function is to bind other tissues together and provide support to various structures of the body.

2. Some connective tissues help in the storage of fat.

3. Fluid connective tissues like blood or lymph help in the transportation of various materials from one part of the body to another.

4. At the sites of infection or injury connective tissues start the process of healing by destroying bacteria and dead cells.

5. Certain types of connective tissues protects the body against the foreign toxic substances. eg. Phagocytes

6. Some connective tissues, because of the gelatinous matrix act as shock absorbing cushions.

7. Connective tissues join bone to bone or bone to muscle bringing about flexible motion.

Classification of connective tissue:

Connective tissues

- Connective tissue Proper
  1. Areolar tissue
  2. Fibrous tissue
  3. Adipose tissue
- Skeletal Tissue
  1. Cartilage
  2. Bone
- Fluid tissue
  1. Blood
  2. Lymph

Connective tissues are basically divided into three categories viz., *connective tissue proper, Skeletal tissue* and *Fluid tissue* (Vascular tissue)
The cells of the connective tissue proper are immersed in a soft jelly like matrix chemically composed of carbohydrates linked to proteins. The connective tissues proper connect one tissue to another and form protective sheaths around organs.

Connective tissue proper is divided into five categories namely *Areolar tissue, Adipose tissue, white fiberous tissue, Yellow elastic tissue* and *reticular tissue.* In some cases only three are recognized namely Areolar tissue, fibrous tissue and adipose tissue.

Areolar tissue: This is the most widely distributed connective tissue in the body of the animals. The main function of this tissue is connective. It usually fixes skin with the muscles and then attaches blood vessels and nerves with the surrounding tissues and fastens the peritoneum to the body wall and the viscera. The other functions of Areolar tissue includes formation of the dermis of the skin and submucosa in the wall of the alimentary canal. It forms packaging material in all organs between the muscles. It also permits considerable stretching of the tissues.

Structurally the areolar tissue has a large amount of transparent sticky matrix enclosing numerous fibrous cells and large quantity of mucin. The fibres in the areolar tissue are of two types *white* and *yellow*. The white fibres occur in fasciles (bundles) and are very fine and wavy. They are bound together by mucin. They are nonelastic and unbranched and are formed of a protein called *collagen.* The collagen fibres are composed of numerous microfibrils which provide tenacity to the tissues.

Fig. 25.7 Aerolar tissue

The Yellow fibres are less in number than the white fibres. They are also much thicker and occur singly. The yellow fibres are branched and elastic with the branches getting interwoven to form a network. Chemically they are composed of a protein called *elastin.* The spaces left between the fibres are called areolae hence the name areolar tissue.

The cells found in the matrix are of three kinds namely *fibroblasts* or *fibrocytes*, *macrophages* or *histocytes* and *mast cells.*

The fibroblasts are elongated flat cells with oval nuclei. They secret the material from which fibres are formed. Those secreting the white fibres lie along with them at the initial stages and later move away. Those producing Yellow fibres are never adjacant to the fibres.

The macrophages are large amoeboid cells with ovoid nuclei. They freely wander in the matrix and engulf foreign particles microbes damaged cells etc. Mast cells are oval shaped cells or irregular shaped cells and have abundant dense granules in cytoplasm . It is these cells that produce the matrix. They also produce haparin, histamine, and serotonin. Haparin prevents clotting of the blood. While histamine is involved in allergic reactions, Serotonin plays a role in control of the blood pressure.

Adipose tissue: The Adipose tissue is mainly concerned with the storage of fat. The cells are spherical or oval and are packed in a matrix. The cells are called fat cells or adipocytes. In the beginning a fat cell develops small droplets which coalesce to form a large fat globule. This globule pushes the cytoplasm and nucleus to the periphery of the cell. The matrix of the tissue consists of fibroblasts, macrophages, collagen fibres, and Elastin fibres. Normally the adipose tissue is white in colour but in some hibernating (Inactive)animals the adipose tissue may be brown in colour. The usual location of the adipose tissue is subcutaneous. It is found around the kidneys, around blood vessels and also in covering of the heart. It is also known to form the yellow bone marrow.

The primary function of the adipose tissue is to store the natural fat which will be utilised whenever necessary . The subcutaneous fat prevents heat loss from the body and also gives a rounded contour. It forms a shock absorbing cushion around the kidneys and eye balls.

White fibrous tissue: This tissue is densely fibrous with the amount of matrix being reduced. The tissue is specially rich in white collagen fibres. The fibres are similar to those found in the areolar tissue but thicker than them. This tissue is very tough and nonelastic. There are two types in white fibrous tissues namely *cords* and *sheets*. The cords are white fibres which run parallely forming what are known as tendons. These are arranged in bundles bound together by areolar tissue. Among the cells, only fibroblast are present. The tendons play an important role in connecting the skeletal muscles with the bones.

Sheets are fibres that lie in a criss cross pattern in one plane. They are found in the pericardium of the heart, duramater of the brain and spinal chord , cornea of the eyeball capsule of the kidney etc.,

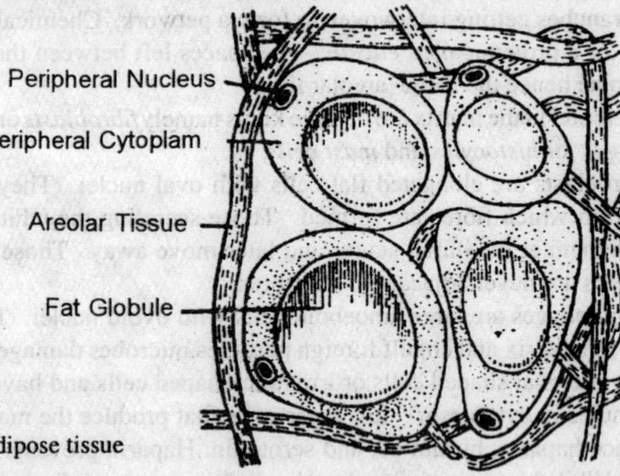

Fig. 25.8 Adipose tissue

Yellow elastic tissue: This tissue mainly consists of elastic fibres. They are similar to those of the areolar tissue but much thicker in size. The fibroblasts among the fibres are scattered. The fibres are elastic. They combine strength along with flexibility. The yellow elastic tissue is also of two types - cords and sheets.

Cords formed by yellow fibres are called ligaments. They serve to bind the bones together. Sheets of yellow fibres are present in the walls of blood vessels, lungs and bronchioles . The expansion and contraction of these vessels is due to the presence of this tissue in their walls.

Differences between tendons and ligaments.

	Tendons	Ligaments
1.	Composed of white fibrous tissue	Composed of yellow elastic tissue.
2.	Fibroblasts lie in almost continuous rows	Fibroblasts scattered.
3.	Tough and non elastic	Strong but elastic
4.	Connects skeletal muscles with bones	Connects one bone with another bone.

Reticular tissue: The tissue is made up of stellate (Star shaped) cells whose protoplasmic processes join together to form a network. Reticular fibres are found superimposed on these cells. These fibres are composed of a branched nonelastic protein called reticuline. Spaces between the cellular network are filled with a fluid matrix and other types of cells. The reticular tissue forms a supporting frame work of lymph glands, spleen and bone marrow. It is also found in the gutwall. The reticular cells are Phagocytic in function and constitute the defence mechanism of the body.

SKELETAL TISSUES:

The function of Skeletal tissue is mainly supportive. The tissues form rigid frame work offering support to the body and provide an infrastructure for the attachment of muscles and bring about locomotion and co-ordination with the muscles. The skeleton found in the higher animals is refered to as endoskeleton as it is found internally.

Skeletal tissues are mainly of two types. Cartilage and Bone.

(1) ***Cartilage***: Cartilage is a form of connective tissue which is solid and semirigid. It has a very firm intercellular matrix called the chondrin. Chemically the chondrin is made up of protein chains connected to long chains of disaccaride hyaluronic acid. Embedded in the matrix are found a number of cartilage cells called chondroblasts or chondrocytes. It is these chondrocytes which secret the matrix. The chondrocytes are somewhat oval and have large nuclei. The cells may remain in groups and are located in a fluid filled space called lacuna. Externally the cartilage is bound by a stiff sheath called perichondrium and it consists of white fibrous tissues. Cartilages lack blood vessels and nerves. The cartilages grow by the addition of new layers of matrix to the outside. Injury to the cartilage generally takes a longer time to heal as there is no blood supply.

Based on the type of matrix and its structure cartilages are classified into three types viz., hyaline, fibrous and calcified.

Hyaline Cartilage: This is the most common cartilage. The matrix is a clear homogenous translucent substance. Very often a net work of collagen fibres are dispersed in the matrix. The interweaving of the collagen fibres is very fine and very close and often difficult to observe. The name hyaline is given to this cartilage because of a translucent hyaline appearance of the matrix. The matrix consists of many lacunae and a number of chondrocytes. The cartilage is covered by a membrane called perichondrium having blood vessels. It is from these blood vessels that nutritive substances difuse into cartilage. In the body, hyaline cartilage is found in trachea , Bronchial rings, Larynx, Nasal septum, Sternal ribs etc.

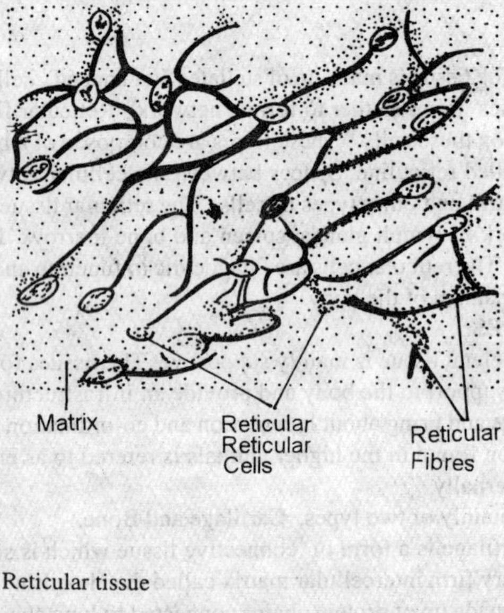

Fig. 25.9 Reticular tissue

Fibrous cartilage: The fibrous cartilage has a number of fibres traversing in the matrix. Based on the type of fibres the fibrous cartilage is divided into white fibrocartilage and elastic cartilage. The white fibrocartilage consists of rigid non-elastic white fibres. This type of a cartilage has a firmness of the fibrous tissue and the elasticity of the cartilage. This tissue is found in intervertebral disks.

The elastic cartilage has more quantity of elastic yellow fibres in the matrix in addition to white fibres. This gives added flexibility to the tissue and can get back its original shape after an injury. The elastic cartilage is somewhat yellowish in appearance and is found in the pinna and external auditory canal of the ear. Elastic cartilage is also found in epiglottis, eustachian tubes etc.

Calcified cartilage: As the very name indicates in this type of cartilage the matrix gets impregnated with calcium salts. A network of coarse collagen fibres traverse in the matrix. It is hard and non-elastic. It is found in the intervertebral disks, pubic symphysis of pelvic girdle etc.

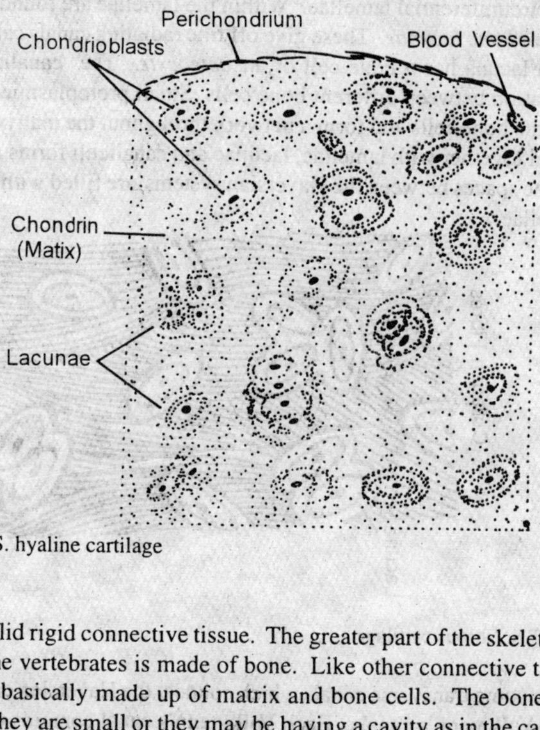

Fig. 25.10 T.S. hyaline cartilage

BONE

This is a solid rigid connective tissue. The greater part of the skeletal system in most of the vertebrates is made of bone. Like other connective tissues the bone is also basically made up of matrix and bone cells. The bones may be solid when they are small or they may be having a cavity as in the case of long bones such as humerus and femur. The substances of the bone are distinguishable into three regions surrounding the central cavity. These three regions are periosteum, matrix and endosteum.

The central cavity is usually refered to as the marrow cavity. It consists of the bone marrow. The bone marrow stores the reserve food material and also is the place from where new blood cells are produced.

External to the marrow cavity is the endosteum. It consists of white fibrous tissue and contains the bone forming cells which produce new bone material.

Matrix: The matrix of the bone occupies a very large portion. It is very hard and made up of a protein called ossien.. Collagen fibres are also found in the matrix but difficult to observe. The matrix is loaded with salts like calcium phosphate, calcium carbonate, sodium carbonate and magnesium sulphate. Of these the major quantity is of calcium phosphate. The organic and inorganic constituents of the matrix are in the ratio of 30:70. Within the matrix are found a number of layers called lamellae. These are arranged in concentric rings around narrow longitudinal cavities called *Haversian canals*. The Lamellae that surround the haversian canals are called haversian lamellae. Some lamellae however are arranged concentrically around the marrow cavity. These are called circumferential lamellae. Within the lamellae are found small fluid spaces called *bone lacunae.* These give off fine radiating canals called *canaliculi.* Each lacuna has a bone cell or an *osteocyte.* The canaliculi help in communication between different bone cells. Small protoplasmic strands extend from the bone cells and form a network throughout the matrix. The Haversian canal together with lamellae, lacunae and canaliculi forms a Haversian system. The spaces between the Haversian systems are filled with small interstitial lamellae.

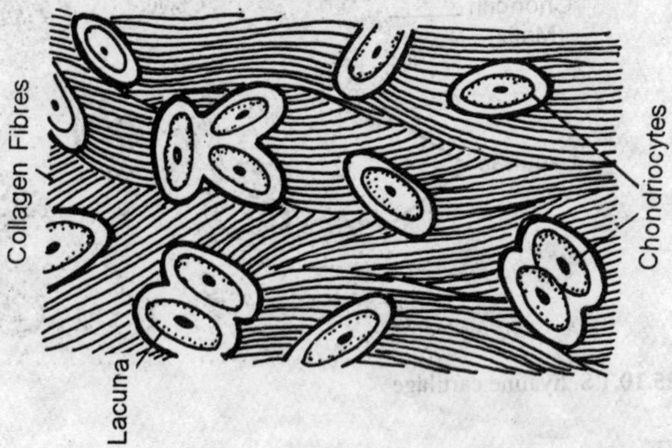

Fig. 25.11 T.S. fibrous cartilage

There are some transverse canals which connect the Haversian canals. These are called Volkmann's canals. Each Volkmann's canal contains the supply of blood vessels, nerves, a lymph vessel and some bone cells. The blood vessels of the Haversian canal are connected to the blood vessels of periosteum and bone marrow. Nourishment and oxygen to the oestocytes is provided through these blood vessels.

External to the endosteum (lining the inner margin of the matrix) is found a layer of cells called the inner layer of osteoblasts. Similarly the outer margin of the matrix also has a layer of cells called the outer layer of osteoblasts.

The periosteum is a tough sheath made up of white fibrous tissue enclosing the bone externally. It is to this sheath that tendons of the muscles are attached. The periosteum has blood vessels which sends branches into the bone to provide nourishment.

There are some differences between the bone of a frog and that of a mammal. In the case of the frog bone the lamellae are concentrically arranged around the bone marrow cavity where as in the case of mammals the lamellae surround the Haversian cannals.

Fig. 25.12 T.S. elastic cartilage

Ossification: The process of bone formation is called ossification. It is of two types *endochondral* and *intramembranous*. In Endochondral ossification the cartilage gets transformed into bone. These are called replacing bones. Eg. the limb bones, vertebrae and ribs. The intramembranous ossification produces bones directly in the connective tissue of the dermis. Bones formed in this manner sink below and get attached over the underlying cartilaginous skeleton. Eg. skull bones and clavicles of girdles.

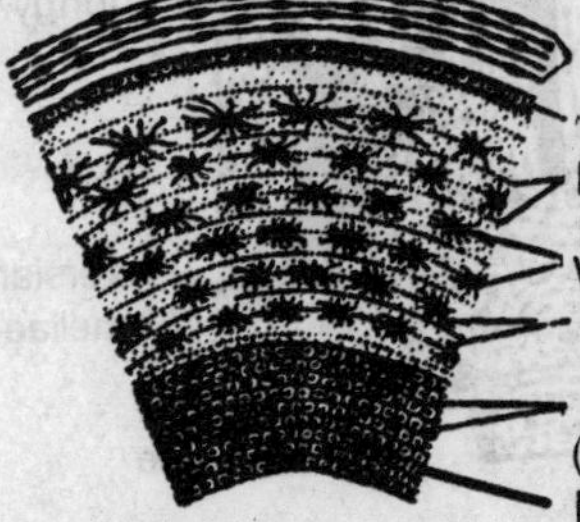

Fig. 25.13 A portion of T.S. of a limb bone of frog

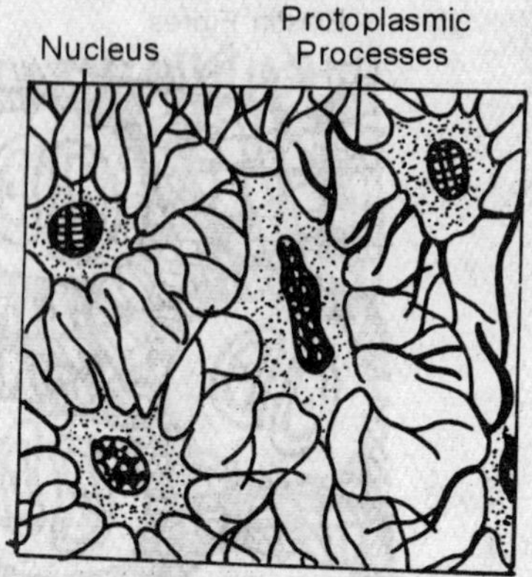

Fig. 25.14 Bone cells and their processes

Fluid tissue (Vascular tissue): The fluid tissues have a liquid matrix and free floating cells. They circulate all over the body. The matrix generally lacks fibres and is not produced by the cells enclosed in it.
The vascular tissues are of two types. Blood and lymph.

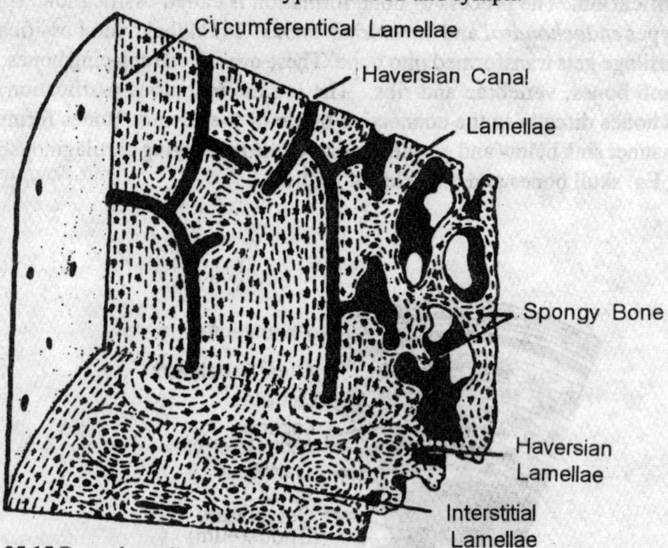

Fig. 25.15 Bone-three dimensional view

BLOOD

Blood is a liquid connective tissue. It is referred to as a tissue because it has a number of cells and it is called a connective tissue as the cells are present in a liquid matrix which travels all over the body.

An average adult human being has about 6.8 lts of blood constituting 6 to 10% of the body weight.

Composition of blood: Basically the blood has two portions namely the plasma and the blood cells. The blood cells are of various types.

Blood

- Plasma
- Formed substances
  - Corpuscles
    - Erythrocytes
    - Leucocytes
      - Granulocytes
        1. Basophils
        2. Eosinophils
        3. Neutrophils
      - Agranulocytes
        1. Monocytes
        2. Lymphocytes
  - Platelets

Plasma: The plasma is the fluid portion. It is a light yellow, slightly alkaline liquid. The plasma comprises of 90 to 92% of water and 8 to 9% of solids. The solids include inorganic salts, proteins and a number of other organic compounds.

(i) The proteins of the plasma occur in colloid form and are produced in the liver. The proteins are globulins, albumins, fibrinogen and prothrombin.

(ii) Salts: These are various kinds like sodium chloride and bicarbonates, sulphates and phosphates of calcium potassium and magnesium.

(iii) Miscellaneous substances: These includes glucose, vitamins, fats, aminoacids, enzymes, hormones etc.,

(iv) Waste products: These includes ammonia, urea etc.,

The plasma also consists of heparin-an anticoagulant substance. This prevents clotting of blood in normal blood vessels. Heparin is produced in the liver.

Blood cells: These are of basically two types. *Corpuscles* and *platelets*. The corpuscles are again classified into two types- *Erythrocytes* and *Leucocytes* (See classification above)

Erythrocytes: These are commonly called the red blood corpuscles (R.B.C). As the name indicates they are scarlet red in colour due to the presence of an oxygen carying pigment called Hemoglobin.

The shape of the RBCs varies in different groups of animals. In frog the RBC is biconvex and nucleated; the same is the condition in fishes, reptiles and birds. The RBCs of mammals are circular and biconcave and do not have any nucleus. The central part of the RBCs is thinner and helps in flexibility.

The size of the human RBCs are 7 to 8 mm in diameter. In human beings on an average there are about 4 to 5 million RBCs per cubic mm of blood. The RBC's are formed in bone marrow by cells called *megaloblasts.* Reduction in the number of RBCs causes anemia.

As has already been said the red colour in RBCs is due to hemoglobin which is a conjugated protein. This is basically made up of protein joined together with a heme group hence the name hemoglobin.

Structurally each RBC is bound by a differentially permeable membrane. The cytoplasm gradually loses all the usual components of the cell during the coarse of maturation. As a result it has lot of space to hold hemoglobin. The process of the formation of RBCs is called *Erythropoiesis*. Proteins and other iron components stimulate the production of RBCs.

Usually human RBCs have a life span of about 120 days and worn out RBC's are destroyed by phagocytosis in the blood itself and also in the spleen and liver.

Leucocytes : Leucocytes are commonly called White Blood Corpuscles (WBC). These cells lack haemoglobin and are generally far less in number when compared with RBCs. A cubic millimeter of blood contains between 5 to 10 thousand WBCs. In abnormal conditions the number may increase or decrease. Unlike RBCs WBCs are nucleated and are irregular in shape. The cytoplasm has the organelles like mitochondria, golgi bodies etc., Leucocytes are formed by the process known as *leucopoiesis*. It takes placé in lymph nodes, spleen, thymus and red bone marrow. The Leucocytes have a life span of three to four days. The worn out ones are destroyed in blood liver and lymph nodes.

Classification of WBCs: Basically there are two types- namely agranulocytes and granulocytes.

(i) **Agranulocytes**: These do not have any granules in cytoplasm or occasionally small granules may be found. The cells have a single nucleus and are generally rounded in shape. They are produced in Lymphoid tissues. There are two types of agranulocytes viz, *monocytes* and *lymphocytes*.

a. *Lymphocytes*: These cells are larger than RBCs. They have a large spherical nucleus and thin cytoplasm. They are non motile and also nonphagocytic. They are produced in lymphnodes and secrete antibodies which destroy invading microbes. The lymphocytes are about 1300 to 3000 in number per cubic milimeter of blood.

b. *Monocytes*: These are the largest of all leucocytes and have a large bin shaped nucleus. They are motile. They come out of the blood vessels and engulf bacteria and other cellular debris.

(ii) **Granulocytes**: These leucocytes have a number of granules in the cytoplasm and have a distinctly lobed nucleus. The Granulocytes are produced in the red bone marrow. They are also motile and squeeze out of the blood vessels and destroy invading cells. Three kinds of granulocytes have been identified based on their staining reaction. These are basophils, eosinophils and neutrophils.

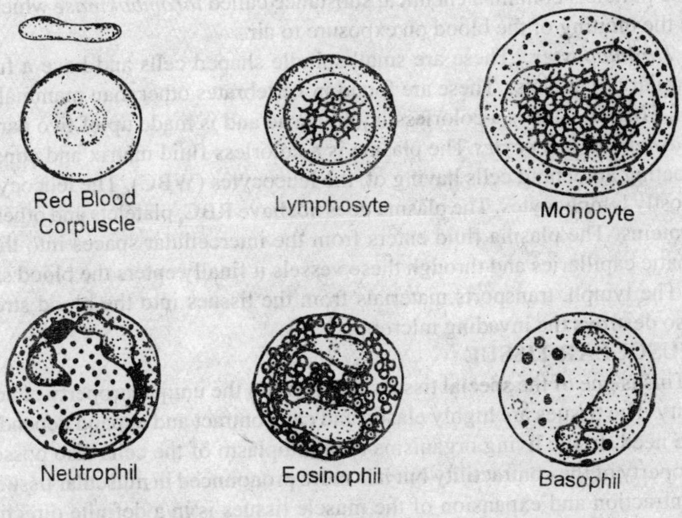

Fig. 25.16 Different types of blood corpuscles

a. *Basophils*: These react positively to the basic stains. They are fairly large in size, have a S shaped nucleus and a few coarse granules. A cubic milimeter of blood consists of about 50 cells averaging to 0.4%. The Basophils release the anticoagulant substance.

b. *Eosinophils*: These leucocytes take up acidic stains like eosin appearing red in colour. They are fairly large, have bilobed nucleus and abundant coarse granules. On an average a cubic milimeter of blood consists of about 60 to 100 cells. Eosinophils destroy toxic substances produced by microbes. In case of infections and allergic reactions the number of eosinophils increases and this condition is called *eosinophilia*.

c. *Neutrophils*: These exhibit a neutral response to both acidic and basic dyes and stain equally well. These are large, have a *manylobed* nucleus and *manyfine* granules in the cytoplasm. They are phagocytic in action.

Platelets:

Platelets are rounded or oval or disklike bodies and lack haemoglobin. Usually a cubic milimeter of blood has about 2,50,000 platelets. Increase or decrease in this number is an abnormality and is known as *thrombocytosis* and *thrombocytopenia* respectively.

Structurally the platelets are flat and bounded by a membrane. The cytoplasm consists of a few organelles and of secretory granules. The platelets are formed in the red bone marrow and survive for about three to seven days only and are disposed off by phagocytosis. The platelets do not have a nucleus. The platelets contain a chemical substance called *thrombokinase* which helps in the clotting of the blood on exposure to air.

Thrombocytes: These are small spindle shaped cells and have a function similar to platelets. These are found in vertebrates other than mammals.

LYMPH: This is a colorless liquid tissue and is made up of two parts viz., *plasma* and *leucocytes.*The plasma is a colorless fluid matrix and consists of floating amoeboid cells having of the leucocytes (WBC). The leucocytes are mostly lymphocytes. The plasma does not have RBC, platelets and other blood proteins. The plasma fluid enters from the intercellular spaces into the lymphatic capillaries and through these vessels it finally enters the blood stream.

The lymph, transports materials from the tissues into the blood stream. It also destroys the invading microbes.

MUSCULAR TISSUE

This is one of the special tissues that possess the unique property of contractility. The tissues are highly elastic and can contract and expand depending on the need. In the living organisms the protoplasm of the cells also possess the property of the contractility but it is more pronounced in muscular tissues. The contraction and expansion of the muscle tissues is in a definite direction ie., along their long axis.

The muscle tissue is made up of elongated thin cells called *muscle fibres* and they are held together by areolar tissue. There is no intercellular tissues between the muscle cells and there is also no matrix. The cytoplasm of muscle fibres is called *sarcoplasm.* Embedded in the cytoplasm are a number of long, fine, parallely arranged protein fibres called *Myofibrils.* These are arranged along the long axis of the fiber. The spaces between the myofirbrils are occupied by a large number of mitochondria (sarcosomes) and glycogen granules. These granules provide the energy necessary for contraction. In some instances the muscle fibres may be bounded by a special membrane called *sarcolemma*. The muscle fibres may have one to many nuclei.

General functions of muscle tissue

1.They support bones and other structures.

2.They are mainly responsible for locomotion and movement of body parts.

3.They are also responsible for heart beat, for flow of the blood through the vessels, passage of food through alimentary canal, flow of air through respiratory tract etc.

4.Every movement of the body whether it is respiration, digestion, secretion or excretion cannot take place without the participation of muscles.

Classification of muscle tissues: Muscle tissues are classified into three categories-

1.Smooth muscle (involuntary muscle)
2.Striated muscle (voluntary muscle)
3.Cardiac muscle

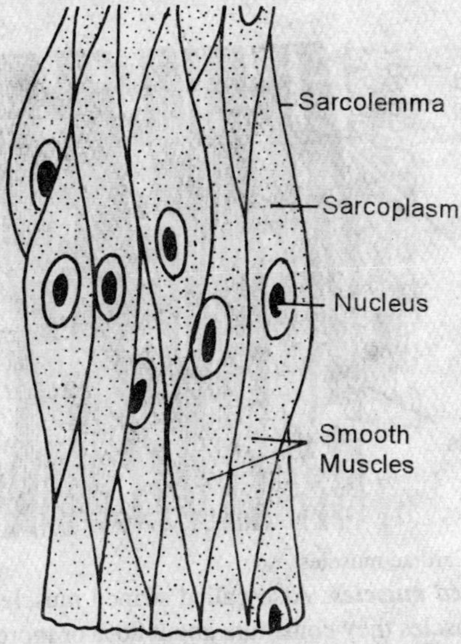

Fig. 25.17 Unstriated muscle fibres forming a sheet

1.***Smooth muscle***: These muscles are also called non-striated or non straight muscles. The muscle fibres are made up of long narrow spindle shaped cells that are generally shorter than striated muscle cells. They average a length of about 0.2 mm. Each muscle fibre has a single nucleus in the central thick portion. The sarcoplasm consists of a number of parallely arranged myofibrils. These are made up of myosin. The muscle cells are not bound by a true sarcolemma; the usual plasma membrane however is present. The cytoplasm consists of mitocondria and endoplasmic reticulum but they are not as many in number as in the striated muscles.

Smooth muscle fibres may occur individually or in the form of bundles of sheets. The muscle fibres of viscera are usually arranged in the form of sheets made up of many layers of fibres. In villi of the small intestine the muscle fibres occur singly.

In some instances, bundles of smooth muscle fibres get surrounded by connective tissue and are represented by *arrector pili* which are present at the base of the hairs. It is these which are responsible for the hair to stand erect at certain times.

Smooth muscle fibres occur in the walls of alimentary canal, genital tract, ducts, blood vessels, urinary bladder etc. Smooth muscles undergo slow and rhythmic contraction regulated by the autonomous nervous system. Hence they are called Involuntary muscles. The contraction of the muscles may be prolonged and it causes shortening of the organ and increase of the diameter of the lumen.

Fig. 25.18 Cardiac muscles

2. ***Striated muscles***: Also called striped muscles or voluntary muscles or skeletal muscles they constitute about 80% or more of the soft tissues present in the body. The stratied muscle fibres are long, cylindrical, unbranched and have blunt ends. Each fiber may be as long as 40mm. Each muscle fibre is bounded by a conspicuous sarcolemma. The sarcoplasm consists of many elongated, flattened nuclei. In addition there are a number of glycogen granules and mitochondria.

The most striking feature of the striated muscle fibre is the presence of alternating dark and light transverse bands called stripes or striations. Hence the name striated muscles. The dark bands are called anisotropic or A bands. Each A band has at its middle a somewhat light zone called the H zone or Henson's line. The lighter band is isotropic and is called the I band. Each I band has at its centre a dark line called the *membrane of Krause* or Z line or Z band. The dark and light bands alternate regularly. The portion of the myofibril between two Z lines is called a Sarcomere and it constitutes a contractile unit. Thus each sarcomere consists of A band and half of each adjacent I band. Electron microscopic studies reveal the sarcomere is a bundle of thin myofilaments of two types- primary and secondary.

3. ***Cardiac Muscles*** :

The cardiac muscles are confined to the walls of heart only. They have a unique function of contracting rhythmically and continuously. They generate a unique wave excitation as they pass from one fibre to another.

The cardiac muscle fibres are made up of short cylindrical fibres joined end to end. These are interconnected by crosslinks forming a network. It is this arrangement that helps in the rhythmic contraction so characteristic of the heart muscles.

The muscle fibres are uninucleate and do not have a sarcolemma. The cardiac muscle fibres are striated but the bands are much paler than those of the striated muscle fibres. The cytoplasm consists of a large number of mitocondira and numerous glycogen granules. There are some speicalised regions of the membrane at the end of the fibres which bring about cris-cross junction - these areas are called *intercalated discs* (rather than Z lines of the striated muscles). These intercalated discs help as boosters of the wave action permitting the motion to be transferred from one fibre to the other. The cardiac muscles like skeletal muscles (striated muscles) contract quickly but do not experience fatigue.

STRIATED	NON-STRIATED	CARDIAC
1. Found in limbs	Found in the wall of alimentary canal, urinary bladder etc.	Found in heart
2. Occur in bundles (flesh of the body)	Form sheet, sphincter or tube	Form a network
3. Long and cylindrical	Long and spindle shaped	Short, cylindrical branched
4. Sarcolemma present	No sarcolemma, bounded by thick plasma membrane	Sarcolemma very thin bounded by plasma membrane
5. Multinucleate (coenocytic)	Uninucleate	Uninucleate or multi nucleate
6. Myofibrils with distinct alternate dark and light cross bands.	Myofibrils in distinct without cross bands	Myofibrils distinct with faint alternate dark and and light bands
7. Intercalated discs absent.	Intercalated discs absent	Intercalated discs present
8. Innervated from central nervous system.	Innervated from autonomous nervous system	Innervated by nerves from autonomous nervous system
9. Contract faster and get fatigued very soon.	Contracts slowly, can remain contracted for	Contracts quickly. Never stops and never get tired
10. Voluntary	Involuntary	Involuntary

NERVOUS TISSUE

The nervous tissues are specialised for receiving and responding to stimuli such as sound, heat, cold chemicals etc. The cells of the nervous tissue help in the transmission of the messages from one to another . The protoplasm of the nerve cells has the property of conductivity through which messages are transferred from one place to another. The nervous system which is composed of nervous tissue has the following three components.

1.Neurons
2.Nerve fibres and
3.Neuroglia

Neurons: Also called nerve cells, the neurons constitute the basic structural unit of the nervous tissue. Each neuron is made up of three parts namely the cyton, dendrites and axon.

(i) *Cyton*: This is also called the cell body, soma or perikaryon. Body of the neuron may be rounded, oval, or stellate. The cytoplasm is abundant and it is called neuroplasm. The cyton has a large nucleus with a distinct nucleolous. The cytoplasm also has mitochondria golgi aparatus, flat globules and fine thread like structures called neurofribrils. There is no centrosome so that the neuron cannot divide. But some zoologists opine that centrosomes are present in the neurons. In addition to this the cytoplasm has numerous basophilic bodies called *Nissle's granules*. The exact functioning of these granules is not understood. But they seem to play an important role in the metabolism of the neurons as indicated by the fact that they undergo changes under different physiological conditions. Electron microscopic studies of the Nissle's granules has shown it to be rough endoplasmic reticulum.

(ii) *Dendrites*: These are small minute processes that arise from the cyton. They show a high degree of branching. The function of these dendrites is to connect one neuron with the another.

(iii) *Axon*: This is a single long process that arises from the cyton. It is usually given off from the small conical region of the cyton which is known as axon hillock or implantation cone. The cone contains ribonucleic acid, protein and iron. The axon at its free end gives rise to minute branches called telodendrites. The cytoplasm of the axon is known as axoplasm.

Nerve Fibres :

The axon along with its enveloping sheath is called as a nerve fibre. The nerve fibres located outside the brain and spinal cord are covered by a thin membrane called *neurilemma* or *sheath of Schwann*. The neurilemma inturn is made up of sheath cells or lemnoblasts or Schwann's cells. These sheath cells are present in a series. Each sheath cell with its cytoplasm and flattened nucleus surrounds the section of the axon. The presence of these cells is essential for the proper functioning of the axons of peripherial nerves.

The nerve fibres are classified into two types. Medullated nerve fibres and Non-medullated nerve fibres.

Medullated nerve fibres: These are having a central core (neuraxis) surrounded by two sheaths - inner thick medulary sheath and outer thin nurilemma. The axis cylinder is nothing but the axon nerve cell. It contains longitudinal neurofribrils in its neuroplasm. The medulary sheath is composed of a shining white fatty substance called *myelin*. This sheath serves as an indicating layer preventing the loss of energy of the nerve impulses during the passage. The medulary sheath is not continuous, at some places it is absent and these regions are known as the *nodes of Ranvier*. The part of the nerve fibre between two nodes of Ranvier is called the internode.

Non-medullated nerve fibres: These nerve fibres lack the myelin sheath and the neurilamma directly surrounds the axon. The nuclei of the neurilamma bring about swellings around the fibre. Electron microscopic studies however have shown a trace of myelin surrounding the axon in the non-medulated fibres also. The fibres appear grey in colour in contrast to the glistening white appearance of the medullated fibres.

The medullated fibres occur in the brain, spinal cord, Cranial and spinal nerves whereas the nonmedullated nerve fibres occur in the autonomous nervous system.

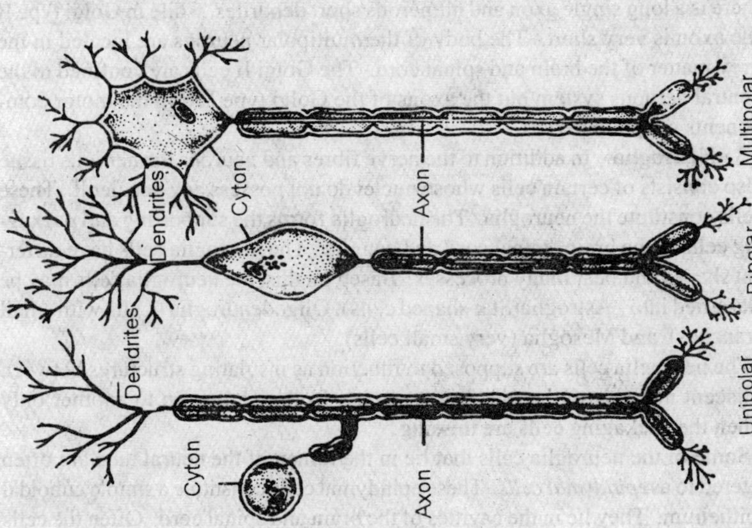

Fig. 25.19 Type of neurons

Types of Neurons

A number of neurons occur in different parts of the nervous system. These are classified as follows:

a. *Unipolar* neurons: In these neurons, a single axon passes from cyton. There are no dendrites. Unipolar neurons are rarely found in the central nervous system of the adults, but they are present during the early developmental stages.

b. *Pseudounipolar neurons*: These cells are first typically bipolar and spindle shaped but as the development proceeds the two processes converge until they meet at one side of the body. This then elongates so that a fine process is formed with a T shaped division at the end. Pseudounipolar neurons are formed in all spinal ganglia and in the ganglia of cranial nerves except the VIII nerve.

c. *Bi-polar neurons*: These cells have a single dendrite and a single axon projecting from opposite ends of the cells. Such neurons are present in the retina of the eye, in the olfactory epithelium and in the ganglia of the VIII cranial nerve.

d. *Multipolar neurons*: These occur most frequently when compared to other neurons. As the very name indicates they have a number of processes. Based on the number, arrangement and length of the processes they are categorised into two types viz., Golgi type I and Golgi type II. In Golgi type I there is a long single axon and numerous short dendrites, while in Golgi type II the axon is very short. The body of the multipolar neurons are located in the grey matter of the brain and spinal cord. The Golgi II cells are confined to the central nervous system but the axons of the Golgi type I form the motor components of spinal nerves.

3. **Neuroglia**: In addition to the nerve fibres and neurons the nervous tissue also consists of certain cells whose nuclei do not possess any neucleoli. These cells constitute the neuroglia. The neuroglia forms the supporting and packaging cells in the brain, spinal cord and ganglia. The neuroglia cells have different shapes and bear many processes. Based on this, the neuroglia cells may be classified into - Astroglia(Star shaped cells), Oligodendroglia (Cells with small branches) and Mesoglia (very small cells).

The neuroglia cells are supposed to function as insulating structures between adjacent neurons so that the impulses pass from one neuron to another only when the packaging cells are missing.

Some of the neuroglia cells that lie in the lumen of the neural tube are often refered to as *epindymal cells*. These epindymal cells constitute a simple cuboidal epithelium. They lie in the cavities of the brain and spinal cord. Often the cells are ciliated.

HISTOLOGY OF THE HUMAN SMALL INTESTINE

The small intestine in human beings is a convoluted tube extending from the pyloric region of the stomach to the ileocaecal valve where it joins with the large intestine. The length of the small intestine is approximately 5mts (16ft). It is made up of three parts namely duodenum, jejunum and ileum.

The duodenum is the first part in the small intestine; it is about 2.5 to 3.0mts long and lies around the head of the pancreas. It is fixed to the posterior wall of the abdomen by means of peritoneum. The jejunum constitutes two-fifths of the small intestine while ileum occupies the remaining portion.

The small intestine, histologically is covered by four coats of tissues. These are-

1. An outer serous coat (peritoneum)
2. A muscular coat
3. A sub mucous coat and
4. Mucous membrane

1. **Outer serous coat**: The serous coat or serous membrane is also called the peritoneum. Like all serous membranes it has a smooth shiny surface composed of a layer of simple squamous epithelium. This is supported by a thin layer of areolar tissue. The membrane forms the outer wall with the epithelial surface facing the cavity which is refered to as peritoneal cavity.

2. **Muscular coat**: The muscle coat is made up of two layers of visceral muscle fibres. These are -a) longitudinal muscle fibres and b) circular muscle fibres. The longitudinal muscle fibres are superficial while the circular muscle fibres form the inner layer (there are no oblique fibres in the small intestine as seen in stomach). The muscle fibres are arranged along the length of the organ. The muscle coat has two important functions. 1. The tone of the muscle coat maintains the intestine in a functional state and 2.The rhythmic contraction of the muscle fibres allows (even without nervous stimulation) the movement of the food in the intestine.

3. **Sub mucous coat**: The sub mucous coat binds the mucous membrane to the muscle coat and constitutes the main area for the circular folds of the mucous membrane. In the duodenum, this coat contains duodenal glands (Brunner's glands) which secrete an alkaline mucous to protect the mucous membrane from the acidity of the food chyme coming from the stomach.

The sub mucous coat is made up of loose connective tissue and contains many elastic fibres. It contains blood vessels to supply the mucous membrane.

4. **Mucous membrane**: This is the inner most membrane of the intestine. The surface layer of the mucous membrane is made up of epithelial cells and it comes in direct contact with the food materials. The cells are of the stratified squamous type where food is rough but columnar where digestion and absorption takes place.

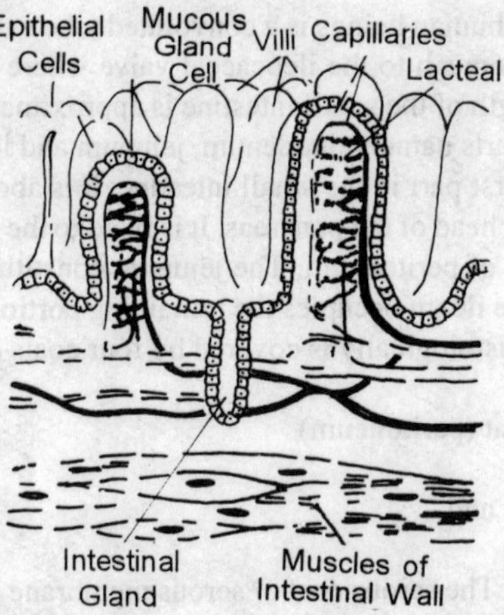

Fig. 25.20 The intestinal villi which increase the absorptive surface of intestine

The mucous membrane is thrown up into permanent circular folds called *Plicae circulares*. These greatly increase the surface area for secretion and absorption. At the sites of absorption the folds of the mucous membrane are filled with minute finger like processes called *villi*. These villi give the surface a soft velvety appearance. The collection of these villi gives a brush border appearance. A microscopic examination of the villi (singular. villus) shows that these are tiny folds of the membrane projecting out from the surface. The surface of each villus is covered by a layer of columnar epithelial cells. These epithelial cells have a brush bordered appearance forming thousands of microvilli. The villi and microvilli greatly increase the total absorptive area of the small intestine. Scattered among the columnar epithelial cells of the surface of the villi are found a number of goblet cells each of which contains a large drop of colourless mucous. The mucous secreted by the goblet cells acts as a lubricant. Each villus has a core area of loose connective tissue and some smooth muscle fibres.In the centre of this is found a lymph capillary with a blind end (lacteal). Into this lacteal some of the absorbed fatty substances get in. Each villus is provided by a small arteriole which feeds the capillary plexus lying close to the epithelium. Food materials absorbed from the lumen by the columnar cells are passed on to the blood through these capillaries.

Between the bases of the villi, epithelium dips into the underlying layer of connective tissue (lamina propria) forming some simple tubular glands called intestinal crypts are crypts of Lieberkuhn. The cells lining these glands secrete the alkaline intestinal juice called Succus entricus. This juice besides being alkaline contains inorganic ions, mucin and many other enzymes. These enzymes complete the breaking up of fats, carbohydrates and proteins to the simple molecular level so that they can be easily absorbed. At the base of the crypts of Lieberkuhn are present the cells of Paneth. These secrete some digestive enzymes.

The small intestine secrets mucous from its entire surface as a protective lubricant and this is very important to protect it against the acidic chyme coming from the stomach. In the first few centimetres of the duodenum are found large mucous glands called Brunner's glands. The secretions from these protect the duodenal mucousa from the powerful effect of the hydrochloric acid and pepsin until the chyme has been neutralised by the pancreatic juice .

Lymphatic tissue of diffused type is scattered throughout the small intestine particularly in the lamina propria. These may occur as solitary nodes or may be present in collected masses (especially in children) called Peyer's patches. These Peyer's patches are mainly found in the ileum.

Twenty six

Digestion

All the organisms require energy to perform various activities. In multicellular organisms every individual cell requires energy. This energy is obtained by organisms by the food they consume. The food that is consumed is digested to its simple state and then ultimately oxidised during the process of respiration to release energy.

The purpose of digestion is to alter food materials and convert them into simple constituents that can be absorbed into the blood and utilised by the various tissues according to their requirements.

Basically the human digestive system consists of the alimentary canal and many associated digestive glands. The accessory organs of digestion are the liver and pancreas.

The alimentary canal is a long tubular structure extending from the mouth to the anus. Although it is a single continuous tube, it is made up of several parts. Most of the alimentary canal lies in the abdomen in a coiled form. The alimentary canal has a lining of epithelium called mucous membrane. The following are the parts of the alimentary canal.

Mouth
Pharynx
Oesophagus
Small intestine and
Large intestine

Digestion of food takes place in the alimentary canal and is aided by the accessory organs of digestion like salivary glands, liver and pancreas. These organs pore their secretions into the canal to help in the process of digestion. Food is physically and chemically a complex compound and two kinds of digestive processes are needed before it can be absorbed by the body cells. These processes are *mechanical digestion* and *chemical digestion*.

The mechanical digestion consists of physically breaking large particles of food into smaller ones. This involves mastication by the teeth, swallowing and the peristalic movements of the alimentary tract which mix the food with the digestive juices and move it along.

The chemical digestion involves breaking up of the bonds of large molecules by involvement of digestive enzymes present in the secretions of glands. A digestive enzyme chemically acts on a large molecule and breaks it up into its individual constituents. For eg. a protein molecule is broken down into amino acids.

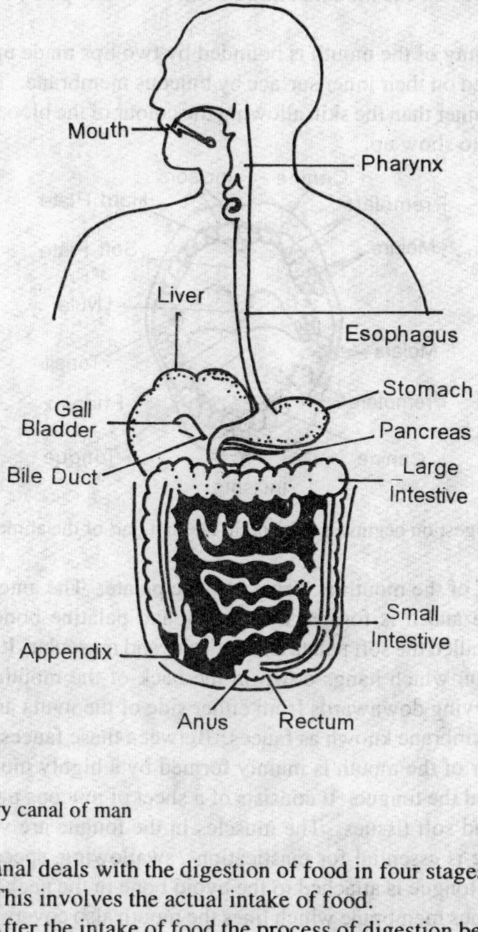

Fig. 26.1 Alimentary canal of man

The alimentary canal deals with the digestion of food in four stages

1. **Ingestion** : This involves the actual intake of food.
2. **Digestion** : After the intake of food the process of digestion begins in the mouth(physical process) and is mainly carried out in the stomach and small intestine (chemical process)
3. **Absorption**: This involves the passage of digestive food substance through the mucosa of the gastrointestinal tract into the blood and lymph.
4. **Egestion** : Food substance that cannot be digested or that are not required by the body are eliminated as excretory material.

We will now study in some detail the various parts of the alimentary canal involved in the process of digestion.

1.**The mouth**: The mouth or buccal cavity is the upper expanded portion of the alimentary canal. It is made up of two parts. The outer part called the vestibule is the space between the teeth and the inner surfaces of the cheeks. The inner part of the mouth communicates with the pharynx and contains the tongue.

The opening of the mouth is bounded by two lips made up of muscle fibres and are lined on their inner surface by mucous membrane. The mucous membrane is thinner than the skin allowing the colour of the blood in the underlying capillaries to show up.

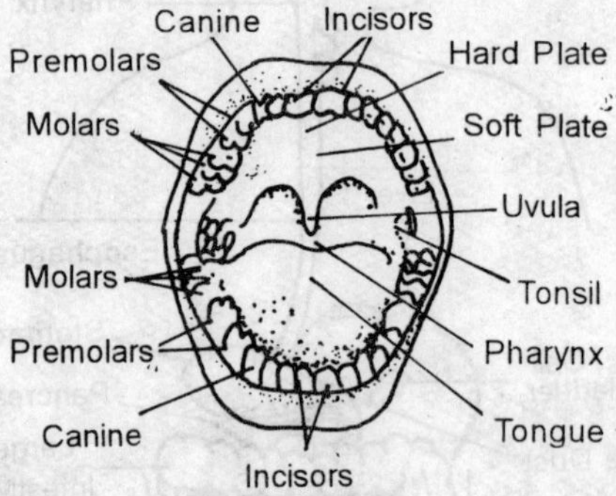

Fig. 26.2 Digestion begins in the mouth, the first part of the alimentary canal

The roof of the mouth is formed by the palate. The anterior portion is the hard palate and it is formed by maxilla and palatine bones. The posterior portion is called the soft palate and is fleshy and muscular. It has a cone shaped prolongation which hangs down in the back of the mouth and is called the *uvula.* Curving downwards from either side of the uvula are two folds of the mucous membrane known as fauces. Between these fauces lie the tonsils.

The floor of the mouth is mainly formed by a highly mobile and muscular organ called the tongue. It consists of a sheet of mucous membrane covering muscles and soft tissues. The muscles in the tongue are voluntary muscles. The tongue is essential for mastication , swallowing, speech and taste. The root of the tongue is attached to the hyoid bone in the neck.

The mucous membrane which lines the mouth also covers the free surface of the tongue. On the upper region of the tongue are numerous minute elevations of the mucous membrane called papillae. Three types of papillae are found on the tongue. These are

a.*Circumvallate papillae.* These are large papillae arranged in a V shaped pattern at the base of the tongue. They contain numerous taste buds.

b.*Fungiform papillae*. These are flattened papillae found on the top and sides of the tongue.

c. *Filiform papillae*. These are small pointed elevations found all over the surface of the tongue.

The tongue has many functions. These are (a) It helps in tasting of food. The taste buds can recognise four tastes salt, sour, bitter and sweet. (b) It helps in shovelling the food properly to be chewed by teeth. (c) It helps in swallowing. Besides, it plays an important role in speech.

Teeth: These are hard structures present in both the jaws. In human beings the teeth are embedded in pits or sockets in the jaw bones. Such teeth are said to be *thecodont*. In human beings there are two sets of teeth. The temporary teeth and the permanent teeth. The temporary teeth are found in childhood and are twenty in number. The permanent teeth are thirty two in number - 16 in the upper and 16 in the lower jaw. According to their shape and position they are classified into the 4 following groups.

Incisors	2 + 2
Canines	1 + 1
Premolars	2 + 2
Molars	3 + 3

The Incisors help in cutting and biting. The canine teeth are used for grasping and tearing. Behind the canines are the premolars and molars which are mainly meant for grinding.

Salivary glands. There are three pairs of salivary glands - the *parotid, submandibular* and *sublingual.* These secret saliva which begins the process of digestion in the mouth. The parotid glands lie below and slightly in front of each ear. They are the largest of the salivary glands. The submandibular glands are situated just under the cover of the angle of the mandible. The sublingual glands lie in the anterior part of the floor of the mouth between the tongue and mandible. They pour their secretions directly into the mouth.

The digestive juices secreted by these glands are commonly called saliva. This is a fluid consisting of 90% water and a digestive enzyme *ptyalin* (salivary amylase). The carbohydrate gets partly digested due to the action of saliva. The saliva also contains mucin and a small quantity of calcium salts. The main functions of saliva are

1.It moistens the mouth and tongue.

2.It lubricates food so that it can be rolled into a soft mass or bolus which can be easily swallowed.

3.Ptyalin acts on the starch and breaks them up into dextrins and maltose.

4.Saliva also dissolves the part of the food which stimulates taste buds.

5.It has a cleansing action in the mouth.

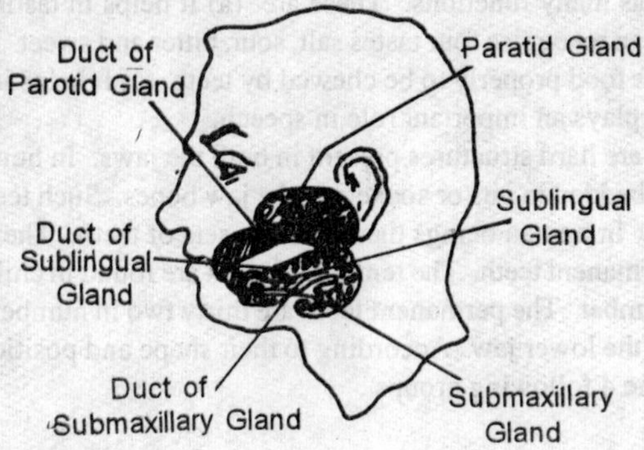

Fig. 26.3 Man, location of salivary glands

Pharynx. The Pharynx is an expanded muscular tube lying behind the nose and mouth. It is about 13 cms long and extends from the base of the skull to the level of the cricoid cartilage and joins with the oesophagus. Anatomically the pharynx is divided into three parts. These are 1. the *nasopharynx* 2. the *oropharynx* 3. the *laryngopharynx.* The nasopharynx lies immediately behind nasal cavities and extends to the level of the soft palate. The oropharynx is situated behind the mouth. It opens to the mouth in the anterior portion and is bounded by the anterior pillars of the fauces. The laryngopharynx is narrower than the other parts and lies behind the larynx. At its lower end it merges into the oesophagus.

The nasopharynx actually does not take part in the digestive process. The oropharynx is a shared entrance for the respiratory and alimentary tracts. Since the air must pass through it to reach the larynx and the food also must pass through it to reach the oesophagus. The oropharynx leads into the larynx in front. The opening to larynx is called *glottis.* This bears a cartilaginous flap called the *epiglottis.* The epiglottis projects into the oropharynx opposite the uvula. During the process of swallowing of food it covers the glottis to prevent the entry of food into it.

The pharynx as such does not take part in the digestive process except it serves as a passage for the food from the oral cavity to the oesophagus. The walls of the pharynx are composed of muscles arranged in thin overlapping sheets. These form the constrictor muscles which contract and help in swallowing.

Oesophagus. The oesophagus are gullet is a collapsible muscular tube. It is about 25 cms long and runs down through the neck, the thorax and the abdominal cavity. It extends from the pharynx to the cardiac orifice of the stomach. The oesophagus leaves the thorax by pearcing through the diaphragm and enters the abdominal cavity before terminating in the stomach. Structurally the oesophagus is composed of four layers of tissue - a mucous membrane made up of stratified epithelium, a submucous layer, a muscle layer consisting of an inner coat of circular muscles and an outer coat of longitudinal muscle fibres and an outer covering of connective tissue.

The muscle fibres of the oesophagus relax when swallowing takes place. They also prevent the contents of the stomach from coming back to the oesophagus.

The food which is in the form of a soft rounded mass called bolus gets into the oesophagus during the process of swallowing. The soft palate rises during the swallowing and the epiglottis moves backwards to cover the opening - glottis which prevents the entry of food in the trachea. The sphincter muscles of the oesophagus relax and the larynx is pulled upwards opening the oesophagus. The bolus is carried by the oesophagus by the contraction of its muscular walls. When the cardiac sphincter muscles of the oesophagus relax the food is pushed into the stomach

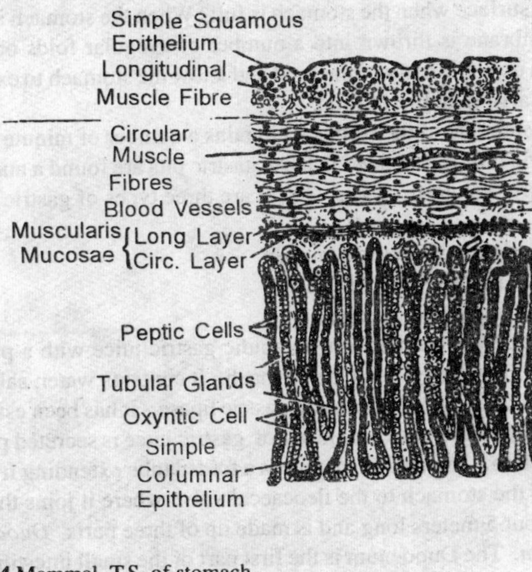

Fig. 26.4 Mammal, T.S. of stomach

The stomach. The dilated portion of the alimentary canal situated between the oesophagus and the beginning of the small intestine is called the stomach. It is a muscular distensible organ and its shape and size varies. It is about 30 cms long and 15 cms wide. There are four regions in the stomach. These are the *cardiac region, fundus, body* and the *pyloric region.*

The *cardiac region* is situated at the upper end of the stomach. The oesophagus opens into this. This opening is called the cardiac orifice. This is guarded by the cardiac sphincter muscles which prevent the regurgitation (moving back) of the food into the oesophagus. The fundus is the dome like upper part of the stomach and it projects above the cardiac orifice. The main part of the stomach is called the body and at its lower end it narrows into the pyloric region. The pyloric region connects itself to the small intestine through the pyloric aparture and is guarded by a valve called the *pyloric sphincter.* This regulates the movement of the food into the intestine.

Structurally the wall of the stomach is made up of four layers of tissue. These are

1. An outer covering of the serous membrane (the peritoneum)
2. A muscle layer which inturn is made up of three layers of visceral muscle fibres namely longitudinal, circular and oblique.
3. A sub mucous layer consisting of loose areolar tissue carrying blood vessels, lymphatics and nerves.
4. The mucous membrane constituting the inner most layer is thick and smooth having a velvety surface when the stomach is full. When the stomach is empty the mucous membrane is thrown into a number of irregular folds or *rugae.* These folds open up at the entry of the food and allow the stomach to expand to accommodate a large quantity of food.

The mucous membrane of the stomach contains a number of minute depressions called gastric pits. Embedded in these gastric pits are found a number of glands which secret the gastric juice. There are three types of gastric glands. These are

1. The peptic cells that produce zymogen
2. The oxyntic cells that secret HCL and
3 Mucous cells which add mucous.

The three secretions together form an acidic gastric juice with a pH of 2. The gastric juice is discharged into the stomach. It contains water, salts, mucin, hydrochloric acid, pepsin, rennin and gastric lipase. It has been estimated that on an average about two to three litres of gastric juice is secreted per day.

The small intestine. The small intestine is a folded tube extending from the p yloric region of the stomach to the ileocaecal valve where it joins the large intestine. It is about 5 meters long and is made up of three parts. *Duodenum, Jejunum* and *Ileum*. The Duodenum is the first part of the small intestine. It is about 25 to 30 cms in length and lies in a C shaped fashion around the head of the pancreas. About half way along, the concave surface of the Duodenum, the bile duct and the pancreatic duct enter the duodenal cavity together as a small papilla called the *ampulla of vater*.

The Jejunum forms 2/5ths of the small intestine while the Ileum constitutes the rest of it. There is very little distinction between Jejunum and Ileum as the former imperceptibly merges into the later. Both the jejunum and the ileum lay coiled in the central and lower abdominal cavity below the liver and the stomach.

Structure of the Small Intestine (see chapter 2.11) The small intestine performs two functions. Completion of the process of chemical digestion and absorption of digested food. It also has the function of moving the undigested food into the large intestine.

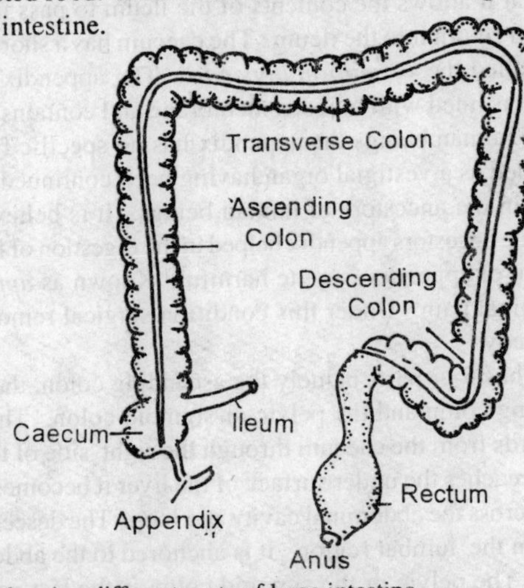

Fig. 26.5 Diagram showing different parts of large intestine

The process of digestion of food in the small intestine: The presence of the acid chyme (food from the stomach) in the duodenum causes the mucosa to secrete mucous and intestinal juice containing the enzyme *enterokinase*. At the same time the mucosa also releases into the blood stream the hormones *secretin* and *cholecystokinin* which are carried to the pancreas. Secretin stimulates the pancreas to secrete the pancreatic juice. The pancreatic juice consists of amylase(for digesting carbohydrates), trypsinogen and chymotrypsinogen (for digestion of proteins), and lipase (for digestion of fats). The pancreatic juice enters the duodenum. The enzymes trypsinogen and chymotrypsinogen are converted into their active state trypsin and chymotrypsin in the presence of enterokinase.

The cholysystokinine hormone also stimulates the gall bladder to empty the bile which is essential for digestion of fat. The bile is emptied into the duodenum via the bile duct.

The small intestine also secrets digestive enzymes from the epithelial cells covering the villi. The food that is digested is absorbed through the epithelial surface of the villi.

Large intestine. The large intestine is shorter than the small intestine and it is called large because it has a wide lumen. It is arranged around the small intestine in the form of a question mark. The large intestine is about 5 feet long and has the following parts Caecum, Colon, Rectum and Anus.

The caecum is a blind dilated sac into which is connected the ileum. The mucous membrane at the point of connection of the ileum and Caecum is so arranged that it allows the contents of the ileum to pass into the caecum but prevents their return into the ileum. The caecum has a short slender worm like projection called the *vermiform appendix*. The appendix is usually about 9 cms long . It is lined with mucous membrane and contains lymphoid tissue in its walls. In human beings the appendix has no specific function to perform and is regarded as a vestigial organ having been continued as an organ which was present in the ancestors of human beings. It is believed that in the non human primate ancestors appendix helped in the digestion of fibrous plant matter. Inflammation of appendix is quite harmful. Known as *appendicitis* it causes acute abdominal pain. Under this condition surgical removal of appendix is the only remedy.

The colon has four parts namely the ascending colon, the transverse colon, the descending colon and the pelvic or sigmoid colon. The ascending colon passes upwards from the caecum through the right side of the abdominal cavity. When it reaches the under surface of the liver it becomes transverse colon. This passes across the abdominal cavity as a loop. The descending colon passes downwards in the lumbar region. It is anchored to the abdominal wall by the peritoneum. The pelvic or the sigmoid colon is the last portion of the colon and it makes an S shaped or sigmoid curve.

The rectum lies in the pelvic cavity and follows the colon. It is about 13 cms long and has longitudinal folds made up of stratified squammous epithelium. The last part of the rectum which is about 2.5 cms long is called the anal cannal. Finally the anal cannal opens to the exterior by anus. The anus is guarded by two anal sphincter muscles namely the internal smooth involuntary muscles and external straited voluntary muscles.

The functions of the large intestine are the absorbtion of water and electrolytes from the chyme as it comes from the small intestine. Another function of large intestine is the storage of faecal matter until it is thrown out as faeces. By the time the contents of the small intestine reach the caecum, the digestion and absorption would have already been over and only the unwanted residue in a liquid form will reach the large intestine. Most of the water and electrolytes are reabsorbed by the first half of the large intestine leaving only a faecal residue between 100 to 200 ml (out of a total of 800 ml) per day.

The faeces is normally a semisolid paste like matter coloured brown by stercobilin, a pigment derived from the bile pigments. Water forms (65 to

70%) the bulk of the faecal matter. The remainder consists of undercomposed cellulose, fatty acids, protein residues etc.

In addition to the above organs two other organs play an important role in the digestion of food. These are the liver and the pancreas **Liver**. The liver is the largest gland in the body weighing about 1.6 kgs. It is reddish brown in colour and it is covered with a connective tissue capsule. The liver is heavier in the male than in the female. It is a wedge shaped organ lying immediately below the diaphragm in the right hypochondrium and epigastrium. The liver has two lobes namely the large right lobe and the small left lobe. Besides this it has superior, inferior, anterior and posterior surfaces.

The right lobe is much larger than the left and division between them is marked by falciform ligament on the anterior surface. The under surface of the right lobe is further subdivided into two smaller lobes called the *quadrate lobe* and *caudate lobe.* A thin walled pyriform sack like structure called the gall bladder is present in a groove in the inferior surface.

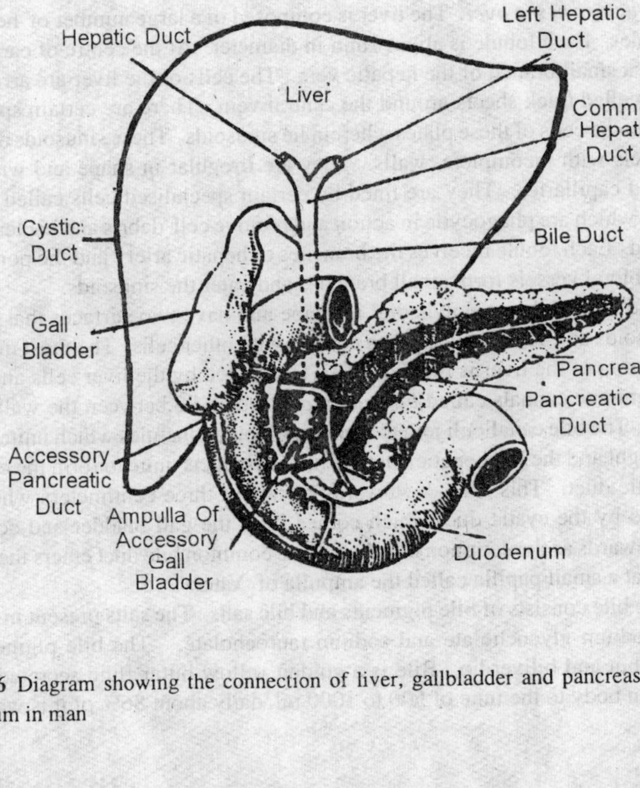

Fig. 26.6 Diagram showing the connection of liver, gallbladder and pancreas with duodenum in man

The surfaces of the liver. As has already been pointed out the liver has four surfaces. The superior surface is in touch with the undersurface of the diaphragm, while the inferior surface is related to other abdominal viscera including the kidney. The anterior surface is separated from the right lower ribs and costal cartilages by the margin of the diaphragam. The posterior surface crosses the vertebral column in the midline and is also related to the aorta.

In the centre of the interior surface lying between the quadrate and caudate lobes is the helium or *porta hepatis*. All the vessels and nerves that enter into or go out of the liver (with the exception of hapatic vein) pass through the hilum.

Blood supply to the liver. The liver receives blood from two sources and is rich in blood and has about 1/5th of its volume made up of blood. The two sources that supply blood are the *hepatic artery* and the *portal vein*. The hepatic artery conveys the oxygenated blood to the liver cells, while the portal vein conveys venous blood from the stomach and the intestines. The hepatic vein drains the venous blood from the liver and empties it into the inferior venacava.

Structure of the liver. The liver is composed of a large number of hexagonal lobules. Each lobule is about 1mm in diameter. At the centre of each lobule runs a small branch of the hepatic vein. The cells of the liver are arranged in one celled thick sheets around the central vein. There are certain spaces between the cells of these plates wherein lie sinusoids. These sinusoids are blood vessels with incomplete walls. They are irregular in shape and wider than blood capillaries. They are lined by certain specialised cells called *Kupffer cells* which are phagocytic in action and remove cell debris and bacteria in the blood. Each lobule receives the branches of hepatic artery and the portal vein. The blood vessels form small branches and enter the sinusoids.

The liver cells are polyhedral in shape and have two surfaces that face the sinusoids and several surfaces in contact with other cells. The most important secretion of the liver is the bile. This is secreted by the liver cells and drains into minute cannals called *bile canaliculi* which lie between the walls of the cells. The bile canaliculi join to form intralobular ductules which unite to form the right and the left hepatic ducts. These two ducts unite to form the common hepatic duct. This runs downwards for about three centimeters where it is joined by the cystic duct which comes from the gall bladder and continues downwards as the common bile duct. This common bile duct enters the duodenum at a small papilla called the ampulla of Vater.

The bile consists of bile pigments and bile salts. The salts present in the bile are sodium glycocholate and sodium taurocholate. The bile pigments are bilirubin and biliverdin. Bile is a golden yellow bitter fluid secreted by the human body to the tune of 500 to 1000 ml. daily about 86% of it is water.

Functions of the liver.

1.During embryonic development R.B.Cs are formed in the liver.

2.Liver stores Vitamin A, B and C.

3.It stores the iron from the destroyed R.B.C.

4.It produces heparin for preventing coagulation of blood in the body.

5.Fibrinogen, blood protein is produced for blood clotting.

6.1/5th of total volume of blood of the body is stored in the liver.

7.Liver plays an important role in carbohydrate metabolism. It stores glycogen and releases it as glucose into blood at times of need, like fasting.

8.It deals with fat metabolism, converting extra carbohydrates into fat.

9.Bile helps in saponification of fat.

10.It is the graveyard for old, worn out R.B.Cs. Bile pigments are formed from haemoglobin.

11.Liver brings about deamination of proteins. The unwanted amino acids are converted into urea, thus toxic substances are rendered harmless and are sent into blood to be expelled through the kidneys.

12.Liver excretes toxic substances.

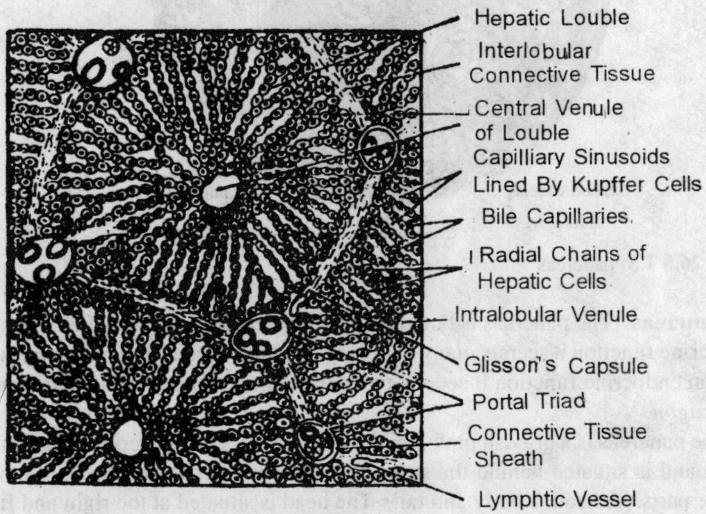

Fig. 26.7 T.S. liver showing histological detailed structure of a hepatic lobule

Gall Bladder: This is a pear shaped sac having approximately a capacity of sixty ml and is attached to the under surface of the right lobe of the liver. Structurally the gall bladder consists of three coats. These are an outer peritoneal coat, a middle muscular coat and an inner coat of mucous membrane. The neck of the gall bladder gives out the cystic duct which joins with the common hepatic duct.

The gall bladder mainly acts as a reservoir of the bile. The bile juice after it leaves the liver through the common hepatic duct moves up the cystic duct into the gall bladder. The mucosal lining of the gall bladder absorbs fluid and electrolytes thus increasing the concentration of other bile constituents. When the food reaches the duodenum, the duodenal walls secret an enzyme- cholecystokinin which enters the gall bladder along with the blood and causes the walls to contract and release the bile.

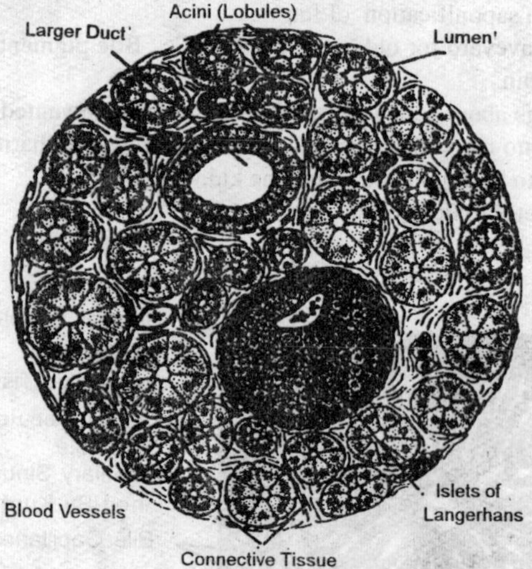

Fig. 26.8 T.S. pancreas

Pancreas: The pancreas is both an exocrine and an endocrine gland. As an exocrine function it secrets many digesting enzymes such as pancreatic lipase. As an endocrine function it secrets two very important hormones *insulin* and *glucagon.*

The pancreas is a grayish pink gland that lies across the posterior abdominal wall and is situated behind the stomach. The pancreas is supposed, to have three parts. A head, a body and tail. The head is situated at the right and fits into the curvature of the duodenum. The body lies behind the stomach and tail of the gland is in contact with the gastric surface of the spleen.

Structurally the pancreas consists of a number of lobules supported by fine connective tissue. Each lobule contains groups of secretory cells called acini. From the lobules arise small ducts which unite to form large ducts untill they reach the main pancreatic duct. This duct extends from left to right and connects itself to the common bile duct to enter the duodenum at the ampulla of vater. Embedded between the acinar cells lie clusters of special cells called the *islets of Langerhans*. There are two types of cells in the islets of langerhans namely the *alpha cells* and the *beta cells*. The alpha cells secret an enzyme called *glucagon* which is concerned with the break down of the glycogen in the liver. It increases the glucose level in the blood. The beta cells secret *insulin* which controls the sugar metabolism and maintains the normal sugar level in the blood.

Digestion of Carbohydrates

The food of human beings consists of mainly two types of carbohydrates namely- polysaccharides and disaccharides. As both these are complex molecules they require to be broken down into simple substances before they can be absorbed by the cells of the small intestine. The normal polysaccharides in the human diet include starch, cellulose, dextrin and glycogen. Among the disaccharides, the most important ones are the ordinary sugar ie., sucrose and lactose or milk sugar. In the human digestive system there are no enzymes for the digestion of the cellulose hence it passes out of the large intestine along with the faecal matter. Many of the herbivorous animals like cattle, have enzymes to digest cellulose. The carbohydrates are digested by enzymes such as carbohydrases to monosaccharides. The process of carbohydrate digestion is initiated in the mouth itself and is completed in the small intestine.

In the buccal cavity, at the contact of food materials, the salivary glands secrete the enzyme salivary amylase or ptyalin. Ptylain breaks starch, glycogen and other carbohydrates into disaccharides like maltose. Further digestion of carbohydrates does not take place in the buccal cavity as food is retained here for a very short time. The digestion of carbohydrates in the buccal cavity is more mechanical than chemical. However it has been estimated that about 30% of starch is hydrolysed in the mouth.

The process of breaking up of the carbohydrates is called hydrolysis because the bonds between the units of sugars are broken down by the addition of water (*hydro*=water; *lysis*=breakdown).

There is no digestive process in the pharynx or the oesophagus. Further digestion begins when the carbohydrates reach the stomach. The salivary digestion however continues until the ptyalin enzyme is denatured by the acid of the gastric juice. The hydrochloric acid present in the gastric juice helps partly in the digestion of carbohydrates. But there are no carbohydrate digesting enzymes in the gastric juice. The food then reaches the intestine. In the intestine two juices act on the carbohydrates. These are the pancreatic juice and intestinal juice. The pancreatic juice produced in the pancreas by the influence of a harmone called secretin consists of carbohydrate digesting enzymes. The main enzyme is the pancreatic amylase. Under the influence of this enzyme starch, glycogen and others get broken down into maltose and other disaccharides.

The intestinal juice has six carbohydrases. These are

1.*Intestinal amylase* which digests the remaining starchy material to disaccharides.

2.*Maltase.* This splits the disaccharides maltose to monosaccharide, glucose.

3.*Isomaltase* breaks isomaltose to glucose.

4.*Sucrase* breaks sucrose into glucose and fructose.

5.*Limit dextrinase* breaks up dextrines into glucose.

6.*Lactase* hydrolyses lactose into glucose and galactose.

As a result of this, the carbohydrates get completely digested into their monosaccharide units such as glucose, fructose and galactose.

There is no mechanism for the digestion of cellulose in the human intestine. As a result of this, cellulose in unaltered form goes out through the faecal matter.

Digestion of Proteins

The diet of human beings consists of different types of proteins. As a result of digestion,these protein ultimately get broken down into their simplest units namely amino acids. The digestion of the protein begins in the stomach and is completed in the small intestine. Proteins remain unaltered in the buccal cavity as the saliva does not have any protein digesting enzymes.

In the stomach, the food undergoes both mechanical and chemical digestion. During the periodic muscular contractions of the stomach, the food is churned and is mixed thoroughly with the gastric juice to form soft mass called bolus. The gastric juice in the stomach is secreted by the gastric glands. This juice contains mucous hydrochloric acid and two proteases viz,. pepsinogen and prorennin. Hydrochloric acid does not digest the proteins but it activates the inactive enzymes pepsinogen and prorennin to pepsin and rennin. Hydrochloric acid also helps in maintaining the proper acidic pH for the action of the protein digesting enzymes. The enzyme pepsin hydrolyses the large molecules of proteins into peptones and other polypeptides. Pepsin also can digest collagen. The enzyme rennin hydrolyses the milk protein casein into paracasein. Paracasein in the presence of calcium gets converted into calcium paracaseinate. Pepsin converts this also into peptones. Further digestion of proteins takes place in the small intestine into which also flows the pancreatic juice. The semidigested proteins are acted upon by the pancreatic juice which consists of three protein digesting enzymes namely trypsin, chymotrypsin and carboxypeptidase. These enzymes are actually in their inactive stage but they become activated under the influence of the intestinal juice. The enzyme trypsin breaks basic proteins into peptides. Chymotrypsin splits proteins into polypeptides and it also hydrolyses milk proteins. The carboxypeptidases break the polypeptide chain into a number of dipeptides.

The intestinal juice also contains two digestive proteases viz., amino peptidases and dipeptidases besides a non digestive protease called enterokinase. Enterokinase activates the proteases in the pancreatic juice. The amino peptidases break up polypeptide chains to dipeptides and free aminoacids. The dipeptidases break up dipeptides into individual ammino acids. The amino acids are finally absorbed by the villi present in the inner wall of the small intestine.

Digestion of Fats

The human diet consists of number of fatty substances and oils. These are digested basically by the enzyme lipases (lipid = fat). The fatty substances are not influenced by the saliva nor is there any fat emulsifying agent in the buccal cavity. The digestion of the fat begins in the stomach where it is partially broken down and it is fully digested in the intestine.

The stomach also does not have any fat emulsifying agent. The gastric juice however consists of gastric lipase which can convert some fats into monoglycerides and fatty acids. This digestion of fats is highly limited in the stomach and most of the fat is digested only in the intestine. In the intestine three secretions of alkaline nature are combined together. These are the bile, the pancreatic juice and the intestinal juice.

The bile juice is secreted by the liver but it does not contain any fat digesting enzyme. The bile salts like *sodium glycocholate* and *sodium taurocholate* have the capacity to reduce the surface tension of large droplets of fat and bring about emulsification. As a result of emulsification, fine droplets of fat called micelles are produced in the intestine. The action of bile salts is to break down large droplets of fat into smaller droplets of fats. This makes the function of fat digesting enzymes like lipases easy.

The pancreatic juice consists of the pancreatic lipase which is the main fat digesting enzyme. Under the influence of lipase, fats are digested in several stages ultimately to yield glycerol and fatty acids (Fats are defined as esters of alcohol like glycerol and fatty acids.

Usually three molecules of fatty acids combine with glycerol to form a molecule of fat. Hence fats are generally called triglycerides) In the first stage, the triglyceride breaks up into a diglyceride with the elimination of one fatty acid; in the second stage a diglyceride breaks up to release another molecule of fatty acid and a monoglyceride; in the third stage the monoglyceride is hydrolysed to the trihydric alcohol glycerol and a fatty acid molecule.

The intestinal juice also consists of a fat digesting enzyme *intenstinal lipase*. This enzyme is produced by the epithelial cells of the intestine and can hydrolyse fats into glycerol and fatty acids just like the pancreatic lipase.

The end products of fat digestion are fatty acids+glycerol and in some cases monoglycerides (fatty acid + glycerol). The glycerol is soluble in water and will be absorbed directly. But the fatty acids are water insoluble. Hence their absorption by the small intestine poses some problems. Two theories have been proposed to account for the absorption of fats. These are the *lipolytic theory* and the *partition theory.* According to lipolytic theory fatty acids which are released from fats due to the action of lipases combine with the bile salts and form soluble complexes which are then absorbed by the intestinal walls. After absorption the complex is broken down to bile salts and fatty acids. According to the partition theory, fats are not completely hydrolysed and these partially hydrolysed fats form emulsions called micelles under the influence of bile salts. These emulsions are absorbed by the intestinal wall.

Summary of Chemical Digestion

DIGESTIVE JUICES	ACTS UPON	RESULTING PRODUCT AND ENZYMES
Saliva (Amylase (Ptylain)	Starch (polysaccharide to complex sugar .	Dextrins (short polysaccharides) and maltose (a disaccharide or double sugar)
Gastric juice Protease (Pepsiongen + HCL)	Proteins, including casein.	Protesoses and peptones (partially digested proteins).
Bil e (contains no enzymes)	Large fat droplets (unemulsified fats)	Small fat droplets or emulsified fats.
Pancreatic juice Protease (Trypsin and chymotrypsin)	Proteins (either intact or partially digested)	Proteoses, peptides, or amino acids
Lipase	Bile-emulsified fats	Fatty acids and glycerol
Amylase	Starch	Maltose
Int estinal juice		
Amino peptidases	Peptides	Amino acids
Sucrase	Sucrose (cane sugar)	Glucose and fructose (simple sugars or monosaccharides)
Lactase	Lactose (milk sugar)	Glucose and galactose (Simple sugars)
Maltase	Maltose (malt sugar)	Glucose (dextrose)
Lipase	Emulsified fats	Fatty acids and glycerol

Hormone	Source	Action
Gastrin	Formed by gastric mucosa in presence of partially digested proteins	Stimulates secretion of gastric juice rich in pepsin and hydrochloric acid
Enterogastrone	Formed by intestinal mucosa in presence of fats	Inhibits gastric secretion and motility
Secretion	Formed by intestinal mucosa in presence of acid, partially digested proteins and fats	Stimulates secretion of pancreatic juice low in enzymes and high in alkalinity (bicarbonate)

Cholecystokinin pancrozymin (CCK-P2)	Fromed by intestinal mucosa in presence of fats, partially digested proteins and acids	Stimulates secretion of bile by liver. Stimulates ejection of bile from gall bladder and secretion of parncreatic juice high in enzymes

A brief survey of Jaundice, Hyperacidity and Ulcer

These are the three important diseases that directly attack the digestive tract or glands associated with digestion

Jaundice : This is called the yellowing disease as people infected with jaundice have yellow coloration of the body tissues including skin, eyeballs etc. The urine also will be yellow in colour. The yellow coloration is because of the presence of excessive amount of the bilirubin yellow pigment in the blood. Strictly speaking jaundice is not a disease in itself but a morphological symptom of the disease of the liver and the biliary tract. Different kinds of infections of liver show signs of jaundice. Infact symptoms of jaundice appear after diseases have set in affecting the function of the liver.

Three different types of jaundice's have been identified mainly depending upon how the excess quantity of bilirubin accumulates in the blood stream. These are -

1. Haemolytic jaundice.
2. Obstructive jaundice. (Cholestatic jaundice.)
3. Hepatocellular jaundice.

1. *Haemolytic jaundice*: This kind of jaundice is due to the increased quantity of bilirubin reaching the liver cells. This additional load of bilirubin is due to the destruction of RBCs due to haemolysis. This kind of destruction of RBC's is generally more common in new born babies and young children as their liver is unable to cope up with the excessive quantity of bilirubin. However jaundice due to haemolysis is usually a mild one.

Causes :

1. Hereditary factors leading to fragile RBC's .
2. Thalassaemia .
3. Spherociposis.
4. Malaria.
5. Erythro bilastosis foetalis.
6. Incompatible blood transfusion.
7. Certain inflammatory and neoplastic diseases.

Clinical features:

In haemolytic anaemia the jaundice is mild. Serum bilirubin is usually less than 4mg/ 100ml. Due to the excessive activity of the reticulo endothelial system the spleen may get enlarged. Stools are brown or orange in colour.

Diagonisis : Examination of the blood will show macrocytes and reticulocytes in peripheral film. Bone marrow shows erythroid hyperplasia.

2. *Obstructive jaundice* : Also known as Cholestatic jaundice, it is caused due to a block in the pathway between conjugation of bile in liver and the entry of bilirubin into the duodenum .The failure of bile flow increases the jaundice progressively because conjugated bilirubin is unable to enter the bile canaliculi and also because there is a failure of clearance of conjugated bilirubin arriving at the liver cell.

Causes: Cholestatic jaundice may occur due to extra hepatic causes or intrahepatic causes. Extra hepatic causes are due to the obstruction in the bile duct preventing the normal flow of bile. This obstruction may be due to gallstones in the common bile duct, carcinoma of the head of pancreas, ampula of vater or rarely in the bile duct. Other extra hepatic causes of obstructive jaundice are fibrosis of the bile duct or obstruction on the bile duct from outside due to the presence of the tumours or enlarged lymph glands.

Interhepatic causes of cholestatic jaundice are as follows.:

1. Drug and Alcohol
2. Primary disease of liver: Viral hepatitis, chronic active hepatitis, cirrhosis.
3. Primary biliary cirrhosis.
4. Secondary carcinoma in the liver.
5. Cholestasis of pregnancy.
6. Pericholangitis or ulcerative colitis.
7. Severe bacterial infections
8. Hodgkin's lymphoma
9. Following surgery.

Symptoms of Cholestatic jaundice :A deep green yellow coloration of the skin, pale or clay color stools, dark brown urine, sever itching of the skin are the common symptoms of the desease. Failure of bilirubin to reach the intestine will result in non formation of stercobilin which is the cause for clay color stools. Excessive bilirubin imparts a dark coloration to the urine. When bile salts do not reach the bowls vitamin K cannot be absorbed which ultimately results in increased clotting time during haemorrhagic conditions. Sometimes there is metallic taste in the mouth associated with upper abdominal pain.

Clinically obstructive jaundice has a gradual onset with serum bilirubin reaching 30mg/100ml.

Obstructive jaundice can be diagonsed by the following -

1. Raised serum bilirubin with high concentration of conjugated proteins.
2. Serum alkalin phosphatase is raised.
3. Urine contains bile.
4. Stools are pale in color.

3. *Hepatocellular jaundice* : This kind of jaundice is produced due to damage to the liver cells. Liver cell damage may be due to many reasons such as -

a. Infective hepatitis
b. Intake of substances toxic to liver, like carbon tetrachloride, alchohol etc
c. Cirrhosis of liver
d. Deficiency of methionine, cystine and choline
5. Liver cell damage due to long standing biliary obstruction

Clinical features

The onset is very rapid; liver and spleen may get enlarged. In advanced cases bleeding from different sources may occur and hepatic encephalopathy may develop.

Diagnosis :

1. High bilirubin content
2.Elevation of the levels of transaminase enzyme
3. Rise in beta and gamma globulins
4. Urine with excess bile pigments

Infective hepatitis leading to jaundice is caused by viruses. This has been categorised into three types namely hepatitis -A , hepatitis -B and hepatitis non-A and non-B.

Hepatitis-A virus is the one that causes infective hepatitis. The disease is characterised by nausea, vomiting and jaundice. The mode of transmission of virus is through the oral or faecal route . In few instances the virus may enter the body through the infected blood. Children are more commonly affected by this disease.

The disease may be diagnosed by increase in the bilirubin content and presence of albumin in the urine. By antigenic analysis also the, disease may be diagnosed.

Hepatitis-B, also known as Serum hepatitis is caused by the hepatitis-B virus. It is more dangerous than the A disease sometimes leading into fatality. The symptoms of the disease include excessive headache, fever, skin rashes, joint pains etc. The disease may be diagnsed in the presence of the antibodies for hepatitis-B in the blood of the infected patient.

The hepatitis-B virus is usually transmitted through contaminated blood during blood transfusion or by the usage of common syringe by drug addicts. Doctor's and others who usually treat the patients are also easily susceptible for the disease.

Hepatitis non-A and non-B virus. Among the several categories which belongs to this group hepatitis-C is the most common one.

Treatment: Jaundice, particularly of the hepatitis type can be prevented by following hygenic methods. Viruses enter the body usually through contaminated water and food. A diet free of fat should be prescribed. Best diet includes fruit juice, sugar cane juice, tender coconut, fresh vegetables etc.

Medicines to toneup the liver- like Liver 52, Analiv and such other medicines may be prescribed.

In the case of obstructive jaundice the reason for obstruction has to be investigated and cleared. If gallstone is the cause of jaundice it has to be removed by surgery.

Hyperacidity and Ulcers : Some amount of hydrochloric acid is always present in the gastric juice to help the digestion of food. But when it is present in excess quantity in addition to digesting the food it can also eat away the membrane of the digestive tract leading to the formation of wounds called Ulcers. Hyperacidity may be defined as production of excessive quantity of acid in the stomach.

Hyperacidity, even though a physical condition, has been mostly described as a psychosomatic disease. When people are unduly anxious having high mental tension and subject to stresses, acid secretion in the stomach is increased exceeding the normal quantity and this is said to be hyperacidity.

Mild hyperacidity may be treated with antacids(chewable tablets as well as suspensions) or medicines that act on the nervous system and supress the production of acid in the stomach. In addition to this the best method to treat hyperacidity -

1. Proper intake of food at regular times
2. Diet with large quantity of fibres
3. Avoiding hot,spicy,oily food
4. Keeping a proper mental balance and avoid tension as far as possible

There is a proverb which says hyperacidity is caused by *hurry*, *worry* and *hot curry*. It is best to avoid all these to prevent hyperacidity.

Ulcers: Ulcer is a term which refers to a wound or a sore injuring the inner lining of the organs mostly of the digestive tract. Ulcers are usually refered to as Peptic ulcers. A peptic ulcer may occur in the lower oesophagus, stomach or duodenum, jejunum or rarely in the ileum also. Based on the location, ulcers may be classified into two types- gastric ulcers and duodenal ulcers.

Gastric ulcers usually occur singly and lie in the lesser curvature of the stomach. Due to hypersecretion of acids, the protective mucosa covering gets digested and the inner blood capillaries are exposed often leading to perforations (holes) . Many a time when perforations are deep, it leads to internal haemorrhage (bleeding).

Duodenal ulcers are wounds that occur in the first part of duodenum(duodenal cap).

Causes of Ulceration - The following may be said to be the causes for ulcer formation.

1. Heredity - patients with ulcers often have a family history of the disease. This is specially so in gastric ulcers. It has been noticed that patients with duodenal ulcers have a high frequency of blood group 'O'.

2. *Occupation* - people who take up professions requiring high mental activity are prone to stress and strain and worry, having to meet a target within a specified time are likely to develop ulcers. In otherwords anxiety seems to be the main cause for ulcers.

3. *Age* Ulceration is usually absent in children. Adults above the age of 20 are prone to ulceration.

4. *Mucosal resistance* - Ulcer is caused due to the digestion of mucosa by the acid contained in the gastric juice. In normal human beings the epithelial cells of the stomach mucosa secrete bicarbonates which create alkaline conditions and prevent injury to the mucous membrane. Many a time either due to drugs or due to other reasons this barier is broken and the mucosa gets damaged.

5. *Smoking* - There is a clear connection between smoking and duodenal ulcers. Smoking also retards wound healing. Further, smoking is known to inhibit the bicarbonate secretion which can prevent the action of acids.

6. *Microorganisms* - Recent researches have shown that certain bacteria (*Helicobacter pylori*) which are present in the gastric mucosa can cause duodenal ulcers.

Symptoms of Ulcer:

1. Abdonimal pain within half an hour of eating is an indication of gastric ulcer. But in duodenal ulcer, pain occurs between meals ie., when the stomach is empty, hence it is called hunger pain.
2. The ulcer pain may spread to the epigastric region or some time to interscapular area also.
3. Heart burn and flatulence(feeling of gas) is common.
4. Vomiting and nausea is also noticed.
5. Apettite may not be very good.

Diagnosis of Ulcer: Peptic ulcer may be confirmed by barium meal X-ray. Sometimes stool examination may also help in diagnosing the disease. Endoscopy of the upper gastrointestinal tract is also useful in detecting duodenal ulcers.

Stool examination very often may reveal the presence of blood indicating bleeding from ulcers. In extreme cases where bleeding is severe the stools become black and it is called *malena*.

Treatment: This involves several approaches. As the disease is a psychosomatic one both the body and mind have to be treated. Treatment includes proper diet, drugs and avoiding harmful agents like tobacco and alchohol. If all these fail, surgery has to be resorted to.

1. Diet: Usually bland diet including lots of fibrous vegetables, fresh fruits, juice, butter milk etc., are known to be of some help. In additon to this, patient should avoid hot spicy snacks. Meals should be taken at regular intervals. For those suffering from duodenal ulcers they should not fill their stomach completely; instead eat small quantities at regular intervals.

2. *Drugs* : Aspirin and other inflammatory drugs should be avoided. Symptomaic releif may be obtained by use of antacid tablets and gels. Certain specific acid suppressing drugs which block the secretion of acid in the stomach can be safely used. These act on the vagus nerve and supress the inducement to produce acid. Some of these drugs which are popular in the market are ranitidine, cimitidine, famotidine etc. Extracts of the plant *solanum nigrum* are known to be helpful in ulcer healing.

3. *Surgical treatment* : When drug treatment is of no use surgical treatment seems to be the only method. There are three differnt kinds of surgical methods; these are - *partial gastrectomy, truncal vagotomy* and *gastrojejunostomy*. In partial gastrectomy, the part of the stomach which produces the acid is removed. In truncal vagotomy the vagal nerve endings that promote the secretion of acid are cut off so that acid secretion is stopped. In gastrojeunostomy the flow of the gastric juice has to be diverted from the stomach to the jejunum.

4. *Psychological treatment* : Anxiety, depression and stress are the three main causes that should be basically treated. Patients should be advised to have a calm mind and take things as they come. Development of hobbies apart from their main work to forget the stress is a useful method. Meditation will also help. In short, patients should be adivsed not to make mountain of a mole hill.

Homeostatic functions of liver and pancreas in the regulation of sugar level in the blood

Homeostasis refers to the maintanance of balance or adjustment to a particular situation in order to see that the metabolic pattern and the rate will not be altered. In the case of maintanance of the level of blood sugar liver and pancreas play an important role.

Blood normally contains glucose in the range of 80 to 100 mg /ml of blood. This might go up to 130 immediately after consumption of food but then again after sometime it comes down to the normal level. These levels of glucose in blood are held extraordinarly constant, as it is very essential for the nourishment of individual tissues. The brain for example has practically no reserves of oxidisable carbohydrates and depends for its proper funciton on the blood sugar. Moreover one must consider that sugar is not taken up continuously but rather intermettiently with food during meal times. Sugar is at first stored in the liver as glycogen. When it is necessary glycogen is broken down and is released into the blood stream to meet the requirements.

The control of sugar level involves both glucose producing and glucose consuming processes. These process are :

Glucose releasing	***Glucose storing***
Glycogen breakdown	Glycogen formation
Conversion of galactose and fructose	Oxidation
Glgenolysis	Fat formation

Numerous hormones participate in the control of blood sugar level. As been pointed out already liver and pancreas play a key role in the maintanance of blood sugar level.

The lowering of blood glucose is effected by insulin a hormone secreted by the beta cells of the islets of langerhans in the pancreas. Insulin acts predominantly in the peripheral tissues by promoting the influx of glucose into the tissues. In muscles it enhances the oxidation of glucose. In the liver it inhibits the release of glucose and stimulates the conversion of carbohydrate to fat. In short insulin acts in a manner as to reduce the content of glucose in the blood.

The normal secretion of insulin maintains a proper level of sugar in the blood and any excess is delivered to the liver where it gets converted into glycogen. In the same way insulin promotes the conversion of glucose to fats in the adipose tissue.

Any hypersecretion of insulin more than the normal, effectively removes even the necessary amount of glucose that has to be mandatorily maintained in the blood (80 to 100mg/100ml). This condition is called *hypoglycemia* and if not checked leads to complications. When blood lacks the normal sugar level it cannot provide enough sugar to the brain cells.

Another kind of cells namely Alpha cells present in the islets of langerhans of the pancreas can retrive the situation when the blood sugar level falls below the normal. These alpha cells secrete a hormone- glucagon which reaches the liver and breaks down the stored glycogen into glucose. This glucose reaches the blood and maintains the normal level. The process of breakdown of glycogen is called *glycogenolysis*. Under exterme conditions(like starvation) liver can convert even proteins and fats into glucose.

In short, the coordianted functions of liver and pancreas are necessary to maintain the normal level of glucose in the blood.

Diabetes: Usually the combined action of liver and pancreas can take care of diffeciency or increased sugar content in the blood and maintian it at the normal level. But in certain situations either due to disease or due to genetic disorders the secretion of insulin becomes less. This hyposecretion of insulin will not be able to bring down the sugar level in the blood to the normal. As a result there is accumulation of sugar in the blood. To begin with however liver tries to rectify the situation by converting excess sugar into glycogen. But with the continued diffeciency of insulin it will be impossible for the liver to convert all of the excess blood glucose into glycogen. Hence excess amount of glucose remains in the blood and if it goes beyond a certian level it will be excreted through the urine. This condition is known as *diabetes mellitus*(sweet urine disease).

Diabetes melliuts results due to insulin defeciency that may be caused due to defective beta cells or it may be only relative when the insulin antagonists may be found in excess.

Symptoms of Diabetes melliuts:

1. Hyperglycemia- presence of excess of glucose level in the blood.
2. Thirst- due to loss of fluid there will be increased thirst to replensih the lost water.
3. Frequent urination- urine is produced in large quantities and very frequently there is urge for urination. This condition is called *polyuria.*
4. Presence of sugar in urine- excess quantities of sugar is excreted through the urine, hence the name sweet urine. This condition is called *glycosuria.*
5. In some cases there is increased hunger.
6. Excessive production of ketone bodies(acetone, acetoacetic acid) will result due to disturbance in fat metabolism. Some times patients develop acidosis due to the production of acetoacetic acid.
7. Nocturia- usually patients get up in the night for urination.
8. Excessive tiredness is experienced.
9. Wound healing is normally delayed.

Diabetes may be primary or secondary. Primary or idiopathic diabetes is of two types- insulin dependent diabetes mellitus(IDDM) and non-insulin dependent diabetes mellitus(NIDDM). IDDM is common among young people and it is also called juvenile diabetes. IDDM is probably due to the beta cell destruction of pancreas induced by a virus infection in genetically suscepetible persons. NIDDM has a strong family history and is noticed in middle age and elderly people. Overeating coupled with inactivity and obesity leads to insufficient insulin production.

Secondary diabetes is purely for pathological reasons of the pancreas like pancreatitis, malignancy, excess of growth hormones etc.

Diagnosis : Disease can be easily diagnosed by measuring the blood glucose level at the fasting and the post prandial level (after eating food). A fasting level above 120mg/100ml and a postprandial level above 140mg is indicative of the disease.

Treatment: There are three kinds of treatment of the disease 1.diet, 2.diet plus oral hypoglycemic agents and 3. diet and insulin. If the disease is detected early about 60% of the cases can be controled by diet alone.

Diet planning is very essential in the control of diabetes. Carbohydrate content of the diet should be regulated depending on the activity. Direct sugars to be avoided. Increased quantity of fibrous vegetables and fruits should be added to the diet. Oral hypoglycemic drugs reduce the level of sugars in the blood. These are available in different brands in the market. The drugs basically belong to two groups- sulphonyl ureas and blguanides.

Direct injection of insulin is recommended for those pantients who do not respond to drug and diet therapy. Insulin injection is given subcutaneously in special syringes.

Twenty seven

Circulation

In higher organisms the body is highly specialised with different kinds of tissues and organs. As a result of this, transportation of materials between them is necessary to maintain their functions. In addition, each and every cell must receive oxygen for its respiration and also the waste materials like carbon dioxide have to be carried away. The hormones of endocrine glands also must reach their targets to co-ordinate their activities. This function of distribution is carried out by the circulation of blood.

It was William Harvey who in 1663 demonstrated that blood makes a complete circuit of the body being pumped out of the heart through one set of vessels and returning to the heart via a different set.

In the circulation of blood, heart plays a very important role as it is this organ which pumps the blood to various parts of the body and also receives blood from various parts of the body. Therefore it can easily be said that the blood vascular system begins and ends in the heart.

The human heart is basically a specialised hollow muscular structure that pumps blood to the various parts of the body. The size of the heart varies in different animals but in human beings the size of the heart is about 0.43% of the body weight in males and about 0.4% in females. Its weight is approximately about 10 ounces.

The human heart is a pyramid shaped structure and is approximately of the size of the clenched fist. It has a base an apex, two surfaces, two borders and also several grooves. The apex of the heart projects downward and forward and to the left lying above the central depression of the diaphragm. Its base projects upward and backward and lies just below the second rib.

Structure of the human heart

The heart is covered by a wall called *pericardium* which is made up of three layers. These are the *epicardium*, the *myocardium* and the *endocardium*. The epicardium is a serous membrane and is the outer most layer. This secrets a serous fluid into the space between the visceral and the parietal pericardium. This fluid called the *pericardial fluid* allows smooth movement when the heart beats. It has been estimated that there will be normally about 15 ml of the pericardial fluid.

The myocardium is the intermediate layer. It is composed of specialised muscle tissues called the cardiac muscles. These muscles are striated fibres. The myocardium is thin in the auricles but thick in the ventricles specially so in the left ventricle.

The endocardium is the inner most layer. This is a thin membranous structure in contact with the blood. It is composed of flattened endothelial cells. The endocardium is thicker in the auricles than in the ventricles.
Internally the heart is composed of 4 chambers namely

1. Right auricle (atrium)
2. Left auricle (atrium)
3. Right ventricle and
4. Left ventricle

There are two auricles - the right and the left. These are separated by a thin interauricular septum. The ventricles are also two in numbers and they are separated by the interventricular septum. The right auricles open into the right ventricle by the right auriclo ventricular openings guarded by a tricuspid valve having three flaps. The left auricle opens into the left ventricle by left auriclo ventricular aperture guarded a bicuspid valve having two flaps. This valve is also called the *mitral valve*.

The right auricle receives venous blood through superior and inferior vena cave. The left auricle receives oxygenated blood through the pulmonary vein returned form the lungs.

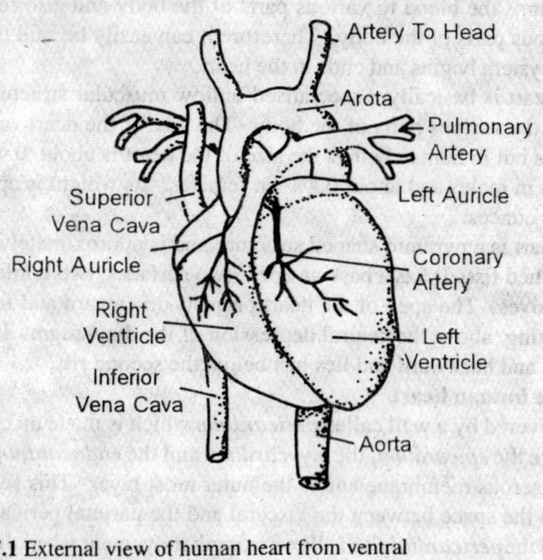

Fig. 27.1 External view of human heart from ventral

Functioning of the heart: Working of heart can be studied under the following heads:

Heart beat: The heart muscles contract and expand (relax) in a regular rhythmic fashion. The phase of contraction is known as *systole* and the phase of relaxation is called *diastole*. One diastole and one systole together constitute a heart beat. The contraction and relaxation of the auricles and ventricles take place in a definite order. The auricles and ventricles do not contract simultaneously.

The functioning of the heart may be divided into three phases.

I phase: The auricle contracts and the blood is pumped into the ventricle which is relaxing. The auricle is said to be in systolic stage while the ventricle is said to be in diastolic stage.

II phase: In this phase the ventricles contract while the auricles are relaxing. The blood is forced into the arteries.

III phase: In this stage, both auricles and ventricles are relaxing and the blood enters into the auricles.

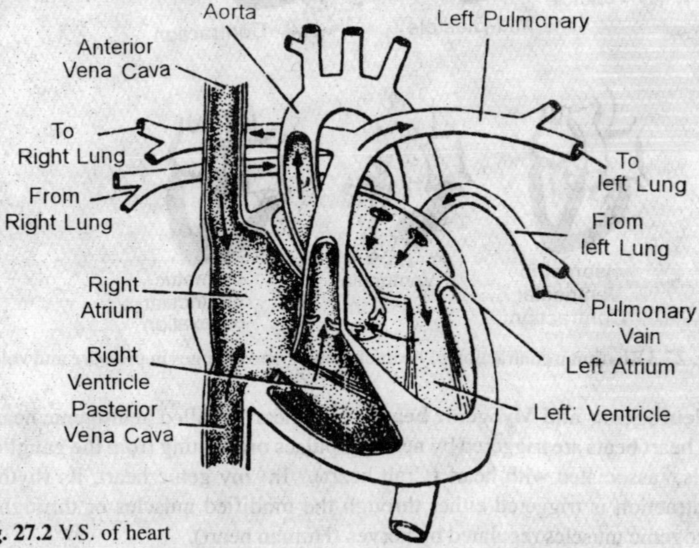

Fig. 27.2 V.S. of heart

Cardiac Cycle: The heart contracts rhythmically throughout life. Each heart beat consists of a systole and a diastole. The total duration is about 0.8 seconds. The total heart action involving all the three phases mentioned above is called the cardiac cycle.

The volume of blood that is ejected during a single heart beat is called the stroke volume (SV). The cardiac output (CO) is the total volume ejected over one minute and is therefore the product of the stroke volume and the heart rate (the heart beats about 70 times per minute in a healthy man and it is known as the heart rate). The cardiac output may be calculated as follows.

$$CO = SV \times HR$$

Where CO = Cardiac output, SV = Stroke volume and HR= Heart rate.

Each heart beat is accompanied by 2 heart sounds which are called the first heart sound and the second heart sound. This usually sound like *lub* and *dup*. The first heart sound results from the closure of the atrioventricular valves while the second heart sound is due to closure of the exit valves of the heart in the roots of the aortic and pulmonary arteries.

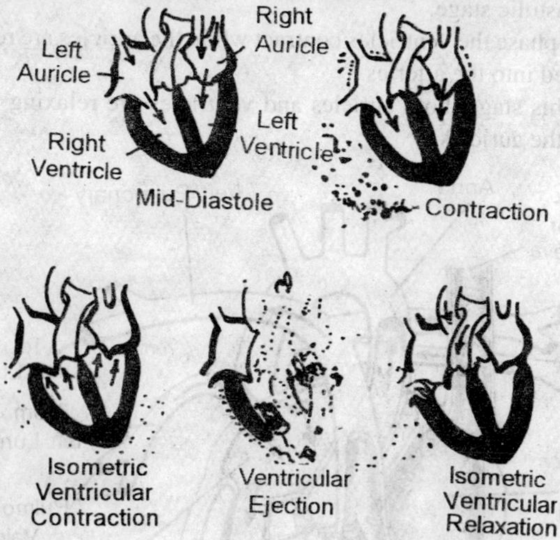

Fig. 27.3 Rhythmic contraction of heart chambers and changes in pressure and volume

Neurogenic and Myogenic heart: The heart is called neurogenic heart, if the heart beats are triggered by nerve impulses originating from the ganglionic cells, associated with heart (Crab heart). In my genic heart, its rhythmic contraction is triggered either through the modified muscles or through the myogenic muscles regulated by nerves (Human heart).

Origin and Conduction of heart beat :

The heart beat starts in a tiny island of tissues present in the upper region of right auricle, called the **sino-auricular node** or **sino-atrial node (S.A.).** It is located near the opening of large vein. S.A. acts as the pace maker of the heart. The heart impulses generated here radiate out from the heart as a wave and cause each muscle fibre, through which they pass, to contract. A similar node is present in the partition dividing the right and left auricles. A bundle of fibres originates from the auriculoventricular node(AV). This is known as ***Bundle of His*** or ***Purkinje fibres***.

The impulse after passing through the AV node passes along bundle of HIS, which happens to be the only direct muscular communication between auricles and ventircles. High in the interventricular septum the bundle of HIS divides into right and left bundle branches which are distributed respectively to the right and left ventricle. The bundle branches terminate in thin fibres called the Purkinje fibres through which the impulses reach the ventricular muscles.

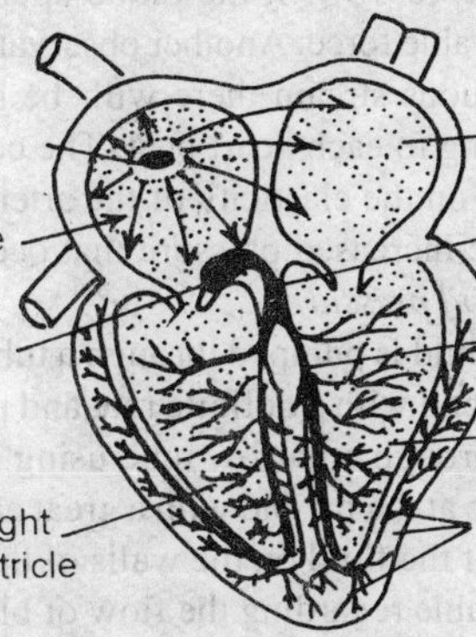

Fig. 27.4 Diagram showing the location of pacemakers in the mammalian heart

The sino-auricular node sends out intermittent impulses. Each impulse initiates a wave of contraction, which passes through both the auricles and causes their contraction. When this wave of contraction alongwith the impulses reaches the auriculo-ventricular node, it is also stimulated. The impulses from it make the ventricles contract. While the ventricles contract the auricles relax. This sequence is repeated again and again.

The whole system- the SA node, the AV node, bundle of HIS and the purkinje fibre systems constitute the conducting system of the heart. These are composed of modified cardiac muscle cells. The stimulus for the contraction of auricles and ventircles is electrical. This sequnece of electrical discharge and the phases of the cardic cycle may be studied with the help of an instrument called the *Electrocardiogram*. This instrument can measure the electrical impulses of the heart muscles and give out the wave forms on a strip of paper and this is called the *Electrocardiograph*.

The Electrocardiogram measures the electrical activity of the heart through the electrodes placed at the surface of the body. After amplification this can either be displayed on a cathode ray tube or for a permanent record can be recorded on graph paper. Analysis of electrocardiograms enables the physician to measure the function of the heart and take appropriate measures. If there is damage to any portion of the heart muscles the electrical transmission through the damaged area is prevented which is recorded on the ECG as variations from the normal.

Blood Pressure

Blood pressure may be defined as the force or pressure the blood exerts on the walls of the artery in which it is flowing. The artial blood pressure is a result of discharge of blood from the left ventricles into the aorta. This can be seen when an artery is cut accross when the blood spurts out from the end nearest the heart with considerable force. Another observation that can be made is that along with the continuous stream there will be regular spurts of increased pressure corresponding to each heart beat. The continous pressure of the blood is partly dependent on the elasticity of the arteries and is called the *Diastolic blood pressure.* The increased pressure that occurs with each heart beat is called the *Systolic blood pressure.*

When a definite amount of fluid is pumped through a tube with a large diameter it will flow out of the far end with a definite rate and pressure. If the same amount of fluid is pumped through a narrow tube using the same amount of force the liquid leaves the tube at the far end with a great amount of force. This is due to the greater friction of the fluid on the walls of the tube.

The same factors are applicable regarding the flow of blood and its pressure within the artries. The force used by the heart may be constant but as the arteries become narrower, due to resistence the flow of blood will be at a higher pressure. The following factors are responsible for the maintenance of blood pressure-

1. The force of the heart beat.
2. The resistance to blood flow in the arteries and arterioles (this depends on their diameter).
3. The volume of the circulating blood.

Measurement of blood pressure: The blood pressure is clinically measured with the help of an instrument is called *Sphygmomanometer***.** Stephen Hales (1733) was supposed to be the first man to have measured the blood pressure of mare by inserting a needle into the carotid artery. The Sphygmomanometer consists of a rubber bag which is placed around the arm. The interior of the bag is connected by rubber tube to a mercury pressure guage. When the pressure in the bag equals the pressure in the artery the latter is compressed flat and the flow of blood through it is temprorarily halted and at this stage the pulse disappears. When a Stethoscope is placed on the brachial artery at the bend of the elbow the systolic heart beats can be heard very clearly. The height to which the column of mercury in the manometer has been forced at this moment is called the *Systolic blood pressure.* In normal human adults this will be about 120mm Hg. As the column of mercury is allowed to fall there comes a stage at which the stethoscopic sounds suddenly disappear. The height of the mercury column at this point is taken as the *diastolic blood pressure.* In the normal human adults this is 80mm Hg. This diastolic pressure indicates the pressure in the arteries (in the diastole) where as the systolic pressure indicates the pressure of the heart beats.

Abnormalities in blood pressure: The maintanance of normal blood pressure is very important for the proper maintenance of health. Any value above 150mm Hg for systolic pressure and any value above 95mm Hg for diastolic pressure is regarded as abnormality. Similarly a systolic pressure of less than 100 is also regarded as an abnormality. Therefore abnormalities in blood pressure are of two types- *High blood pressure* and *low blood pressure.*

High blood pressure: It is also called Hypertension and indicates an abnormality in the flow of the blood through the arteries. Temporarily however there may be an increase in blood pressure due to reasons such as exercise, emotional upset etc. But when the blood pressure is constantly high even when an individual is resting it is an indication of abnormality and has to be immediately attended to medically.

There are two main reasons which cause hypertension. These are- *arterioselerosis* and *nephritis.* Arteriosclerosis is a disease of the blood vessels. Due to various reasons the blood vessels loose the elasticity and offer more resistance to the blood flow with the resultant increase in the pressure. The walls of the blood vessels lose their elasticity as a result of hardening which again is due to the absorption of calcium salts. Generally as advanced age sets in the walls of the blood vessels gradually loose their elasticity and become hardened. This is the reason why generally in aged persons the blood pressure is higher than in the youngsters.

Nephritis is a disease of the nephrons of the kidney which also has an effect on blood pressure. Some other causes of blood pressure are overeating, smoking, herediatry conditions and psychological conditions like continued stress brought about by various reasons.

Initially an increased blood pressure may be symptomless but generally the external symptoms that may indicate increased blood pressure are - headache, throbbing of the head, tiredness, giddiness and palpitation of the heart. The organs that are most likely to be affected by high blood pressure are heart, brain and kidneys. Continued neglect of blood pressure may even cause the blood vessels to burst leading to internal haemmorhage specially in the brain. This is called *Apoplectic stroke*.

Hypertension if detected early can be completely kept under check if not cured. Apart from medication there has to be an alround change in the mental attitude of the patient. While medication may give a temporary relief ultimately it is the mental attitude that will keep the blood pressure under check. Medication includes specific antihypertension drugs like *reserpine* obtained from the roots of a plant *Rawolfia serpentina.* The extracts from this plant have been in use to control blood pressure in the Indian system of medicine since time immemorial.

In addition to medication proper rest, mild exercise, giving up consumption of alcohol, giving up excessive use of salt, moderation in the drinking of coffee, tea etc. are recommended. Smoking should be given up altogether as it has an effect on the narrowing of the arteries. Meditation and yoga are also of great help in controlling hypertension.

Low blood pressure: A fall in the systolic blood pressure below the normal range is reffered to as low blood pressure or *hypotension.* A systolic reading below 100mm Hg indicates a very feeble circulation and a reading below 80mm Hg is indicative of an impending death. A low blood pressure results in case of severe haemorrhage due to accidents or extensive injuries when there is severe loss of blood. A sudden shock also might lower blood pressure. Hypotension may also result due to the burning of the tissues, failure of heart, distrubances of endocrine functioning like hypothyroidism and Addison's disease of the adrrenal glands etc.

Low blood pressure can be treated by medication as well as by adopting healthy habits. Proper mental attitude, mild exercise, sufficient rest etc., will help the blood pressure to come to normal. Some times use of mild intoxicants like alcohol may also give a temporarily relief.

ABNORMALITIES IN THE HEART

Heart enlargement :

The normal healthy heart usually will be the size of a closed fist. Some times various factors like infections, malfunctioning, inflammation etc. lead to enlargement of the size of the heart and this affects its functioning.

There are three kinds of inflammation leading to the enlargement of the heart. These are - 1. inflammation of the inner lining of the endocardium of the heart (endocarditis) 2. inflammation of the myocardium of the heart (myocarditis). and 3. inflammation of the pericardium of the heart (pericarditis).

Endocarditis- The inflammation or swelling of the endocardium of the heart leads to a condition known as endocarditis. Usually endocarditis in patients with heart problems is caused due to bacterial infection and is refered to as bacterial endocarditis. Risk of bacterial infection is high among patients with valve diseases such as mitral valve disease, aortic valve disease etc.

Bacterial infection is usually caused by *Streptococcus viridans, Enterocci, Pneumococci* etc. In some drug addicts endocarditis is caused by yeast and fungal infection also after a cardiac surgery.

Myocarditis - Also called cardiomyopathy in this there is deranged function of myocardium . Cardiomyopahty is of three kinds- congestive cardiomyopathy, hypertropic cardiomyopathy and restrictive cardiomyopathy. In congestive cardiomyopathy the myocardium becomes fibrous and lacks the contractile power. Hence heart is dialated. Hypertropic cardiomyopathy is hereditary generally. Septum is hypertrophied and may cause obstruction of left ventricular flow. In restrictive cardiomyophaty the diastolic function is deranged but systolic function is normal. The rigid myocardium restricts diastolic filling of the blood. Restrictive myocardiopahty is mostly caused by fibrosis.

Pericarditis - This is an inflammation of pericardium and is caused by a variety of causes such as viral and bacterial infections, neoplastic diseases, myocaridal infraction and metabolic autoimmune disordes. The pericardium in inflammed conditions becomes thick and sticky resulting in pain. Often a fluid accumulates between the two layers of the pericardium.

Treatment : Treatment of enlarged heart condition depends on the diagnosis and the type of infection that causes heart enlargement. Treatment includes prescribing tempororialy pain relievers and treatment with specific antibiotics depending on the type of infection. Antibiotics like pencillin, streptomycin or gentamycin are useful.

Myocardial Infraction

This refers to the death of a small area of the tissue of the heart due to lack of blood supply. The death of the heart tissue is called myocardial necrosis or myocardial infraction. This leads to heart attack and the patient exhibits symptoms of excessive chest pain, difficulty in breathing etc.

Within a short time medicaiton should reach the patient or otherwise fatality would result. Proper medicare should primarily include identifying the causes for lack of blood supply and to rescue the area of necrosis and preventing the extension of necrosis to other areas.

Depending on the extent of necrosis funcitonal abnormalities appear. Decreased ventricular contraction results in increase of end diastolic pressure and may bring cardiogenic shock. Healing of infected myocordinum occurs over a period of weeks. Within 24 hours cardiac enzymes are released from the cells and the necrotic cells are removed and replaced by scar tissues. A stable scar is established within 3 to 6 months.

Myocardial infraction may be diagnosed by ECG findings. In addition to this, Echocardiography will also help us to understand the problem.

Approximately 60% of the deaths from myocardial infraction occur within 2 hours after the onset of chest pain. Ventricular fibrillation and asystole account for sudden cardiac deaths. Hence medical assistance should be provided as early as possible. Patient should be put under oxygen to remove breathing difficulties. Proper physical rest is essential. Treatment for myocardial infraction includes the removal of block in the vessels which results in the blockade of blood supply to the heart. This may be identified by coronary angiography where in a small catheter is inserted through an artery in the leg and guided into the coronary artery. A dye injected will show the path of movement and indicates obstruction if any. Depdending on the number and extent of blocks treatment is decided. If the block is small a coronary angioplasty is performed where a small balloon is inserted into the artery and inflated at the region of the block to dilate the walls of the artery. If this treatment is not satisfactory a bypass surgery is performed to provide an alternate pathway for the flow of blood.

In bypass surgery usually a vein from the leg is removed and is attached to the aorta at one end and the other end is connected to the diseased artery above the block. This will provide an easy pathway for the flow of blood.

Defects in the valves (valvular diseases of the heart)

The valves in the heart regulate proper flow of the blood in specific directions. A healthy valve is necessary to perform these functions. But in a number of cases, defects occur in the valves either due to infection or due to congenital (along with the birth) problems. The following are the basic defects in the valves. These are-Mitral stenosis, Mitral incompetence, Aortic stenosis, Aortic incompetence, Tricuspid incompetence, Pulmonary stenosis and Pulmonary incompetence.

Among the infections that bring about the defects in the valves endocarditis and rheumatic fever are the most important. Rheumatic fever is caused by an autoimmune phenomenon following throat infection with betahaemolytic streptococci.

Mitral stenosis : If for some reason the valves are inflammated the valve leaflets fuse together causing obstruction of blood flow. Mitral stenosis is the usual aftermath of a rheumatic heart disease. In Mitral stenosis there is obstruction to forward flow of the blood. The normal circumference of the Mitral opening is 10 m. Anything less than 7 m is regarded as a block. As a result of block the left atrium dilates and pressure in the pulmonary vein increases. Cardiac output is decreased. The valves are thickened, distorted and often heavily calcified.

Mitral incompetence (Mitral regurgitation) : In Mitral incompetence the valves donot get closed completely as a result there is back flow of blood from left ventricle into left atrium during ventricular systole. The back flow of blood passes across the Mitral valve again into to left ventricle during diastole. As a result the work of left ventricle is increased and it gets dialeted.

Aortic stenosis : This is caused by calcification of aorta or endocardities or due to congenital defects. As a result of the block the forward flow is decreased and the pulse is weak.

Aortic incompetence (Aortic regurgitation) : In this the ejected blood from aorta leaks back into left ventricle hence, left ventricle has to handle more quantity of blood and gets dilated.

Tricuspid stenosis : This causes dialatation of right atrium and high right auricular pressure. This high pressure results in dilatation of the right auricle. In some instances it may lead to right heart failure.

Tricuspid incompetence : Insufficency of tricuspid valve is caused due to dilatation of valve ring and may occur after mitral stenosis. Some times it may occur as an after effect of rheumatic fever. Tricuspid incompetence results in right auricular dilatation and right heart failure.

Pulmonary stenosis : This is caused by congenital defects, rheumatic fever or it may be associated with Fallot's tetralogy. As a result of pulmonary stenosis there is right ventricular hypertrophy.

Pulmonary incompetence : Pulmonary incompetence may be due to mitral stenosis and may be due to conginetal defects also. Effects of pulmonary incompetence include right ventricular hypertrophy with dilatation of its cavity.

Septal holes

In a normal human heart the left side of the heart is completely separated from the right side by means of a partition called Septum. In abnormal conditions holes appear in this septum leading to complications such as mixing of the venous blood with the oxygenated blood. Most of the septal defects are congenital and usually occur from the child hood itself. But some times septal holes may be caused due to rheumatic fever also. The congenital defect is related to the disorder of embryonic development during 5th to 8th week in the uterus of the mother. Congenital heart diseases of the septum may be classified into Acyanotic and Cyanotic. In acyanotic defects there is generally no total mix up of venous blood with arterial blood while it is so in cyanotic defects. The following are some of the diseases of septal holes -

1. Ventricular septal defect.
2. Artial septal defect
3. Patent ductus arteriosus.
4. Tetralogy of Fallot

1. Ventricular septal defect - These are generally small openings or holes present in the interventricular septum. If the hole is small it closes spontaneoulsy by 3 years of age of the child. If the defect is larger more blood passes from left to right; thus burdening the right ventricle and increasing pulmonary circulation. Children with this defect get tired soon. Defects can be diagnosed by systolic murmur (heart murmur). The defect can be closed by putting a dacron patch under surgery.

2. Atrial septal defect - This results from a defeciency in the interartial septum. Due to the presence of the hole, blood passes from left atrium to right atrium increasing the blood flow in right side of the heart. Children suffering from this get tired and contact frequent respiratory infections. The defect can be treated by a dacron or pericardial patch under surgical procedure.

3. Patent ductus arteriosus (PDA) - In this disease, the duct connecting aorta to pulmonary artery remains open which otherwise should be closed within 48 hours of birth. As a result of this blood, flow to the lungs is increased and left ventircle has to exert to pump extra blood. Premature babies have increased incidence of PDA.

4. Tetralogy of Fallot - This is a cynatoic heart defect. As has already been said in cynatoic heart disease there is mixture of venous blood with arterial blood. As a result arterial oxygen saturation is reduced and tissues get less oxygen. Children suffering from this disease become blue during feeding and defaecation. This disease is also called a *Blue hole* disease.

Twenty eight

Respiration

The respiratory system in human beings consists of lungs and air passages. In its course to the lungs the air passes through the nose, mouth, pharynx, larynx, trachea, bronchi and ultimately the lungs.

The nose has the external(anterior) nares, nasal chambers and the internal (posterior) nares.

External nares : The paired external nares are situated at the anterior end of the upper jaw slightly behind the mouth. These are rounded in form and communicate posteriorly with paired nasal chambers.

Nasal chambers : The nasal passages are paired elongated chambers situated dorsal to the buccal cavity and separated from the latter by bony palate. The nasal passage opens anteriorly by external nares to the outside and posteriorly it leads into the pharynx through the nostril. The nasal passages-

a. Prevent the entrance of dust into the lungs.

b. Hold and sweep minute organisms or bacteria,

c. Incoming air is warmed.

d. Moisten the dry air with the help of mucous secreted by glandular respiratory epithelium.

The internal nares : These are situated at the back of the nasal cavity and constitute the enterance to the nasopharynx.

The Pharynx : There are three parts in the pharynx. These are the nasopharynx (nasal part), the oropharynx (oral part) and the laryngeal part.

The nasopharynx lies immediately behind the nasal cavities and opens into it *via* the internal nares. The oropharynx is continuous in front with the buccal cavity and below with the laryngeal part of the pharynx. The laryngeal part of the pharynx is continuous with the oesphagus.

The Larynx : The larynx is situated in the anterior part of trachea and communicates with the pharynx through glottis. Besides acting as a part of the air passages, the larynx is modified in structure to help in voice production.

The larynx consists of a framework of the following hyaline cartilages.

1. *Thyroid cartilage*: This is the largest and consists of two side flaps united in the centre to form an angular projection sometimes refered to as Adam's apple.

2. *Cricoid Cartilage*: This is situated below the thyroid cartilage and forms the lowest part of the larynx where it connects to the trachea. It is circular in shape with an expansion at the back giving it a signet ring shape.(The lobes of the thyroid gland lie on each side of the cricoid and thyroid cartilages)

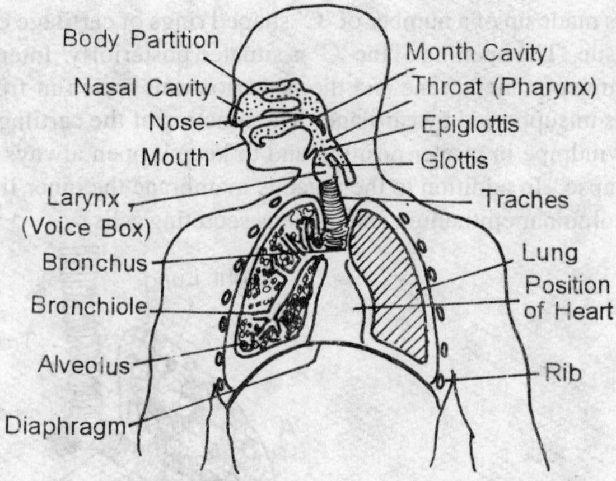

Fig. 28.1 The organs concenred with breathing and external respiration of man

3. *Arytenoid Cartilages*: These are two small structures located on the upper surface of the cricoid cartilage. These are pyramidal in shape and provide attachment to the posterior ends of the vocal cords. In front, the vocal cords are attached to the posterior surface of the thyroid cartilage. The muscles present in the cords vary the tension and rotate the arytenoid cartilages thereby vibrations are produced resulting in the alterations of the pitch of the voice.

4. *Epiglottis*: This is a leaf shaped plate made up of yellow elastic fibrocartilage and is situated between the base of the tongue and the upper opening of the larynx. The lower portion of the epiglottis is attached through a ligament to the thyroid cartilage. The main function of the epiglottis is to prevent the entry of food to the larynx during swallowing.

5. *Hyoid*: This is a horseshoe shaped bone lying between the mandible above and larynx below. It is located at the base of the tongue and provides attachment to this. This however does not take any part in the formation of the larynx. Internally the larynx is lined by the mucous membrane which is covered with ciliated columnar epithelium except over the vocal cords.

The vocal cords are fibroelastic bands extending from the posterior part of the thyroid cartilage in front and to the arytinoid cartilages at the back. These are the true vocal cords. In addition to this, there are two false vocal cords which are nothing but the folds of the mucous membrane situated above the true vocal cords.

The Trachea: Commonly called the windpipe the trachea is about 12 cms long and about 2.5 cms in diameter. The upper half of the trachea is situated in the midline of the neck while the lower half is in the superior mediastinum of the thorax. It lies in front of the oesophagus and ends opposite the fourth thoracic vertebra. Here it divides into two branches called bronchi.

The trachea is made up of a number of 'C' shaped rings of cartilage connected by fibrous tissue. The opening of the 'C' is situated posteriorly. Internally it is lined by the mucous membrane and the posterior wall which in front of the oesophagus is unsupported by cartilage. The function of the cartilage rings is to keep the windpipe in proper position and to keep it open always so that it does not collapse. In addition to the mucous mambrane the inner lining also has ciliated columnar epithelium and mucous secreting cells.

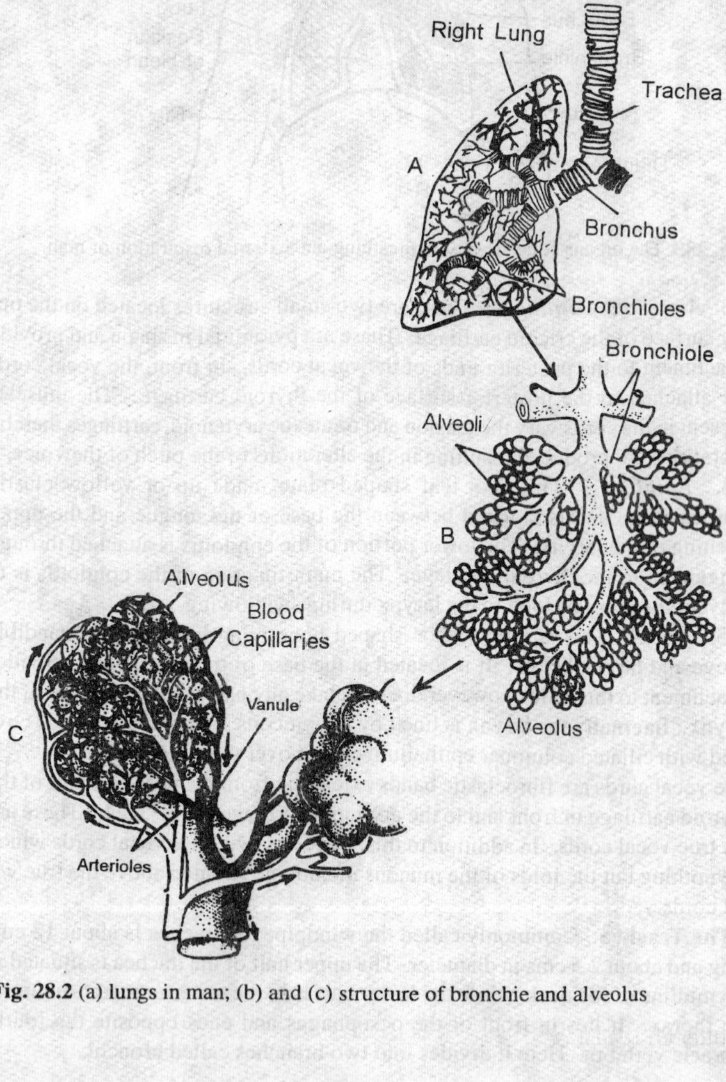

Fig. 28.2 (a) lungs in man; (b) and (c) structure of bronchie and alveolus

The Bronchi: The trachea bifurcates into two main branches called the bronchi (right and left) opposite the fourth thorcsic vertebra. Each bronchus enters to respective lung (left bronchus to left lung; right bronchus to right lung). Each main bronchus divides into numerous smaller branches repeatedly and the ultimate small tubes of the bronchus are called *bronchioles*.

In structure, the bronchi are similar to trachea; they are surrounded by incomplete hyaline cartilages lined with mucous membrane made up of ciliated columnar epithelium. The walls of the bronchi also have some smooth involuntary muscles. The bronchioles have a structure similar to bronchi but there are no cartilaginous structures; instead they consist of muscle fibres.

Copius mucous covers the epithelial lining and the action of the cilia pushes the mucous up to the larynx and pharynx from where it is either swallowed or thrown out. The mucous is secreted by the goblet cells which are present between the epithelial cells.

The Alveoli or Air sacs: The Bronchioles ultimately terminate in irregular sac like structures made up of a number of air packets called *alveoli* or *air sacs*. The alveoli are lined with a delicate layer of epithelial cells and are surrounded by numerous capillaries. It is through the walls of these capillaries that the exchange of air takes place.

The wall of the air sacs is made up of three layers, namely a surface epithelium, a supporting connective tissue layer and blood vessels. The connective tissue layer is absent in most of the alveolar valve and only a shared basement mambrane lies between the epithelial cells and the endothelial cells of the blood capillary. Thus a very thin valve separates the air in the alveoli and the blood in the capillaries. Some of the epithelial cells of the alveoli secret a detergent like substance called surfactant which reduces the surface tension and allows to maintain the shape. The air spaces in the alveoli and alveolar macrophages (dust cells) help in engulfing the debris and dust particles that reach the alveoli.

The Lungs : In human beings these are the main respiratory organs. These are a pair of conical organs enveloped by a serous membrane - the *pleura*. The lungs occupy a greater part of the thoracic cavity. The two lungs are separated from each other by the mediastinum. Each lung is divided by deep fissures into lobes. The right lung has three lobes - the upper, middle and lower while the left lung has two lobes the upper and the lower. The lung is also described as having a mediastinal and a costal surface, an apex and a base.

The outer(costal) surface is in contact with the wall of the pleural cavity while the medial surface of the lung is applied to the mediastinum. The main feature here is the presence of hilum where the main bronchus and pulmonary artery enter and the pulmonary veins leave the lung. Lymph nodes are present at the hilum which may get enlarged by diseases.

The apex of the lungs rises into the root of the neck for about an inch above the clavicle. The base is concave and is related to the upper part of the diaphragm.

Structurally the lungs are spongy and each lobe of the lung is made up of a number of small lobules consisting of alveoli and bronchioles which form larger bronchi. In addition to this there is a framework of interstitial or fibrous tissue in which are found the blood vessels and lymphatics.

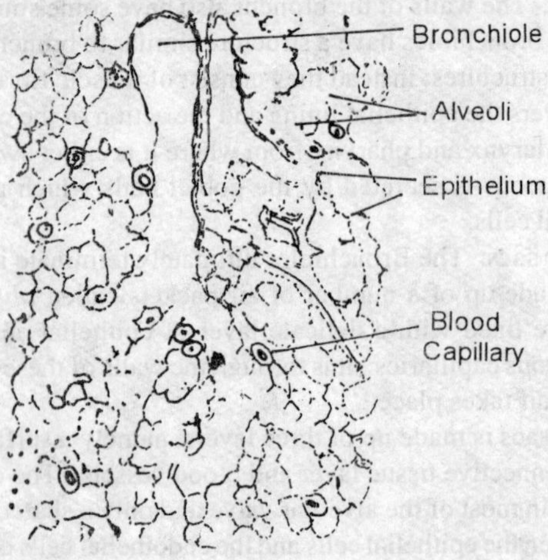

Fig. 28.3 T.S. Human lung

The pleura: This is a serous membrane and is made up of two layers - the *parietal* and the *visceral.* The visceral layer forms the outer cover of the lung which it encloses completely except at the region of the hilum. The parietal layer lines the interior of the chest wall and upper surface of the diaphragm. Both the layers are smooth and shiny and are always kept moist by a small amount of serous fluid acting as a lubricant. This helps the two surfaces to glide smoothly over each other during the process of breathing.

Mechanism of breathing: The intercostal muscles associated with the ribs and the radial muscles of the diaphragm bring aobut the process of breathing. The process can be accomplished in two steps.

1. *Inspiration*: The intake of fresh air is accomplished with the enlargement of the thoracic cavity. It is brought about by the contraction of muscles. As a result, the ribs are drawn outward and forward; sternum; is carried forward and downward and the dome shaped diaphragm flattens down. Thus the volume of thoracic cavity increases in all directions. Due to this increase in thoracic cavity, the pressure inside the lungs decreases and the fresh atmospheric air rushes in through the respiratory passage and gets filled in the alveolar spaces.

2. *Expiration*: During expiration the radial muscles relax to assume the normal form and the diaphragm acquires the original dome shaped appearance. Simultaneously, the external intercostal muscles relax and the internal intercostal muscles contract so that the ribs are pulled back to their normal position. Hence the volume of the thoracic cavity is reduced.
The reduction in the volume of thoracic cavity exerts pressure on the lungs and air is squeezed out of the lungs to the exterior through the respiratory passage.

Regulation of Breathing:

1. *Respiratory* centre: A respiratory centre located bilaterally in medulla oblongata and pons varolii of the brain, serves as the controller of respiration. It sends signals for contraction of diaphragm and other respiratory muscles for respiration, but remains dormant during expiration.

2. *Hering - Breurer reflex*: Located in walls of bronchi and bronchioles these are stimulated when lungs are overstretched and transmit impulse through vagus nerves to the respiratory centre. This reflex serves as a protective mechanism for preventing excess lung inflation.

3. *Chemical control*: A chemosensitive area, located close to the respiratory centre in medulla, maintains the proper concentration of it oxygen, carbon dioxide and hydrogen ions in the blood and body fluids. It is highly sensitive to changes in CO_2 and pH in blood.

Volume of Gases in Man:

1. *Tidal volume*: Volume of air inspired or expired in each normal breath is called tidal volume of air. It is about 500ml of air.

2. *Inspiratory reserve volume*: It is the volume of air that we can inspire over and above the tidal air by the deepest possible inspiration (it is about 3000 ml of air).

3. *Expiratory reserve volume*: It is the volume of air that we can expire over and above the tidal air by the most forceful expiration. (it is about 1100 ml).

4. *Residual volume*: The volume of air that remains in the lungs after the most powerful expiration is called residual volume of air. (it is about 1200 ml of air).

5. *Vital Capacity*: This is the maximum volume of air a person expires from the lungs after filling with a maximum deep inspiratory reserve volume plus the tidal volume plus the expiratory reserve volume. It is about 4600 ml.

6. *Total lung capacity*: This equals the maximum volume of air that our lungs can normally hold. It includes total of tidal, inspiratory, reserve, expiratory reserve and residual volumes of air. It is about 5800ml.

Physiology of Respiration:

This can be studied under two headings:

a) **Gaseous exchange in the alveoli**: The uptake of O_2 and release of CO_2 by the blood of the alveolar capillaries is by diffusion, when the gases pass from the regions of high pressure to those of low pressure. The partial pressure of O_2 in the blood of the alveolar capillaries is considerably lower (40mm of Hg). The oxygen diffuses into the capillaries; its O_2 pressure has been raised to about 100mm of Hg, the same as alvelolar O_2. In the similar manner diffusion of CO_2 out of the blood takes place.

b) **Transport of gases in lungs**: It involves two steps:

1) **Transport of Oxygen**: In mammalian blood, the O_2 carrier respiratory pigment is haemoglobin. The maximum amount of O_2 which the normal human blood can absorb is 20 ml / 100ml of blood. When O_2 passes from the lung alveoli into the lung capillaries, it diffuses into the blood and unites with haemoglobin to form oxyhaemoglobin.

$$Hb + O_2 \longrightarrow HbO_2$$

(Haemoblogin Oxyhaemoblogin)

This reaction is a reversible one. The oxyhaemoglobin reaches the place of diffusion. The liberation of O_2 from the blood to the tissues is just as important as its rapid absorption by the blood during its passage through the lungs.

$$HbO_2 \longrightarrow Hb + O_2$$

(Oxyhaemoglobin Reduced haemoglobin)

The reduced haemoglobin is again transported *via* blood to the lungs and the cycle is repeated.

Oxygen Dissociation curve: Normally the haemoglobin combines with O_2 to form oxyhaemoglobin. Each iron atom can attach one molecule of O_2 and when all sites are occupied, the haemoglobin cannot take on anymore since it is fully loaded or saturated. This saturation varies greatly. In other words, the proportion of oxyhaemoglobin to haemoglobin present in the blood at any time is dependent on the tension of O_2 in the blood. This is known as dissociation curve of oxyhaemoglobin. In this way the actual relationship between the partial pressure of O_2 and the degree of saturation of haemoglobin dissociation can be measured. The following factors affect the oxygen dissociation curve.

1.Partial pressure of O_2
2.Partial pressure of CO_2
3.Effect of temperature
4.Effect of haemoglobin
5.Effect of pH

2. **Transport of CO_2**: With the result of various metabolic activities CO_2 is evolved which diffuses into the blood. The total amount of CO_2 in the venous blood is about 60nk / 100ml of blood and the arterial blood contains about 50ml CO_2 per 100ml. Thus, a relatively small amount of CO_2 is carried in the blood, yet blood reaction does not become acidic. This proves that in CO_2 transport the blood buffers play a very important role. CO_2 is carried in the following 3 ways.

(i) *Transport of CO_2 in Physical solution*: Under normal temperature and pressure only 2.7 ml of CO_2 is carried in physical solution as H_2CO_3. As CO_2 enters the blood, it combines with H_2O of Plasma forming carbonic acid.

$CO_2 + H_2O$ —— H_2CO_3 (Carbonic acid)

(ii) *Transport of CO_2 as Bicarbonates*:

Nearly 85% of the total CO_2 is carried as bicarbonates in both plasma and RBC's.

$NaHCO_3$ in Plasma	$KHCO_3$ in RBC's
Arterial Blood = 33.1%	Arterial Blood = 9.8%
Venous Blood = 35.2%	Venous Blood = 10.5%

As CO_2 enters the blood cells from the tissues, it combines with H_2O to form H_2CO_3 which dissociates to hydrogen ions and bicarbonate ions. The latter diffuses into the plasma forming sodium and potassium bicarbonates.

H_2CO_3 —— $H+ \ +HCO_3^-$

$Na+ \ + HCO_{3-}$ —— $NaHCO_3$

$K \ + \ HCO_3^-$ —— $KHCO_3$

(iii) *Transport of CO_2 as Carbamino Compounds*: About 10% of CO_2 is transported as carbamino compounds

Carbino-Haemoglobin in RBCs	Carbinoproteins in plasma
Arterial Blood = 2%	Arterial blood = 1%
Venous blood = 2.6%	Venous blood =1.1.%

$CO_2 \ + \ HbNH_2$ ———— HbNHCOOH Carbohaemoglobin

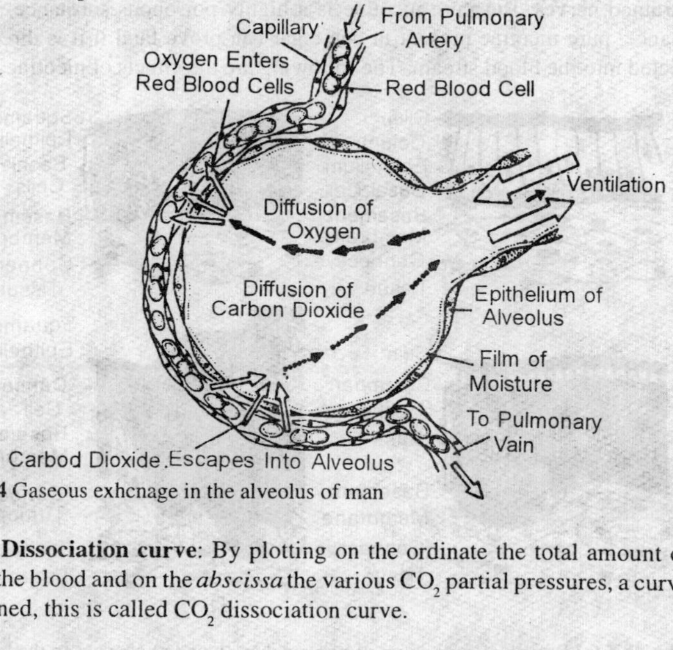

Fig. 28.4 Gaseous exhcnage in the alveolus of man

CO_2 Dissociation curve: By plotting on the ordinate the total amount of CO_2 in the blood and on the *abscissa* the various CO_2 partial pressures, a curve is obtained, this is called CO_2 dissociation curve.

Effect of smoking and pollutants on the respiratory system :
Effects of smoking :

Tobacco smoking is one of the oldest evil habits that man has developed since time immemorial. The practise of inhaling vapours of toabcco in the form of cigarattes, beedies or cigars is defined as smoking. Smoking is an all pervading habit that is practised by people of all countr·'s all age groups and all economic strata (both rich and poor).

Besides inhaling the smoke of tobacco, other habits of tobacco use include chewing of tobacco leaves, inhaling the powdered leaf in the form of snuff etc. While smoking and other forms of tobacco use may give a momentary happiness, they also bring different types of diseases affecting the health of man. Tobacco primarly contains *nicotine* besides about 300 other chemical compounds all known to be harmful for man. Some of these compounds are carbon monoxide, hydrogen cyanide, polycyclic aromatic hydrocarbons, tar, many irritant substances and atleast about 16 dangerous chemical compounds known to induce cancer.

Tobacco is primarly smoked for inhaling nicotine which is supposed to be a stimulatory compound. Researches have shown that in small quantities nicotine can activate the nervous system and also have a soothing effect for the strained nerves. But pure nicotine is a highly poisonous substance. For instance, pure nicotine present in one cigar can prove fatal if it is directly injected into the blood stream. The following are the effects of nicotine.

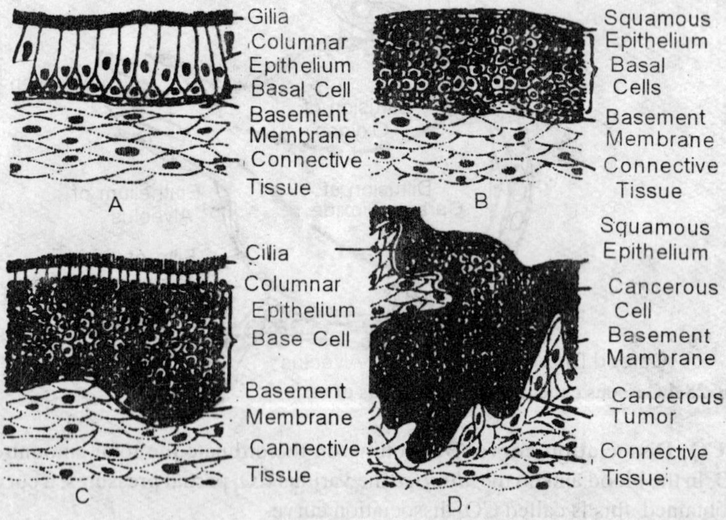

Fig. 28.5 (a) Bronchial epithelium of non-smoker, (b and c) changes in the bronchial epithelium of smokers, (d) Cancerous growth as a result of smoking

1. It stimulates the nervous system. 2. Causes relaxing of muscles. 3. Induces the release of adrenalin. 4. Increases heart beat and contraction of peripheral blood vessels leading to increased blood pressure. 5. Causes retardation of the growth of the foetus in pregnant women.

The following are the diseases that are associated with long term and heavy smoking.

1. *Lung cancer*-This is a most serious effect found in people who are long time smokers.

Even passive smokers(persons who are closely associated with smokers ie., who inhale the smoke released by the smokers) are prone to this disease. Polycyclic hydrocarbons are harmful to lungs. (for details see later in the same chapter)

Chewing tobacco leaves or gutkas or pan masalas mixed with tobacco is known to cause cancer of the mouth.

2. *Bronchitis* - This is an inflamation of mucous membrane of throat and bronchial tubes.(for details see later in the same chapter)

3. *Emphysema* - Collection of phlegm in the lungs makes breathing difficult and results in the streching of alveoli and bronchioles. This brings about persistant cough and weakining of the walls of alveoli. Ultimately the surface of the alveoli for gaseous exchange is reduced and more amount of carbon dioxide is retained in the blood.

4. *Heart diseases* - Smokers have a high rate of heart diseases because smoking constricts the blood vessels increasing the blood pressure.

5. Though smoking does not cause tuberculosis it weakens the system and helps in the spreading of the disease.

6. Gastric irritation leading to ulcers are known to be aggravated due to smoking.

Effects of pollutants :

Air pollutants like carbon dioxide, carbon monoxide, oxides of nitrogen, oxides of sulphur, photochemical oxidants, hydrocarbons, particulate matter, aerosols etc. constitute the major air pollutants which are known to have an effect on the human health. These pollutants also affect the respiratory system. The following are the diseases/abnormalities caused by air pollutants.

1. Carbon monoxide is a very poisonous gas produced by the incomplete combustion of fuel in automobiles. When inhaled it combines with the haemoglobin of the blood and reduces its oxygen carrying capacity. This leads to inactivity, reduced vision and finally death.

2. Oxides of nitrogen which are released by automobiles and chemical industries are also harmful as they can also combine with haemoglobin reducing its oxygen carrying capacity.

3. Oxides of sulphur cause damage to mucous membrane, drying of the mouth and scratchy throat. It is a pungent gas which may even lead to chokeing of the respiratory tract.

4. Hydrocarbons and pollutants like asbestos fibres act as carcinogens inducing cancer of the lung.

5. Cotton dust (particularly in workers of textile mills) leads to bronchitis and asthama.

6. Breathing of mercury vapours causes damage to central nervous system in industrial workers. Cadmium is another respiratory poison.

7. Photochemcial oxidants are formed when the primary pollutants react with unburnt hydrocarbons in the presence of sunlight. This photochemical reaction produces peroxyacyl nitrate (PAN), ozone and other compounds in the air. Of these, ozone causes dryness of mucous membrane of mouth, nose and throat. It also causes headache, pulmonary congestion and odema.

DISORDERS OF RESPIRATORY SYSTEM

LUNG CANCER - Techincally known as bronchial carcino, lung cancer is the leading cause of cancer death in men of all countries. It is four times more common in men than in women. This is possibly because smokers are more in men than in women. Of all the causes bringing about lung cancer the one that has been established beyond doubt is smoking. 90% of lung cancer patients are smokers. The increased risk of lung cancer in smokers is directly proportional to the amount of smoking per day and the tar content in cigrattes. Death rate from lung cancer is 40 times more in smokers than in non smokers. Those who give up smoking have lesser chance of contracting lung cancer.

Pathology of lung cancer- Lung cancer affects the body in three ways:

1. Local invasion- Carcinoma of the bronchus arises from the epithelial cells that lie in the bronchus. Subsequently the tumours invade the deeper layers of the bronchial wall and spread into surrounding lung tissue. It may cause ulceration and bleeding from the bronchus. The malignant (tumours) cells spread to the lymphatic system causing the glands to enlarge.

2. Maliganant cells from the lungs may enter the blood stream and form metatasis leading to secondary deposists in bones, liver and brain. Tumours develop in all these parts.

3. Toxins produced by malignant cells lead to certain extra pulmonary (outside the lung) manifestations like functional aberrations such as damage to nervous system, clubbing of fingers, hypertrophy of glands and several other complications.

Symptoms of lung cancer :

At early stages, lung cancer is symptomless. Persistant cough which does not yield to any treatment is the earliest symptom. Shortness of breath, chest pain, wheezing, hoarsenses of the voice are all early symptoms of localised lung cancer. When lung cancer is in the metastasis stage symptoms include severe headache, unsteadiness of gait, tumours in the bone, enlargement of liver, jaundice, tender skin nodules etc.

Diagnosis: Progressive loss of weight, weakness associated with persistant cough must arose, suspicion of lung cancer. Presence of enlarged glands in neck and clubing of fingers provide further supportive evidence on physical examination.

X-ray of the chest is the most important step in the diagnosis of lung cancer. A dense opacity (non clear) of the lungs with irregular enlarged hilar glands suggest bronchial carcinoma.

Bronchoscopy is another important method of diagnosis. Inspection of intrabronchial portion of the tumour and biopsy (removal of a small amount of tissue) for histological examination are very important. Other methods of diagnosis include CT scan, magentic resonance imaging etc.

Treatment: The treatment depends on the state of the disease. These are as follows:

1. Surgical treatment- Surgical removal of the tumour is the best method if the tumours are detected at an early stage and are not spread to other organs. If only one lobe of the lung is affected it can be removed by lobectomy.

2. *Radiotherapy*- Small tumours can be destroyed by subjecting them to radiation.

3. Chemotherapy- Some of the cytotoxic (drugs that are toxic to malignant cells) drugs in combination with raditation therapy are known to be beneficial. Some of these cytotoxic drugs are nitorgen mustard, cyclophosphamide, methotrexate, vincrystine etc. Treatment with these drugs prolong the survival of the patients for upto one year. It should be pointed out however, both radiotherapy and chemotherapy also have so many side effects.

4. Immunotherapy- Patients suffering from lung cancer very frequently have their immune system retarded. Immune stimulating drugs are helpful in preventing secondary infections.

5. Palliative treatment include nutritions diet, vitamin suppliments and analgesics to suppress the pain.

Prevention :

It is best to prevent the disease rather than trying to cure it. As it has been clearly established that smoking is one of the prominent causes of lung cancer it should be avoided at any cost. Proper education on these lines is very essential that should be undertaken by doctors, nurses and other health and social workers. Non smokers should avoid inhaling the passive smoke let out by the smokers as otherwise they have to suffer for others sin.

In many of the countries educative measures are undertaken to discourage people from smoking. In all most all countires legislation has made it mandatory that cigaratte packs should carry the warning that *smoking is injurious to health*. Many countires have banned smoking in public places, offices, buses, trains etc. All these will go a long way in the prevention of the disease.

Allergic disorders : Allergy may be defined as sensitivity to certain things or substances. These may be gases, solids, liquids, pollen grains, dust etc.People who are allergic, on coming into contact with these substances develop harmful side effects. These include rashes on the skin, sneezing, cold, cough, etc. Physiologically on coming into contact with allergic substances the body reacts by releasing compounds called *histamines.* These histamines cause many side effects.

Substances that cause allergy are called allergens. Allergens are of various kinds. As has been pointed out already a number of substances which are biotic as well as abiotic (living and non-living) are known to cause allergy.
There are three main allergic disorders of the respiratory system. These are - asthma, bronchitis and hay fever.

Asthma - Bronchial asthma is an allergic disorder of the bronchial tube which gets narrowed due to allergic reactions. It is accompained by wheezing and coughing. The narrowing of the bronchus is due to reasons such as contraction of the muscles of smaller bronchi (broncho spasm), thickening of the lining of the bronchus or increased secretion of mucous resulting in the blocking of the smaller bronchi.

Pateints suffering from asthma experience difficulty in breathing accompained by wheezing(noisy breathing) .

Factors that cause asthma: A number of factors precipitate the attack of asthma . These are

1. Presence of allergic substances like dust, cotton dust, animal hairs, pollen etc. In Bangalore particularly it has been noticed that during winter season a number of plants produce allergic pollen that induce allergic asthma in susceptible persons.

2. In some persons ingestion of analgesics and antiinflamatory drugs may induce asthma.

3. Many environmental factors such as air pollutants from industries also cause asthma.

4. Industrial workers particularly those working in chemical, plastic, pharmacetucial, wood and other industries have a higher chance of suffering from asthma.

5. Bacterial and other infections of the respiratory tract may also aggravate an asthmatic attack.

6. Sometimes physical exertion also may cause asthma.

7. Emotional stress and hereditary factors are other causes that precipitate asthma.

Treatment :

This may be classified into two - home treatment and hospital treatment. Home treatment includes administration of oxygen in a portable cylinder as asthmatic attack may severely restrict breathing. Bronchodilator drugs (drugs that increase the lumen of the bronchioles) should be given. Hospital treatment includes oxygen via oxygen mask or nasal catheter. To compensate fluid loss oral fluids are to be given. The most important treatment includes the use of bronchodilators which give immediate relief. Some of these are -

a. Subcutaneous injection of adrenaline relieves bronchospasm immediately.

b. Aerosols of salbutamol can be given as inhalations

c. Aminophyline may also be given as an intravenous injection. The drug should be *injected very slowly*.

d. Steriods are life saving drugs to be used in times of emergency. Hydrocartiozone is the most quick acting drug which can be given intravenously.

e. Antibiotics are also helpful as they control respiratory infections that aggravate asthmatic conditons.

In extreme cases mechanical ventilation can be used if the patient does not respond to any of the above therapy.

Prevention : The first step in prevention of asthma particularly in those persons who have a history is to identify the cause of allergy . Sensitivity to allergens may be identified by several tests which includes skin contact test. When once the cause of allergy has been established, patients should try to keep away from it.

Bronchitis : This is one of the diseases of the lower respiratory tract including the bronchi, bronchioles and some times lungs also. Usually bronchitis refers to the inflammed condition of the bronchi (air ways). There are two types of bronchitis- *Acute bronchitis* and *chronic bronchitis.*

Acute bronchitis : This is followed either by a viral or bacterial infection of the upper respiratory tract. Bacterial agents such as *Streptococcus pneumoniae, Haemophilus influenzae* and others cause secondary infections. Acute bronchitis is common in winter months.

Symptoms of acute bronchitis are- headache, mild fever (38-39°C), cough with mucous, pain below the sternum, fast and laboured breathing with whistling sounds that can be heard through a stethoscope.

Diagnosis of the disease includes blood test (raised WBC count) besides physical examination.

Treatment of the disease includes bed rest in a warm well ventilated room, antiobiotics such as ampicillin or oxytetracycline, inhalation of medicated steam (menthol), exprectorants (which bring out the mucous) and bronchodilators if necessary.

Chronic bronchitis *:* The attack of chronic bronchitis is not sudden. It is persistant over a long time. Usually persistant cough over 3 months (some times up to two years) is an indication of chronic bronchitis. Condition becomes aggravated due to smoking and atmospheric pollutants. Infection by bacteria increases the disease already existing.

Clinical symptoms include persistant cought severly increasing at times, breathlessness on exertion, oedema of legs, strain on the right ventricle, impairment of the functioning of the lungs etc.

Treatment : The first step in the treatment is to persuade the patient to give up smoking (if he is a smoker). The patient should also avoid dusty or smoky environment. Chemotherapy with specific anitbiotics to tide over the acute condition must be followed. If wheezing is present, bornchodilators are to be used. In some cases of suspected allergy steriods also may be used.

Hay Fever : (Allergic rhinitis or atopic rhinitis)

This is an allergic disease of the upper respiratory tract caused due to exposure to certain substances called *allergens.* Allergens are of various types and include dust, smoke, pollen from a number of plants (grasses, members of Asteraceae), animal hair, fungal spores, house dust, wool, feathers, fabrics etc. It should be noted however that these do not cause allergic reactions in all people. Only certain people are sensitive to any one or all of the above mentioned allergens. Many a time sensisivity is due to hereditary factors also.

Hay fever is technically called *paroxysmal rhinorrhoea* and is seen in people who are hypersensitive to allergens. Symptoms of hay fever appear after coming into contact and inhaling the allergens. Hypersensitivity is beleived to be inherited hence hay fever is also known as *atopic rhinitis.*

Symptoms : In the initial stage, the nasal mucosa becomes red due to inflammation, becomes odematous and swollen. The mucosal surface in bathed by a watery or mucoid fluid which becomes mucopurulent (infected with microbes) at later stages. These symptoms are followed by continuous sneezing, and watery discharge from the eyes and nose. Eyes also become red. There is however little or no rise in body temperature.

Treatment : There are two steps in the treatment - to treat the symptoms of the disease and secondly to identify the source of allergy and to keep away from it. Symptoms of allergy can be treated by prescribing antihistamines.

Identification of the allergen is the most important aspect of treatment. All the possible allergens from the surrounding are collected and the patient is tested for allergic reactions. This is usually done by skin test. In skin test the suspected allergen is applied to the skin to test the reaction. If rashes or pustules appear, then the role of the allergen is confirmed. After the identification of the allergen, the patient is adivsed to keep away from it. If for some reason the person cannot keep away from the allergen, then it is necessary to build up resistance against the allergen. This is called *desensitization* in which small doses of allergen (extract) is injected over a period of time gradually increasing the quantity of the allergen. This method builds up natural immunity in the body over a period of time and the symptoms of disease do not appear even if the individual is exposed to the allergen. This method however is to be followed only under the supervision of physicians who are specialized in allergy treatment.

Twenty nine

Excretion

Excretion may be defined as the separation and elimination of the metabolic wastes from the body usually in an aqueous solution. By the metabolism of proteints, nitrogenous wastes like ammonia, urea, uric acid are formed. These waste products are harmful to the tissues and hence have to be removed from the place of their production and taken to places for expulsion from the body. This process of removal of the waste products is termed as Excretion. In our body and other vertebrate animals, kidney, skin, lungs, large intestine and liver are concerned with excretion.

Excretory organs of the body.

Exrcretory organs	**Substances excreted**
Kidneys	Nitrogenous wastes from protein, catabolism, Toxins from bacteria etc Water from ingestion and from catabolism; Mineral salts
Skin (Sweat glands)	Water and small amounts of nitrogenous wastes
Lungs	Carbon dioxide
Intestine	Wastes from digestion, some metabolic wastes like bile pigments and also salts of calcium and other heavy metals

Metabolic waste products :

They may be grouped under the following headings :

1. Respiratory waste products : Catabolic waste products of food stuffs are CO_2 and water. In higher animals CO_2 is excreted with the expired air through lungs. Excess of water is eliminated in the form of urine and sweat.

2. **Nitrogenous waste products :** Nitrogenous waste products are derived from the deamination of the excess of amino acids taken in with the diet and also from the breakdown of animal proteins and nucleic acids. These are of following types:

a. **Ammonia :** It is formed by the deamination of amino acids during protein metabolism. Ammonia is highly toxic and its rapid elimination is essential. It is highly soluble in water and is quickly eliminated into surrounding water in acquatic animals such as Protozoans, *Hydra*, etc.

b. **Urea :** It is less toxic than ammonia. Most mammals, fishes and amphibians excrete their nitrogenous waste products in the form of urea.

c. **Uric acid :** Insects, some snails, birds and reptiles with few exceptions, excrete uric acid. Uric acid is excreted in solid form. As uric acid is insoluble in water, it crystallizes out from the fluid urine and water is reabsorbed, thereby allowing conservation of water.

d. **Amino Acid:** In certain animals like molluscs and echinoderms, the excess of amino acids is removed as such without undergoing any change.

e. **Other nitrogenous compounds :** Allantoin and allantoic acid are other nitrogenous excretory products. These are insoluble in water and are formed during embryonic development by amniotes with shelled eggs.

There are certain other nitrogenous waste products such as guanine and adenine from nucleic acid break down and creatine from the creatinine of muscles that are excreted along with other nitrogenous waste products in the urine.

3. **Mineral Ions :**

Sodium, potassium, calcium, magensium, chloride and ammonia are the essential mineral ions of the animals. The excess of these ions taken in with the diet are excreted by one means or the other.

Classification of animals based upon the types of main nitrogenous wastes:

a. **Ammonotelic animals**: The main nitrogenous waste is ammonia, e.g., certain protozoans, polycheate annelids, crustacean arthropods, molluscs (Aplysia, Sepia and Octopus), fresh water teleost fishes, tadpoles of amphibia and crocodiles.

b. **Ureotelic animals** : Main nitrogenous waste is Urea, e.g., elasmobranch fishes, amphibians and mammals.

c. **Uricotelic animals**: Main waste is uric acid, e.g., insects, some gastropods, lizards, snakes and birds.

EXCRETORY SYSTEM OF MAN (MAMMAL);

Kidneys: These are the most essential organs of man and other mammals. These are 2 in number, and are dark red in colour placed one on either side of the vertebral column.

Each kidney is convex on the outer surface and concave on the inner side termed hilum through which renal artery enters and renal vein and ureter leave. The ureters lead behind to open into the urinary bladder and are continued as urethra which opens at the tip of the penis in the man and in front of the vaginal opening in the female.

STRUCTURE OF THE KIDNEY

In human beings the kidneys are paired. They are bean shaped and are located between the levels of last thoracic and third lumbar vertebrae. The right kidney is placed slightly lower than the left because of the presence of the liver. The average size of the human kidney measures about 11 to 13 cms in length; 5 to 7.5 cms wide and 2.5 cms thick. In adults each weighs about 150 gms and this correspondence to about 1% of the body weight. The medial border of the kidney is slightly concave and has a marked depression.

A long section of the kidney shows two main regions - an outer reddish cortex and an inner pale coloured medulla.

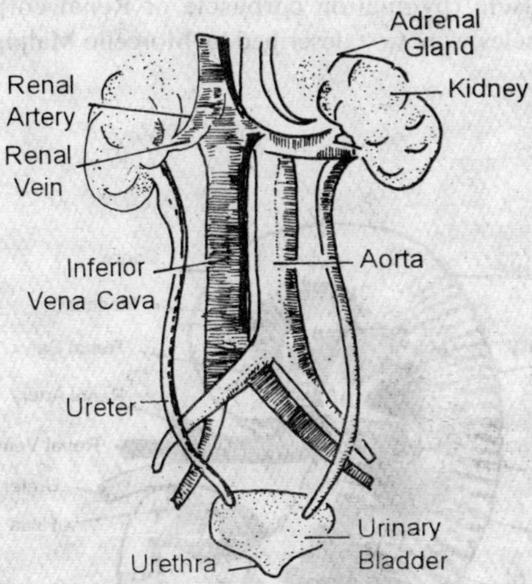

Fig. 29.1 Urinary system of man

Cortex: The Cortex is granular in appearance because of the presence of convoluted uriniferous tubules. The cortex overlies the bases of the medullary apices called medullary pyramids. The cortex is also present between the medullary pyramids. Under a magnifying glass the cortex can be seen divided into alternating radial tracks known as *rays* and *labyrinths*. The rays are radially straight because they contain the straight parts of the uriniferous tubules. The Labyrinths have a granular appearance because the tubules are cut irregularly.

Medulla: The medulla looks striated as the tubules run through it in a straight course. It consists of 10 to 15 conical masses called the renal pyramids. Two or three pyramids usually fuse and form the common apex called papilla. These papillae project into cup shaped channels known as calyces. The calyces join to form the pelvis which is the expanded initial part of the ureter.

NEPHRON (URINIFEROUS TUBULE) This is the basic filtering unit of the kidney. It has been estimated that nearly one million nephrons are present in the human kidney. The nephron basically consists of two parts. The secretory tubule and the excretory portion.

1. ***SECRETORY TUBULE***. The secretory tubule consists of the Bowman's capsule, the proximal convoluted tubule, the loop of Henle and distal convoluted tubule.

Bowman's capsule: At the closed end which is dilated the nephron consists of a baloon shaped structure called the Bowman's capsule. The Bowman's capsule is a cup shaped structure. It lies in the cortex of the kidney and contains an invaginated tuft of capillary vessels called the glomerulus (pl. glomeruli). The Bowman's capsule and the glomerulus together constitute the malpighian corpuscle (pygmalion corpuscle or Renal corpuscle). The Malphigian corpuscles were first described by Morcello Malpighi in the year 1966.

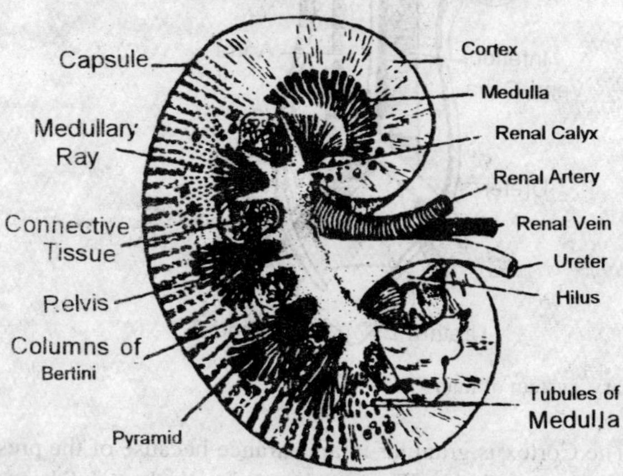

Fig. 29.2 Longitudinal section of the kidney of a mammal

The inner layer of the wall of the Bowman's capsule is a visceral layer and along with the endothelium of the glomerulus forms an endotheliocapsular membrane. This membrane has three parts namely the Endothelium of the glomerulus (porus membrane), Basement membrane (This is between the endothelium of the blood vessels and Epithelium of Bowman's capsule. This is without pores and helps as a dialyzing membrane) and the epithelium of the Bowman's capsule (This layer is made up of epithelial cells called podocytes. The podocytes have foot like pedicels arranged parallel to the glomerulus and cover the basement membrane except for spaces called slit pores or filteration slits).

Proximal convoluted tubule: The Bowman's capsule opens thorough a narrow constricted passage into a greatly contorted tube called the convoluted tubule which also lies in the cortex. This tubule is about 14 milimeters in length. The wall of the tubule is made up of cuboidal epithelial cells which have microvilli. Each cell has about 150 microvilli per square micron. Each microvillus is about 1 micron in length. The microvilli together form a brush border region.

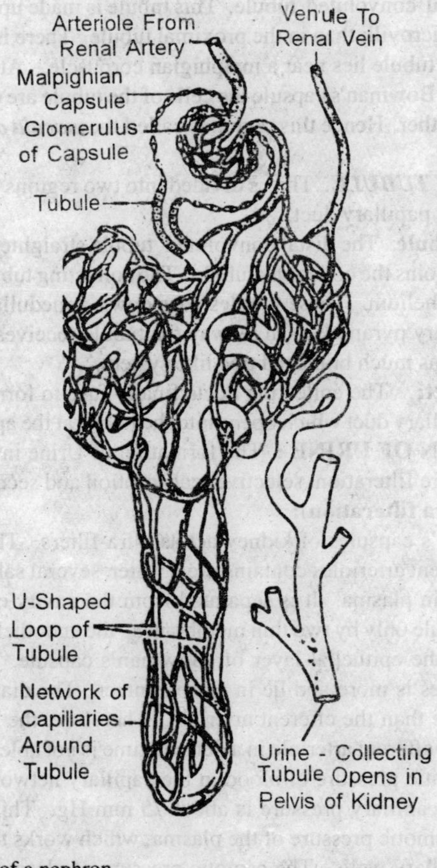

Fig. 29.3 Structure of a nephron

Loop of Henle: The proximal convulutd tubule takes a straight course down into the medulla region. This descending limb is called the loop of Henle. This loop of Henle is divided into three segments namely -

a) Thick descending segment which is the straight portion of the proximal convoluted tubule.

b) A thin segment which takes a direct radial course in the medulla. This segment takes a sharp hairpin bend extending past the tip of the Henle's loop.

c) A thick ascending segment. This is a straight radial tubule. In this region the squamous epithelial cells suddenly change into cuboidal epithelial cells. This segment is relatively impermiable to salts and water.

Distal convoluted tubule : The tubule that comes out of the Limb of Henle leads to the distal convoluted tubule. This tubule is made up of cuboidal cells and have less microvilli than in the proximal tubule. There is no brush border formation. The tubule lies near a malphigian corpuscle. At the region of the contact with the Bowman's capsule the cells of the tubule are columnar and are aggregated together. Hence this region is called the *macula densa* or crowded spot.

EXCRETORY TUBULE: This is divided into two regions namely a collecting tubule and a papillary duct.

Collecting tubule: The distal convoluted tubule straightens and by a short junction tubule joins the collecting tubule. The collecting tubules have a lining of cuboidal epithelium. These tubules pass down a medullary ray and enter into the medullary pyramids. On its way the tubule receives other junctional tubules and forms much branched tree like system.

Papillary duct: The collecting ducts finally fuse to form a large straight tubule - the papillary duct which opens into the pelvis at the apex of the papilla.

FORMATION OF URINE : The formation of Urine involves three processes. These are filteration, selective reabsorption and secretion.

Filtration (Ultra filteration):

The Bowman's capsules of kidney act as ultra-filters. The blood flowing though the afferent arterioles contains urea, water, several salts and blood proteins dissolved in plasma. It is separated from the cavity of renal tubule of Bowman's capsule only by two thin membranes - the endothelial layer of blood capillaries and the epithelial layer of Bowman's capsule. The diameter of afferent arterioles is more and lie in close contact. The diameter of afferent arteriole is more than the efferent arteriole. Therefore, the amount of blood which enters the afferent arteriole in a definite time is not able to be drained out and the hydrostatic pressure of blood in the capillary network of glomerulus increases. The capillary pressure is about 75 mm Hg. This pressure has to overcome the osmotic pressure of the plasma, which works to retain constituents within capillary walls. The osmotic pressure of plasma is about 30 mm Hg. The intestinal pressure on the capillaries themselve, together with the resistance to flow in the uriniferous tubules contributes a pressure about 20mm Hg. This also works against the capillary pressure. Therefore, the net filtration pressure supplying energy for the filtration is about 25mm Hg. As a result, water and diffusible solute molecules are forced across the membrane from the blood plasma into the Bowman's capsule. This process is known as ultra-filteration.

The effective filtration pressure (EFP) can be calculated as follows;

Glomerular pressure = 75 mm Hg.

Osmotic pressure of blood capillaries+ capsular pressure = 25 mm Hg.

EFP = 25 mm Hg.

Tubular Reabsorption: The nephron functions like an exclusive filter and as such it filters all the substances including those that are to be retained in the body and are necessary for metabolism. These substances are glucose, amino acids inorganic salts etc., The filterate also consists of urea, creatin and uric acid which are to be removed from the body. In selective reabsorption the essential constituents are reabsorbed into the body during its passage through the kidney tubules. This is necessary to maintain the fluid and electrolyte balance and also of the alkalinity of the blood. (pH 7.4)

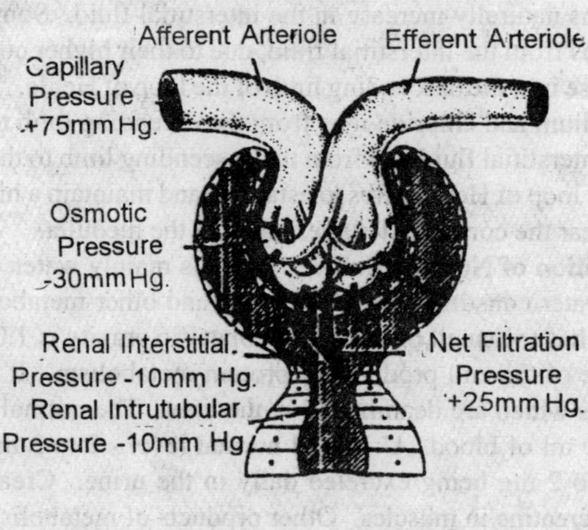

Fig. 29.4 Diagrammatic representation of ultrafiltration

Every minute about 125ml of the filterate enters the tubule of which about 124 ml is reabsorbed. Inorganic salts like sodium chloride, water, glucose, amino acids etc., first difuse passively into the epithelial cell from the tubular lumen. Then glucose and amino acids are actively transported (using metabolic energy) and are reabsorbed. The different parts of the tubule absorb different substances. The Henle's loop and the distal convoluted tubule reabsorb water. It has been estimated that about 80% of the water is reabsorbed in the proximal tubule by a passive process and is referred to as obligatory reabsorbtion.

Secretion (active secretion): While tubular reabsorbtion removes the necessary substances from the filterate into the blood, secretion removes materials from the blood and adds to the filterate. Metabolic products present in the blood but not required by the body and in fact are injurious or secreted out from the blood to the filterate. The tubular epithelium plays an important role in this. The substances that are secreted into the filterate include potassium, hydrogen, ammonia, creatine, uricacid and drugs like pencillin. These secretions mix with the filterate of the nephrons and constitute the urine.

Tubular secretion has two main roles - it rids the body of unwanted materials and it also controls the pH of the blood.

Counter current Mechanism for Urine concentration :

Higher vertebrates, including man, have evolved a counter current mechanism to excrete hypertonic urine (urine is more concentrated than the blood) for conserving body water, necessary for land life. As the filterate passes through the ascending limb of Henle's loop, sodium is actively transported into the interstitial tissue fluid and chloride ions pass out passively. The sodium and chloride ions naturally increase in the interstitial fluid. Some of sodium and chloride ions from the interstitial fluid, due to their higher concentration, passively diffuse into the descending limb of the loop of Henle. The cyclic movement of sodium and chloride ions from the ascending limb to the descending limb via., interstitial fluid and from the descending limb to the ascending limb through the loop of Henle helps to establish and maintain a high concentration is lowest near the cortex and highest deep in the medulla.

Composition of Normal Urine : Urine is mainly water, urea and sodium chloride, water constitutes 96% urea 2% and other metabolic products 2%. Stale urine has the smell of ammonia. Specific gravity is 1.005 to 1.030. Urea is one of the end products of protein metabolism. It is prepared from amino acids which are deaminated in the liver. The normal blood urea is 30 mg per 100 ml of blood. Uric acid normal level - 2 to 3 mg per 100 ml, of blood 1.5 to 2 mg being excreted daily in the urine. Creatinine is a waste product of creatine in muscles. Other products of metabolism include purine bodies, oxalates, phosphates, sulphates and urates. Electrolytes of salts such as sodium, potassium chloride are excreted to balance the amounts taken by mouth.

Disorders of the excretory system

The excretory system (kidneys) throws out the waste products by filtering the blood. The functioning of the kidneys may get impaired due to various reasons. Some of the diseases of the kidney malfuncitoning are - Renal failure, nephritis, nehprotic syndrome, cystitis, urethritis, renal stone formation etc. In this book renal failure and kidney stone formation are explained in some detail.

Renal failure (Kidney failure)

Renal failure refers to non performance of the functioning of the kidneys when they fail to filter the blood of the waste products. As a result waste products accumulate in the blood leading to severe complications.

Renal failure is not a disease in itself; it is the result of one or many diseases that affect the kidneys. There are two types or renal failure- Acute renal failure (ARF) and Chronic renal failure (CRF).

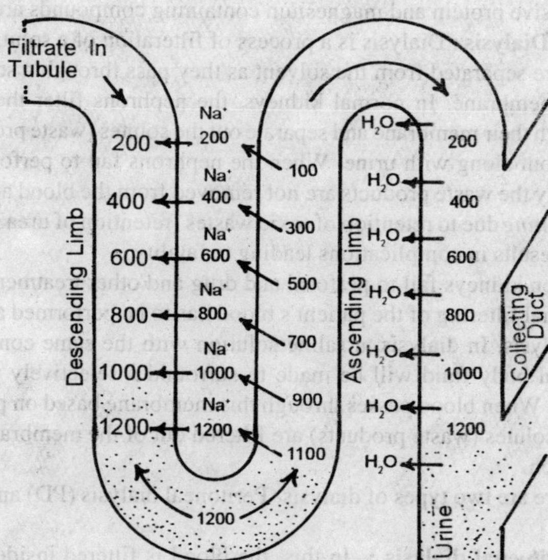

Fig. 29.5 Counter current concentration in loop of Henle

Acute renal failure (ARF) **:** This is characterized by the sudden decline in the filteration rate of the glomerules, which limits the ability of the kidney. Metabolic wastes like urea, creatinine etc accumulate in the blood. ARF occurs in two forms-

(i) *Oliguric ARF* in which urine output is less than 400ml/day.

(ii) *Non olliguric ARF* in which urine output is more than 400ml/day.

Causes of ARF are three types

a. *Pre renal:* includes dehydration, hypotension and heart failure.

b. *Renal :* Nephritis, drug induced injury, mismatched blood transfusion.

c. *Post renal :* Obstruction of the urethra by stones, tumours; bladder obstruction by hypertrophy of prostrate glands.

Treatment of ARF

(i) Hypotension may be remedied by infusion of saline solutions. If hypotension is due to blood loss, blood transfusion has to be undertaken; heart failure also to be treated if it is a cause.

(ii) Blood potassium and other electrolytes have to be estimated regularly in order not to allow their accumulation. Excess potassium if any may have to be removed by intravenous bicarbonates.

(iii) Proper diet is necessary which should include all the essential nutrients. Excessive protein and magnesium containing compounds are to be avoided.

(iv) **Dialysis** : Dialysis is a process of filteration of a solution in which solutes are separated from the solvent as they pass through a selectively permeable membrane. In normal kidneys, the nephrons filter the blood (dialyse) through their membrane and separate out the solutes (waste products) and throw them out along with urine. When the nephrons fail to perform thier filtering activity the waste products are not removed from the blood and result in *blood poisioning* due to retention of toxic wastes (retention of urea is called uremia). This results in complications leading to fatality.

When kidneys fail to perform and drug and other treatments are of no help artificial filtering of the patient's blood has to be performed and this is known as dialysis. In dialysis, a saline solution with the same concentration of the normal body fluid will be made to surround a selectively permeable membrane. When blood passes through this membrane based on principle of diffusion, solutes (waste products) are filtered out of the membrane into the saline solution.

There are two types of dialysis- Peritoneal dialysis (PD) and Haemodialysis (HD).

Peritoenal dialysis : In this, the blood is filtered inside the body of the patient using the serous membrane of the peritoneum for dialysis. This is performed by slowly introducing one or two litres of saline solution (dialysate) over a period of 20 minutes through a catheter. After a period of 20 minutes stay in the peritoneum, the saline solution is drained out slowly for the next 20 minutes. Dialysis (taking one hour approximately) has to be performed several times over a period of 24-72 hours.

In this method of dialysis, the peritoneum acts as dialysis membrane across which waste products diffuse into dialysate fluid. This treatment is simple and should be continued till condition of the patient improves.

Haemodialysis : In this, blood and saline solution are placed across a membrane to remove unwanted materials from the blood and at the same time adding useful compnents to it. A constant flow of blood on one side of the membrane and dialysate on the other side allows removal of waste products in the same way as glomerular filteration takes place in the kidney. Blood from the patient is continuously drained into a haemodialysis machine by surgically creating a fistula between an artery and a vein (either in the hand or leg). The haemodialysis machine functions exactly like a nephron and filteres out waste products while platelets and blood cells remain in the blood. After filteration, the blood is pumped back into the blood vessel of the patient. The temperature of the blood is brought to normal body level before it is pumpted into the body. Haemodialysis takes about 7-8 hours as the entire blood of the body has to come out and get filtered. Haemodialysis about thrice a week is necessary to maintain the normal health of the patient.

Differences between Peritoneal dialysis and Haemodialysis

Peritoneal dialysis	Haemodialysis
1. Blood filtered inside the body	Blood filtered outside the body using a machine
2. Pertioneal membrane (serous) is used for filteration	Artificial membrane (not belonging to patient) is used
3. Saline solution is injected into the body	Saline solution is not injected; it remains in the machine.

Continuous ambulatory peritoneal dialysis (CAPD): This is a technique by which the patient himself can carry out dialysis while moving around doing the normal activity. In this, the patient injects the dialysate fluid into his body (peritoenal cavity) throught the catheter, seals the catheter and goes on to do the normal work. After every 4-6 hours the catheter is opened, fluid is drained out and fresh fluid is introduced. Two litre containers of dialysate are available for this purpose and patient need not be hospitalized.

Chronic Renal failure (CFR): This occurs over a period of time and is the end result of many progressive renal diseases. CRF is defined as a *gradual, quantitative reduction of kidney function to a point when kidneys are no longer able to maintain the biochemical balance in the body.* Waste products accumulate in the blood and affect all organs of the body.

CRF is caused by a variety of reasons- Diabetes (causing nephropathy); nephritis, hypertension over a long period, urinary tract obstruction and infection, toxic drugs etc.

Symptoms appear gradually. These include drowsiness, nausea, twitching of muscles, convulsions, high blood pressure, pericarditis, pulmonary disorders, yellow tinge on the skin, drfective vision, bone pain, muscle weakness etc.

Treatment

(i) Diet with restricted amount of protein.

(ii) High blood pressure is to be treated promptly.

(iii) If there is anaemia, blood transfusion may be necessary.

(iv) Long term dialysis: This has to be done continuously to maintain the life of the patient. (for details see under ARF).

(v) ***Kidney transplantation*** **:** This is the best solution for CRF as the patient's kidney can never function normally. For this, kidney has to be obtained from a relative or an outsider. In bothe cases the *tissue match test* is done to find out the suitability. Kidneys from dead bodies *(cadaver transplantation)* may also be used provided they are found suitable in all respects.

Kidney stones (Renal stones)

Kidney stones or calculi develop as a result of various metabloic disorders which affect the fate of calcium and other mineral elements in the body. Stones may be formed in the kidney, urinary bladder and lumens of ureter and urethra. Kidney stones are solid structures made up of mainly uric acid, calcium oxalate, triple phosphate or cystine. Based on their composition four kinds of stones are identified-

1. calcium stones 2. uric acid stones 3. cystine stones and 4. struvite stones. Stones are formed when the urinary salts in amorphous form are bound to a colloid matrix. These are deposited surrounding a nucleus of organic origin. Mechanism of stone formation is poorly understood. Calcium stones are formed due to excess secretion of calcium in the urine. Other causes of stone formation include- low urinary output, hypoglycemia followed by hypercalcularia, hyper parathyrodism, bone tumours etc.

Symptoms of renal stones include extreme pain in the ureter or abdomen. If the stone gets stuck in the ureter urine is blocked leading to dilatation of the kidney. Some times small stones may come out along with urine. Clinically renal stones can be detected by X-ray, ultrasound scan and magnetic resonance image techniques.

Structure of the stones

1. Uric acid stones : These are composed of uric acid with sodium and ammonium urates. Stones appear smooth, hard; light or dark brown in colour.

2. *Oxalate stones :* These are found in pure form and extremely hard with prickly outline. The stones are dark coloured due to coating of blood which again is due to injury caused by the spiny exterior (of the stone).

3. *Phosphate stones :* These may occur in pure form or in combination with others. These stones are white, smooth and often look like a stag horn hence the name *stag horn calculi.*

Treatment: It is both medical and surgical. Heat formentation, analgesic drugs etc may reduce pain. Surgery is the best method when stones are larger than 1c.m.

A recent technique of stone removal does not involve surgery for stones not larger than 3 c.m. This technique called *Extra corporeal shock wave lithotrypsy* (ESWL) sends shock waves over the kidney to pulverize the stones completely and the powdered particles come out along the urine. This is a safe method (though costly) and surgery is avoided.

Patients (even after removal of the stones) should take lots of liquids including tender coconut, fruit juices (tomato or citrus juice should be avoided) etc., to prevent formation of stones again.

Thirty

Nervous system

The nervous system receives information about external and internal changes, conducts such information between different parts of the body and coordinates the activities of different organs and tissues in the light of those changes. In mammals, the nervous system consists of the central nervous system comprising the brain and the spinal cord, and the peripheral nerves present outside the central nervous system.

The nervous system is divisible into three parts:

1. Central nervous system
2. Peripheral nervous system and
3. Autonomic nervous system

Central nervous system: It includes brain and spinal cord.

Structure and function of human brain :

Brain is the controlling centre of all the body activities. It is protected by cranium. It is covered by three membranes together known as *meninges*. The inner membrane is *piamater* the middle *arachnoid* and the outer membrane is *duramater.* Outer part of the brain has grey matter and its inner part has white matter. Inside the brain there are 2 spaces- subdural space which is the space between duramater and arachnoid and subarachnoid space which is the space between arachnoid and piamater. These are continuous with the spinal canal of the spinal cord and are filled with cerebro-spinal fluid. The human brain weighs about 1,400 gms in volume.

Divisions of Human Brain:

Human brain has three primary divisions (with subdivisions):
Forebrain, Midbrain, Hindbrain

Forebrain: It is the largest part of the brain constituting about 2/3 of the human brain. The forebrain is distinguished into *cerebrum, olfactory lobes* and *diencephalon.*

1. Cerebrum: The cerebrum is the largest and most prominent part of the brain. It is divided into the right and left cerebral hemispheres. Each hemisphere is divided by sulci into distinct areas known as frontal, parietal, temporal and occipital lobes. Ventrally, the two cerebral hemispheres are connected by a transverse band of white medullated fibres the corpus callosum. Its outer wall is the cerebral cortex which is thrown into folds forming elevations called *gyri* with depressions between them termed as *sulci.*

Functions of Cerebrum: It is the dominant seat of highest mental faculties viz. will, conscious sensation, skilled work and other voluntary activities, and intelligence (memory, experience, learning, thinking, reasoning, knowledge and speech). This part also exerts strong control over many such reflexes like laughing, weeping, micturation and defaecation etc.

2. Olfactory lobes: These form the anterior most part of the forebrain and are more distinct on the ventral side.

Functions of Olfactory lobes: These are the centres of olfaction or sense of smell.

3. Diencephalon : It is a small structure covered dorsally by cerebral hemispheres. Its dorsal surface is raised into a pineal stalk bearing a small rounded pineal body. From the floor of diencephalon arises a funnel-shaped pituitary body. The dorasl wall of diencephalon is thin and vascular and represents the anterior choroid plexus. The cavity of diencephalon constitutes the third ventricle. It communicates through the Foramen of Monro with the cavity of cerebral hemispheres.

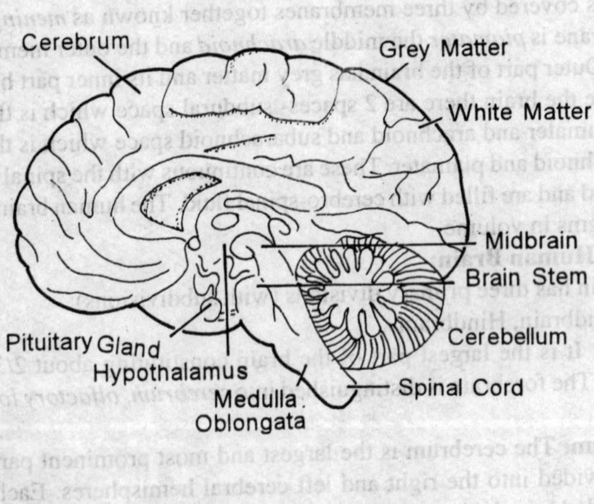

Fig. 30.1 Human brain, side view

Functions of Diencephalon: It is the centre of recognition of heat, cold and pain. The hypothalamus or central part of diencephalon is the control centre of autonomic nervous system thus it controls hunger, thirst, sweating, sleep, fatigue, temperature, sex and emotions like love, hate, satisfaction, anger, pleasure and penance etc. It also controls carbohydrate, fat and water metabolism.

Midbrain: The midbrain consists of optic lobes and cura cerebri.

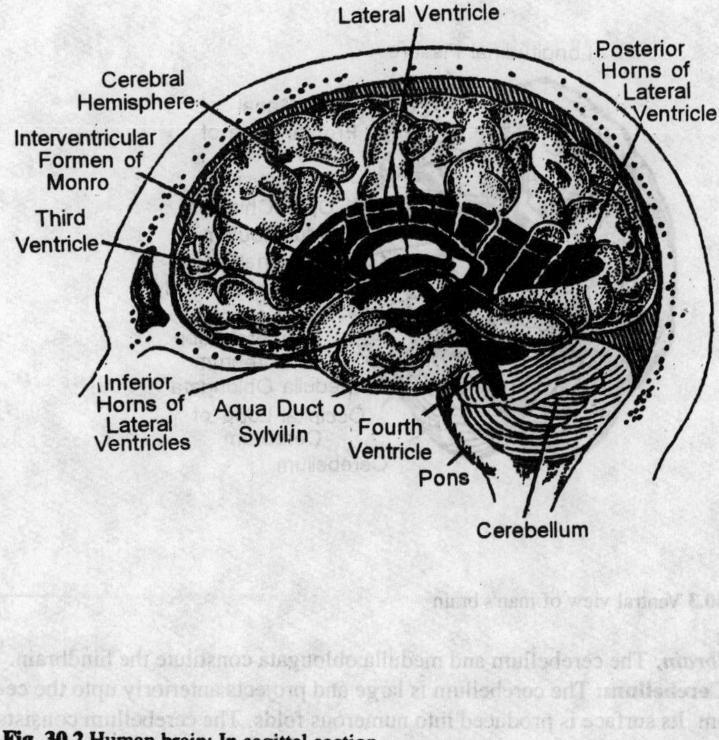

Fig. 30.2 Human brain; In sagittal section

1. Optic lobes: The optic lobes are four solid lobes lying behind the diencephalon. They collectively constitute the corpora quadrigemina. The anterior two lobes are slightly larger than the posterior lobes. The anterior and posterior lobes are connected by means of posterior commissure. They are devoid

of optic ventricles, but between them passes a narrow cavity, the iter, which connects the third and fourth ventricles together.

Functions of Optic lobes: It is the centre of visual and auditory sensation. The crura cerebri of optic lobes serve to transmit motor impulses to limb muscles from cerebrum. It probably helps in control of pigment cells of skin and amplexus during breeding season.

2. Cura cerebri: These are two longitudinal bands of nervous matter situated on the ventral side of the optic lobes which connect the medulla oblongata with the cerebral hemisphere.

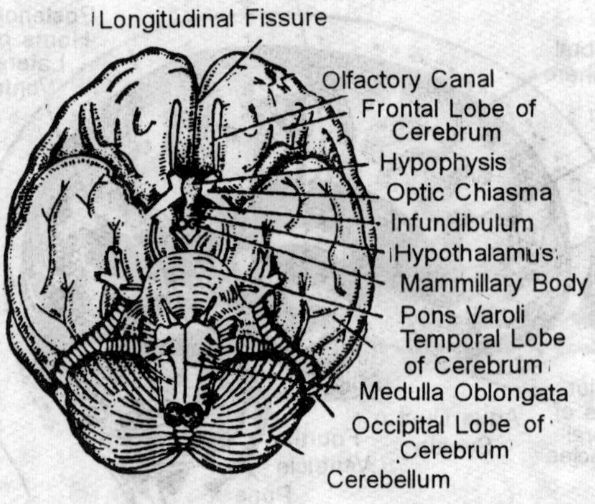

Fig. 30.3 Ventral view of man's brain

Hindbrain: The cerebellum and medulla oblongata constitute the hindbrain.

1. Cerebellum: The cerebellum is large and projects anteriorly upto the cerebrum. Its surface is produced into numerous folds. The cerebellum consists of a central lobe, the vermis, two lateral lobes and two floccular lobes attached to the lateral lobes. The vermis is divided by transverse furrows into anterior, middle and posterior portions. Its grey matter presents a branched tree like structure which is known as arbor vitae.

Function of Cerebellum: Its primary function is to maintain posture, orientation and equilibrium of body by coordinating and regulating tone and contraction of voluntary muscles, mainly according to the command of cerebrum. It also receives sensory impulses from the equilibratory portion of internal ear.

2. Medulla oblongata: It is the posterior most part of the brain and is located underneath the cerebellum. Its floor is beset with a transverse broad band of fibres of white matter which is known as *pons varolii.* The anterior triangular part of medulla possesses thick floor and lateral walls but its roof is exceptionally thin and is formed of highly vascular piamater. It represents the posterior choroid plexus. It is perforated by a small aperture which is known as foramen of Magendie. Its cavity is known as metacoel.

Functions of Medulla oblongata: It is the controlling centre of metabolism, heart beat, breathing, vasodilation and vasoconstriction, blood pressure, gut peristalsis, prey catching, swallowing of food and gland secretion.

Spinal cord:

At the posterior end of Foramen magnum of skull, medulla oblongata of brain is extended as Spinal cord. The spinal cord has the grey matter at its centre and the white matter around it. Groups of nerve cells constitute spinal nuclei in the grey matter. Bundles of nerve fibres ascend or descend along the white matter forming the nerve tracts. The ascending tracts conduct nerve impulses coming from the peripheral tissues to the brain. The descending tracts conduct impulses down from the brain to be relayed ultimately to muscles and the glands. Its cavity is called centrocoel or nurocoel.

Functions of the spinal cord: The spinal cord through afferent nerves passes the impulse from various parts of the body to the brain and the necessary command from the brain to the concerned through the efferent nerves.

Reflex action is a motor response to a sensory stimulus performed by the spinal cord without the brain taking any part, though for all the reactions the brain is the controlling centre and there are reflex centers in the brain. The sensory stimulus is carried by afferent nerves through the dorsal root of the spinal nerve to the spinal cord. The spinal cord gives necessary message which is passed on along the ventral root, motot nerves to the muscles for necessary functioning. The reflex actions are automatic and instantaneous.

Pavlov, a Russian Scientist has the credit of finding the *conditioned reflex.* In normal condition, when the meat is placed on the tongue of a dog, it sends stimulus resulting in production of saliva. Pavlov experimented with a dog, by ringing a bell and giving meat, thus introducing a new factor namely the ringing of the bell. By such a training, the dog associated the ringing of the bell with the availability of food. Finally when the bell was rung and even when food was not given, the salivation was normal. This phenomenon of reaction to a factor not necessarily needed for a response is termed conditioned reflex.

ENDOCRINE FUNCTIONS OF THE PITUITARY (Hypophysis):

The Pituitary gland is one of the most important among the endocrine glands. Very often it is also called the master gland as it controls the functioning of many other endocrine glands. In, size the pituitary gland is about one centimeter and is situated at the base of the brain in the saddle shaped depression in the sphenoid bone known as the *sella turica*. The pituitary gland weighs approximately about 0.5 gms and becomes slightly larger in the pregnant women. It is attached to the brain by a short stalk placed just behind the optic chiasma where the optic nerves from each eye meet.

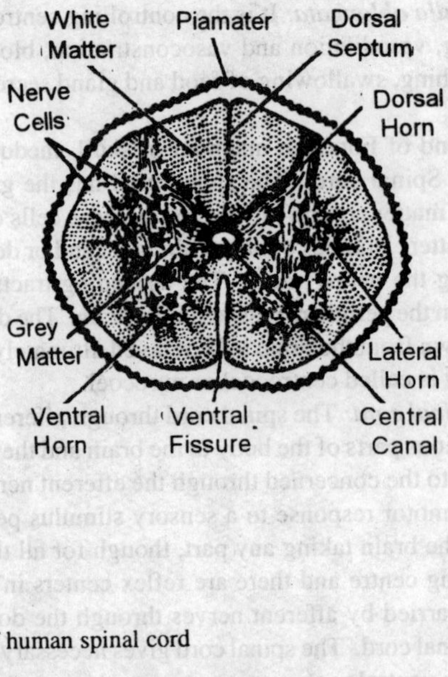

Fig. 30.4 T.S. of human spinal cord

Origin: The pituitary is ectodermal origin. It develops from the roof of the embryonic pharynx.

Structure: Structurally the pituitary gland has two parts viz., adenohypophysis (anterior lobe) and neurohypophysis (*posterior lobe*).

Adenohypohysis : This is a large and compact structure having abundant vasculature. It is also sometimes called pars buccalis. The adenohypohysis consists of the following three parts.

1. Pars distalis: This is the largest part of the adenohypophysis constituting about 75% of the pituitary. Here the cells are arranged in cords and inbetween the cords are found capillaries.

2. Pars tuberalis: This is smaller than pars distalis. The cells here are polyhyderal in shape with a centrally located nucleus. The cytoplasm is full of fine granules. The cells are arranged in clusters or threads. Pars tuberalis is rarely found in man.

3. Pars intermedia: This part of the adenohypophysis is poorly developed in human beings.

Histologically the adenohypophysis consists of cells arranged in clusters or columns surrounded by connective tissue. The cells are basically of two types - *chromophobes* and *chromophils*. The chromophobes are colourless while the chromophils are coloured. On the basis of the staining reaction the chromophils are again divided into *acidophils* and *basophils*. Acidophils take up acidic stain while basophils take up basic stain.

Neurohypophysis:

Also known as the posterior lobe this is further differentiated into *pars nervosa* or *neural lobe* and *infundibulum*. Pars nervosa is massive in mammals. Structurally the posterior lobe consists of non medullated nerve fibres neuroglia and blood vessels.

Hormones of the adenohypophysis : Six hormones are produced by the adenohypophysis. These are

1. Somatotrophic hormone (STH) or Growth hormone (GH)
2. Adenocarticotrophic hormone (ACTH)
3. Thyroid stimulating hormone (TSH)
4. Prolactin or Lactogenic hormone(LTH)
5. Gonadotropic hormone(GTH)
   a. Follicle stimulating hormone (FSH)
   b. Luteinizing hormone (LH)

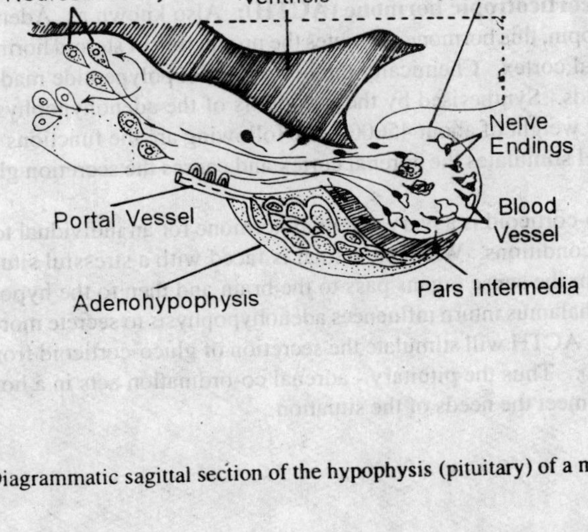

Fig. 30.5 Diagrammatic sagittal section of the hypophysis (pituitary) of a mammal

Hormones secreted by the Neurohypophysis

1. Vasopressin
2. Oxytocin

1. **Somatotrophic hormone (STH) :** This is also known as Growth hormone and it influences growth in human beings by controlling the synthesis of proteins known as somatomedins by liver cells. The hormone is secreted by the acidophilic cells of adenohypophysis. Chemically the hormone is proteinaceous being made up of about 200 amino acids. It has a molecular weight ranging between 21500 to 48000 and precipitates at pH 4.9. In children, the concentration of STH is about 300 micrograms per litre of plasma which drops to about 30% in adult human beings. STH does not have a target reaction instead it shows its influence on all most all tissues of the body by increasing the number and size of the cells.

The growth hormone should be secreted at the normal rate. It should neither be high nor low. Oversecretion of STH in childhood leads to excessive growth in length of the bones and results in a condition known as *gigantism*. In adults hypersecretion of STH results in a condition known as *acromegali.* In this condition there is enlargement of bone and soft tissues resulting in hands and feet getting grossly enlarged, jaws growing abnormally long and broad. In some cases there is less secretion of pituitary during the growth period. As a result of this there is undergrowth resulting in a small sized body. And this is known as *midget.* Over secretion of growth hormone in midget adults will result in an unproportionate growth of the body and it is known as *acromicria.* Chemically the growth hormone increases the release of fatty acids from the adipose tissue and thereby increases the fatty acid concentration in the body fluids. STH decreases the utilisation of the carbohydrates for energy and promotes protein synthesis.

Adenocorticotropic hormone (ACTH): Also known as Adenotropin or Corticotropin, this hormone promotes the production of steroid hormones from the adrenal cortex. Chemically the hormone is a polypeptide made up of 39 amino acids. Synthesised by the basophills of the adenohypophysis it has a molecular weight of about 45000. The following are the functions of ACTH.

1. ACTH stimulates the adrenal cortex and causes the secretion gluco-corticoids.

2. Gluco-corticoid is a very essential hormone for an individual to meet the stressfull conditions. When an animal is faced with a stressful situation, impulses from the sense organs pass to the brain and then to the hypothalamus. The hypothalamus inturn influences adenohypophysis to secrete more quantity of ACTH. ACTH will stimulate the secretion of gluco-corticoid from the adrenal cortex. Thus the pituitary - adrenal co-ordination acts in a homeostatic manner to meet the needs of the situation.

3. Melanocyte pigmentation is also under the control of ACTH and patients in whom there is insufficiency of adrenal secretion are abnormally pigmented as a result of the loss of negative feedback provided by cortisol. This condition is known as *Addison's disease.*

4. There is a diurnal variation in the secretion of ACTH. As a result of this, the cortisol levels in the plasma will be at its peak in the morning (about 6 am) and will reach their lowest level around midnight. In an abnormality known as *Cushing's disease* there is hypersecretion of cortisol and the diurnal variation is lost.

Thyroid stimulating hormone (TSH): Also known as Thyrotropin, this hormone is produced by the basophills of the adenohypophysis. This hormone stimulates the growth and activity of the thyroid gland.

Chemically TSH is a glycoprotein with a molecular weight ranging between 10,000 to 30,000. The concentration of TSH in blood is approximately 25 mili units per ml of plasma. The daily secretion of TSH in normal individual is a little more than 2 international units per day. TSH has the following functions.

1. It co ıtrols the various aspects of the functioning of the thyroid gland.
2. Promotes accumulation of iodine and increases the quantity of intracellular colloids in the epithelial cells of the thyroid gland. This induces the liberation of the hormone thyroxin.

There is a negetive feed back between the level of thyroxin and TSH. If the secretion is decreased the level of thyroxin is also decreased resulting in *hypothyroidism.* This results in reduced growth. On the other hand increased TSH secretion resulting in increased thyroxin production causes *hyperthyroidism.* This is characterised by irritability, increased rate of metabolism, bulging of the eye balls etc.,

The production of TSH is controlled by thyrotrophin releasing hormone (TRH) produced from the hypothalamus.

Prolactin (lactogenic hormone - LTH): This hormone is also known as lactogen or galactin or mammotropic hormone.

Chemically prolactin is a polypeptide consisting of more than 205 amino acids. It has a molecular weight ranging between 25000 to 30000.

This hormone has a variety of activities and according to Nicoll and Bern(1971), more than 75 different activities are ascribed to this hormone. The secretion of this hormone is under the control of hypothalamus which produces a prolactin realising factor and a prolactin inhibiting factor.

The general functions assigned to this hormone are

1. It stimulates the production of milk in the mammary glands.
2. It stimulates the production of Corpus-luteum
3. During pregnancy prolactin helps in maintaining the functional state of corpus-leutem which is necessary for continued pregnancy.

4. This hormone stimulates the maternal instinct and behavior in women which is very essential for the maintenance of the offspring. It is also believed that this hormone induces in the mother and the child a strong emotional attachment.

Gonadotropic hormone (GTH): Also called Gonadotropins, these hormones are secreted by the basophils of the adenohypophysis. There are two kinds of Gonadotropins in human beings. These are follicle stimulating hormone (FSH) and luteinizing hormone (LH).

Follicle Stimulating hormone(FSH): This hormone is found in both males and females influencing the activity of the reproductive structures. Chemically FSH is a water soluble glycoprotein having a molecular weight between 30000 and 67000. The main components of this hormone are Galactose,Glucosamine, Galactosamine,Mannose etc.,

The main function of FSH in females is to stimulate the development and maturation of ovarian follicles up to the stage of ovulation. In males FSH stimulates the process of spermatogenesis including sperm formation. FSH is also known as Gametokinetic factor as it has influence on both male and female gametes.

FSH is not found in childhood. It is secreted in little quantities only in sexually matured organisms.

Luteinizing Hormone (LH): Also known as interstitial cell stimulating hormone (ICSH), this hormone is a conjugated protein having a molecular weight between 26000 to 30000. The following are the functions of LH.

1. It causes ovulation(release of mature ovum from the ovary) and helps in the formation of corpus lutem working along with FSH. The corpus lutem inturn secrets progesterone in women which is essential for the continuation of the pregnancy.

2. In males LH is generally known as ICSH. This stimulates the interstitial cells(cells of Leydig)of testis to secrete the male sex hormone - testosterone. Testosterone is also necessary for the maintenance of secondary sexual characters.

Hypothalamus controls the production of LH through a hormone called gonadotropin releasing hormone (GnRH). Many environmental factors like stimulation of the genitals,light, temperature etc., also have a regulatory effect on the secretion of LH.

HORMONES OF THE NEUROHYPOPHYSIS.

The hormones of the neurohypophysis (Posterior lobe of the Pituitary) are actually synthesised in the hypothalamus by the neurosecretary cells hence the name neurohypophyseal hormones. After secretion, the hormones reach the posterior lobe of the pituitary *via* the neural pathways. These hormones are stored in the neurohypophysis before their release into circulation.

The neurohypophysis secrets two hormones viz vasopressin and oxytocin.

Vasopressin: This is also known as antidiuretic hormone (ADH) and causes the reabsorption of water into the blood from the collecting ducts of the kidney. Chemically ADH is a protein chain made up of 8 amino acids. The following are the functions of ADH

1. It plays an important role in the water balance(osmoregulation) of the body. When the fluid contents of the body become highly concentrated, ADH is secreted. This reaches the kidney via blood capillaries where it increases the permiability of the tubules thereby most of the water is reabsorbed and the urine gets concentrated with reduction in volume.
2. This hormone maintains the proper osmotic concentration of the blood plasma.
3. The hormone is destroyed as soon as its function is over and is excreted out along with the urine.
4. It has the role to play in the maintenance of the blood pressure. The secretion of this hormone is all the more increased during haemorrhage.
5. Alcohol inhibits the secretion of ADH thereby causing diuresis.
6. Insufficient secretion of ADH causes diuresis or *diabetes insipidus.* Such individuals consume large quantity of water to compensate for the excessive loss of water in urination.

Oxytocin (Pitocin): The term oxytocin in greek means "quick birth". Chemically it is a peptide made up of 8 aminoacids. It is extremely short lived in human blood having a half life of about 1 to 4 minutes. After its function is over it is either destroyed by the tissues or excreted by the kidneys.

This hormone is released by the posterior lobe of the pituitary by a neural mechanism bringing about contraction of the smooth muscles of the uterus and breast. The secretion of this hormone is stimulated by suckling and other kinds of stimuli. It is believed that suckling stimulates the neural pathways as a result of which oxytocin is released causing release of milk from the breast of the mother.

This hormone has a powerful stimulating action on the pregnant uterus, particularly during parturition. It has been seen that a normal level of oxytocin in the blood is essential for normal delivery. This hormone also has an influence on the neutrality of the uterus which promotes fertilization during coitus.

Disorders of Nervous system

EPILEPSY : Also called fits, epilepsy is a seizure disorder affecting the central nervous system. Seizures may be defined as sudden changes in sensation, behaviour, muscular activity or levels of consciousness. These are beyond voluntary control and are caused by the overactivity of the brain cells. The exact machinism of seizure is not known.

A variety of causes have been attributed to the onset of epilepsy. These causes may be classified into local causes, general causes, psychological factors and hereditary factors.

Local causes include pathological conditions of the brain like tumours, abscesses, haemorrhage, viral or bacterial infection(meningitis and encephalitis), head injury, cerebral thrombosis, cerebral embolism etc. General causes include consumption of exogenous poisonous substances like alcohol, lead poisoning, chloroform, carbon monoxide poisoning, hypoglycemia, hepatic failure, uraemia, etc. Psychological factors like extreme stressful condition, trauma, shocks may also trigger the onset of epilepsy. Hereditary factors also play an important role in epilepsy.

Diagnosis of epilepsy includes X-ray of skull, CT scan, study of metabolic parameters and many other screening tests.

Epilepsy is generally categorised into two types-*general* and *localised*. General epilepsy is again of two types-*Grandmal* and *Petitmal*. Localised epilepsy is again of two types-*Jacksonian* and *Psychomotor*.

Grandmal Epilepsy(Grandmal seizure): Also called major epilepsy, this involves the total body. The onset of epilepsy includes four stages namely- a. Aura, b. Epileptic cry, c. loss of consciousness and d. convulsions.

a. *Aura* - This consists of premonitory symptoms before the actual season. These symptoms occur usually a few hours or days before attack. Symptoms include visual hallucinations in the form of flashes of light in front of eyes, numbness of the body, strange sounds in the ear, vertigo etc.

b. *Epileptic cry* - This may not be seen in all cases . It is a scream caused due to partly closed vocal cards.

c. *Loss of consciousness* - The patient loses consciousness and is not aware of the surroundings or of himself.

d. *Convulsions* - This includes two phases *tonic phase* and *clonic phase*. In tonic phase, there is tonic spasm of the muscles. The head and eyes rotate to one side; the mouth also is drawn to one side. There is disturbance in respiration. Tonic phase lasts for a few seconds only. In clonic phase, the patient shows sharp jerks of the body, foam comes out of the mouth. The tongue may be caught between the teeth. The patient may also urinate.

The clonic phase gradually comes to an end. This stage is called the post convulsive stage. The muscular contractions gradually become slow. The patient regains consciousness; the period of unconsciousness however might vary. At the end of this, patient has a severe headache.

Petitmal Epilepsy (Petitmal seizure) : This is a minor type of epilepsy characterised by momentary loss of consciousness. The individual may look blank or may stop for a moment while walking. At that moment he has lost contact with the surroundings. After the attack, the individual is back to normal and is totally unaware of having had an attack. There are no spasms of muscles and no loss of bladder control. Petitmal seizure is common in young children. EEG (Electro Encephalograph) is very helpful in the diagnosis of the disease

Jacksonian Epilepsy (Jacksonian seizure) : These are small seizures which occur without loss of consciousness. They usually begin in one area such as thumb or index finger or angle of mouth or great toe and gradually involve adjacent muscles and finally involve the whole body.

Psychomotor Epilepsy (Psychomotor seizure) : This has a characteristic aura(the first phase before the onset of actual seizure) followed by a dream like state in which the individual repeats the same actions several times such as lip smacking, dressing and undressing, rage, uncontrolled laughter and apprehension. At this stage the patient may commit some crimes being totally unaware of what he is doing. In this stage he is totally taken over by ammesia.

Treatment :

Treatment includes three kinds of approach- general management, care during an attack and treatment with drugs.

1. General management includes explaining the patient his condition and the risks involved. He should be advised to avoid stress and fatigue. As far as possible he should keep away from fire, water and busy traffic. Preferably he should not drive any vehicle. It is better for him to carry a card indicating his condition which will help him to obtain assistance during an attack.

2. *Care during an attack* : Proper care should be taken to see that the patient does not injure himself like biting his tongue. A gag should be placed between the teeth to maintain the airway.

3. Drug treatment: This includes a variety of anticonvulsant drugs like phenobarbitones, primidone, carbamazepine etc. Antiepileptic drugs are to be taken for a period of three years from the date of the last seizure. These drugs however are to be used with extreme precaution as some of them may cause several side effects.

PARALYSIS

This refers to a condition in which there is loss of sensation in the entire body or in some parts of the body. The loss of sensation is accompanied with partial or total loss of movements in some parts or the whole body. In paralysis, the patient is totally unaware of the existence of that part of the body.

Paralysis is a disease of the nervous system because of the impaired functioning of the brain. This may be due to a stroke which is caused by either haemorrhage or a blood clot in some part of the brain. Sometimes spinal cord injuries and brain tumors are even brain infections may cause paralysis.

Paralysis is classified into the following types depending on the extent of attack on the parts of the body.

1. **Hemiplegia:** This refers to paralysis of one side of the body. Hemiplegia has two stages of attack. The first stage is the onset in which there are symptoms like shock effect, patient may become unconscious and the angle of mouth will be twisted to one side which is affected. The second stage is the stage of recovery. In this stage there is change of gait where the paralysed side is dragged around in semicircle. The tongue is protruded to the affected side.

2. **Paraplegia** : This condition is characterised by weakness of both lower limbs. There may be partial or complete paralysis of the legs. Paraplegia is caused by spinal cord injuries, vascular syphilis, cervical spondylosis, tumors of the brain etc.

3. ***Diplegia :*** In this type of paralysis, at both sides of the body are affected including the same regions. For instance if right hand is affected left hand is also affected.

4. **General paralysis :** This is a type of paralysis where the entire body is involved. There is total loss of movement or sensation in the entire body.

HAEMORRHAGE

Haemorrhage refers to loss of blood from the body. In the case of the brain, the haemorrhage is refered to as the cerebral haemorrhage wherein due to internal injury there will be rupture of the capillaries and oozing out of the blood which collect internally. This causes a variety of complications leading to stroke and even death. Initially however internal haemorrhage, if it is of a little quantity, may bring headache and nausea. Prolonged bleeding leads to unconsciousness.

Brain haemorrhage may be classified into four types- *extradural haemorrhage, subdural haemorrhage, subarachnoid haemorrhage* and *intracerebral haemorrhage*.

Extradural haemorrhage refers to collection of blood in the space between the skull and duramater. Rupture of the meningeal artery causes oozing of blood. The laceration of the artery may be due to fracture of the temporal bone as during a head injury. As a result of the collection of blood, there is increased intracranial pressure and the patient becomes unconscious. Immediate surgical intervention can save the patient.

Sudbural haemorrhage refers to collection of blood between the dura and the arachnoid mater. This is due to the rupture of cerebral veins which may be caused during a head injury. Several cortical veins bleed due to rupture to produce haematoma. In some instances subdural haemorrhage may be of the chronic type as seen in boxers who are prone to head injury. Minor haemorrhages accumulate and simultaneously blood clotting may also take place.

Sub arachnoid haemorrhage is the collection of the blood between the arachnoid and the piamater. This is due to trauma caused during head injury when blood leaks into the space. Sometimes it could be spontaneous also due to congenital problems.

Intracerebral haemorrhage may be of two types-*massive cerebral haemorrhage* and *punctate cerebral haemorrhage*. Massive cerebral haemorrhage is caused due to tumours, hypertension, trauma and various other reasons. This kind of haemorrhage leads to cerebral stroke or *apoplexy*. In this, the arteries rupture followed by abrupt rise in blood pressure. This may be caused by physical or emotional stress. Punctate cerebral haemorrhages are refered to as multiple, small, pinpoint haemorrhages within the brain substance. This may be due to brain contusion, infection, embolism or even asphyxia.

Treatment: Treatment for all types of cerebral haemorrhages depends on the extent of damage caused to the brain. In most of the instances when detected early, surgical treatment followed by medication is the only remedy.

EFFECT OF ALCOHOL AND NARCOTIC DRUGS ON THE BODY

Human beings since time immemorial are using intoxicants, hallucinogens and other types of drugs mainly to combat with stress or strain. In order to find an easy way out and forget the worries they seek the refuge of these agents with the main hope that they will provide some happiness. Use of alcohol has been recorded even in the earliest writings of human beings. In fact Rigveda mentions the intoxicating *soma* drink. While some people opine that mild drinking may be health invigorating it is a matter of debate. But in any case nothing useful has ever been proved to the body by means of drinking. In the following paragraphs we will discuss two evil habits of human beings namely- alcoholism and drug addiction.

Alcohol and alcoholism

Chemically the alcohol that is normally consumed in various forms is ethyl alcohol (CH_3CH_2OH). Alcohol is produced by the fermentation of the carbohydrates using various types of cereals as source material. Microbes like yeasts help in the fermentation process during their anaerobic respiration.

Varieties of types of alcohol depending on the source of fermentation and percentage of alcohol are available in the market. Some of these are beer, whisky, toddy, arrack, brandy, vodka, gin, rum, saki etc. While beer, whisky and others are the products of barley and other cereals, wine which is also alcoholic is produced by the fermentation of grape juice.

The consumption of alcohol or drinking is regarded as a social custom and is practised by both poor and rich. While mild drinking may be regarded as partly harmless and drinkers may be called social drinkers, excessive consumption of alcohol leads to what is known as alcoholism. Individuals suffering from alcoholism require their drink always. They cannot do without it.

Alcohol intoxication actually begins from the first sip of the drink. With the increase in dose the body loses control and in extreme cases unconsciousness and even death might occur.

People who get into the habit of drinking gradually become so much accustomed to it that they always crave for drinking. This is the first sign of addiction. People who are addicted to drinking, immediately cannot give it up also as they have become victims of alcohol. Their body is so accustomed to the intake of the alcohol that any withdrawal suddenly will have many effects on the body.

Effects of alcohol drinking

It is said very often that modest drinking relives tension and keeps the mind calm. But the good effects of moderate drinking are yet to be proved. It is certain however that alcohol drinking impairs judgement, slows down nerve impulses, causes disturbance in muscular coordination and generally retards the time taken by the body to show response to situations. This will be indeed very dangerous under certain circumstances. In addition to this, alcohol addiction has the following bad effects on the various parts of the body.

1. Alcohol lowers the blood sugar level and is harmful to the brain. It might even cause irrepairable nerve damage.

2. The one organ that is mostly affected by drinking is the liver. Basically liver is capable of converting alcohol into other compounds rendering them harmless to the body. Liver cells synthesise fat from excess alcohol which is deposited in hepatic cells and bile passages. This deposition disturbs protein synthesis and conversion of glycogen. Excessive accumulation of fats disturbs the enzyme production and the condition is known is fatty liver syndrome. In advance stages fatty cells accumulate so much that it causes liver cirrhosis.

3. Alcohol intake results in dialation of peripheral blood vessels. This is due to the oxidation of alcohol and the resultant production of heat. Due to constant dialation the arterial walls get damaged which has an effect on the heart.

4. Alcohol consumption has a harmful effect on kidneys also as they have to do extra work to eleminate excess of water. Dialation of arterioles reduces glomerular filteration pressure.

5. Excessive alcohol intake might induce loss of memory.

Drinking and driving : Many drunkards without realizing the danger that they bring upon themselves and also to others take to driving the vehicles after having had drinks. This is highly harmful as the individual will not be in control of himself , hence the question of his control over the vehicle does not arise. Drinking affects the alertness, retards response to a situation as there is impaired coordination between the mind and the body. Further, alcohol may also blur the vision. *Hence no one who has had a drink should take to driving.*

Treatment : As the alcohol addiction is a social evil the treatment is more psychological than medical. Research studies have revealed that people who resort to heavily drinking are highly tensed up having family and financial problems and they take to drinking in order to escape from the realities of life. They should be convinced first that *drinking is not a solution to a problem but a problem in itself.* Instead of solving a problem a drunkard gets into more problems. Psychotherapy is employed in many centres called Deaddiction centres. In these centres the patient is advised to give up drinking gradually. Because giving up drinking all of a sudden will have some effects on the body known as withdrawal symptoms as the body has been physically dependent on the alcohol. *Withdrawal symptoms* include lack of coordination, confused mind, fear, trembling of the body, sleeplessness, hallucinations etc.Medication for patients (to gradually wean them away) includes nutritive food, vitamins, adequate rest etc. In addition deaddicted individuals should be advised to develop hobbies or to practise yoga and meditation.

Drug addiction

Generally speaking, drugs are chemical substances that are usually prescribed to treat a disease. These interact with the central nervous system either to stimulate it or to calm it down. Drugs acting on the nervous system are usually prescribed to relieve pain in chronic diseases.

Drugs that act on the central nervous system are generally called narcotic drugs. In the same way as in drinking some people are accustomed to taking drugs as they provide an easy escape from the day to day problems. But as in the case of alcohol drinking drug consumption also creates more problems than it solves.

Narcotic drugs have the capacity to make the individual dependent on them. This stage is called drug addiction. The psychotropic drugs (narcotic drugs) have an effect on the brain and bring about alteration in the behaviour, consciousness, power of thinking and even perception. Psychotropic drugs take the addicts on a hallucinogenic trip where initially the head becomes light and the mind becomes very calm. Persons who take to these drugs will become totally dependent on them. Drug addiction is infact more dangerous than alcoholism.

Psychotropic drugs are basically classified into narcotic drugs and stimulants. Narcotic drugs depress the activities of the central nervous system eg. opium derivatives. Stimulants are those drugs which increase the activity of the central nervous system either directly acting on it or by blocking the inhibitory activity of certain other nerves. eg. drugs of amphetamine group, caffeine, cocaine etc.

In another type of classification psychotropic drugs can be classified into sedatives and tranquilisers, pain killers, stimulants and hallucinogens.

Type of drug	**Examples**	**Effects**
1. Sedatives and tranquillisers (depressant)	Benodiazephines (e.g. Valium) barbiturates	Depress brain activity, creates calmness, relaxation drowsiness, and deep sleep under high dose
2. Opiate narcotics	Opium, morphine, heroin, pethidine, methadone	Supress brain activity, provides relief from intense pain
3. Stimulants	Caffeine (very mild), cocaine, amphetamines	Stimulate nervous system; makes a person more alert, wakeful and provides excitment
4. Hallucionogens	LSD, mescaline, psilocybin, charas, bhang , marijuana, hashish	Alerts thoughts, feelings and perceptions; creates visions.

Some of the commonly used psychotropic drugs

Opium and its derivatives: This is a drug extracted from the fruits of the plant-opium poppy(*Papver somniferum* of the family Papaveraceae). The crude drug is obtained in the form of latex from the unripe fruits of the plant. Inciscions are made in the fruit and the latex is collected. It is then further refined. Morphine and codine are the derivatives of opium. While another drug heroin is synthesised from morphine following several chemical procedures. Medically morphine is used as a pain killer while codine is an ingredient of many cough syrups.

Of all the derivatives of morphine the one that is more dangerous and has the ability of severe addiction is heroin. It is one of the highly classified drugs. Use of heroin leads to hallucinations and the addict feels light headedness.

Opium and other derivatives have the effect of reducing the respiratory and cardiac activity. They cause the constriction of the pupil of the eye. For addicts if supply is not available, they suffer from terrible withdrawal symptoms such as muscular cramps, running nose, vomiting and even fits.

Cocaine : This is a stimulant drug obtained from the leaves of the plant Coca (*Erythroxylon coca)*. It is cultivated extensively in south America, particularly Columbia. Pure cocaine is a fine white powder bitter in taste. Taken in mild doses it is a pain killer but in large doses it has the same ability of addiction like heroin and causes the same temporary effect of stimulation and permanent damage to the body functions. Cocaine is taken both orally as well as in the form of intravenous injections. The drug has the effect on the nervous system and also on the cardiovascular system. Cocaine addicts are seized by a feeling of anixety and fear and sometimes even may turn violent. Other bad effects of cocaine are insomnia(sleeplessness) , loss of appetite, hallucinations, delirium and even convulsions.

Ganja (Bhang or charas or marijuana) : The source of this drug is the leaves of the plant *Cannabis sativus* of the family Cannabinaceae. Besides leaves, flowers also are a source of ganja. The different forms of the basic drug ganja are *marijuana* or *charas*. Drug addicts either chew the leaves or more often smoke. Hashish is anothe name for charas and it is the resinous exudate of the plant. This is also smoked.

Chemically the drug contained in Cannabis leaves is tetrahydro cannabinol. This is a very dangerous drug causing dialation of the pupils of the eye, increase in blood sugar level and is also diuretic(increase in the frequency of the urination). Initially like any other narcotic drug, a sense of euphoria overtakes the patient and he feels he is very happy but eventually number of harmful effects on the body will kill the patient.

LSD (Lysergic acid dimethylamide): This is a most potent and most harmful psychotropic drug. The crude drug is obtained from infected barley and rye plants. These plants get the infection of a fungus *Claviceps purpurea*. As a result of infection in the inflorescence hard masses of the fungual mycelia called

Ergot is produced. This contains several alkaloids. The drug LSD in the purified form is a powder and has its principle effects on the nervous system. It affects the mind by changing all the perceptions of sensory organs. The individual will be completely lost and is totally unaware either of himself or of his surroundings. Hallucinations are common for those who take a "trip" on LSD. Addicts eventually suffer mental depression and may even develop suicidal tendencies.

Amphetamines : The effects of these drugs are similar to those of cocaine. Higher doses of these chemically synthesised drugs cause sleeplessness, impairment of vision and also nervousness.

Barbiturates : These are synthetic drugs which are generally used medically to act as sedatives. Given in mild doses they soothe and calm the nerves. Strictly speaking barbiturates are not narcotic drugs but some people have become addicts to this drug to prevent sleeplessness.

Effects of drug addiction

1. Regular use of drugs slows down reflexes and leads to a confused state of mind affecting even the body.
2. Addicts are normally prone to increase the dose which proves to be fatal.
3. Liver and digestive tracts are affected besides the nervous system.
4. Drugs like LSD can even induce chromosomal aberrations and result in congenital abnormalities in their children.

Reasons for drug addiciton

Several reasons may be attributed for a person to become a drug addict. Very often, personal problems, general displeasure with the surroundings, over ambitions but inability to realise them are some of the causes. Besides children who are neglected even from wealthy societies where both the parents are bothered about their business or social activities are also known to become drug addicts. Uncared for street children may also become addicts ultimately turning into criminals when they cannot obtain money to buy the drugs. The following may be cited as some of the general reasons.

1. *Curiosity* : Frequent reference to drugs in cinemas, T.V., newspaper or mentioning of drugs by friends makes a person curious to have an expreience.

2. *Pressure by friends :* Friends who are drug addicts coax a person to have a taste of the drug.

3. *Family history :* If a person in a family is a drug addict others will be influenced to take drugs.

4. *To overcome frustration :* Modern life puts lot of pressure on the individual as there is high competition in every field. In such circumstances individuals may become frustrated when they cannot realize their ambitions and hence seek refuge in drugs.

5. *Looking for a different world :* Some individuals believe that drugs take them to a new world and a new kind of perception.

6. *Desire to do extra work :* Many people, like hard working, intellectuals or students believe that drugs provide them extra energy to do the work. Similarly many athletes also take to some of the endurance increasing drugs.

7. *Relief from pain :* It is true that in some circumstances drugs help in the relieving of pain. But constant usage has many harmful side effects.

Treatment (Efforts to counter the drug menace) :

Drug addiction by youngsters is one of the greatest social problems that many of the developed and developing countries are facing. Indeed it is a formidable task to counter the effect of drugs and protecting the youngesters from its clutches. A variety of methods like legislation, psychiatric counselling (both to the addict as well as to his family members), medication etc are necessary to combat the drug menace. The following are some of the methods-

1. *Legislation* : All the countries of the world have very strict regulations against smuggling and trafficking of the drugs. Both the addict and the trafficker (the seller of the drug) are severely punished. In fact the punishment is more for the trafficker than for the addict. This is true because while an addict is a victim in himself a trafficker is a dangerous person who creates several victims. Some of the countries like Singapore and Malaysia have very stringent legislation so that a person caught having a certain quantity of drug is mandatorily given capital punishement.

2. *Psychiatric counselling* : This is a very important step as it helps the victim to overcome the habit of drug taking. This procedure is a very slow one and has several dimensions. First the background of the individual has to be analysed which includes his social status, personality analysis, analysis of family background, his circle of friends, his social aspirations etc which are all needed to be studied in great detail by counsellers before starting the actual treatment. In this treatment it is very much necessary to involve the parents also if the addict happens to be a teenager or an youngster. Many a time lack of parental attention and a feeling of not being wanted results in youngsters seeking solace elsewhere. The first step in a treatment actually includes a counselling to the parents and not to the addict himself.

The addict has to be advised about the bad effects of the drug on the personal life as well as the harmful influence it will bring to his surroundings. He should be told a drug addict not only spoils his life but of his family members also because the entire family is looked down upon. After preparing his mind, the process of deaddiction should start. If an individual is taking drugs regularly he should not be asked to stop all of a sudden as there are several harmful withdrawal symptoms. The quantity of the drug taken and the duration of taking the doses should be gradually controlled until a stage comes when the victim himself feels that he does not need drugs. Deaddiction centres are established in almost all cities where medical as well as psychological counselling is provided.

Advising the addict to develop healthy hobbies such as gardening, music, yoga, meditation or attending to religious discourses will have a salutory effect in the long run. A deaddicted individual should take a nutritious food to counter the bad effects of the drug on the body.

3. *General education of the population :* The population in general and the youngsters in particular should be educated about the harmful effects of drugs. This is a better method than curing the drug addict as prevention is always better than cure. Use of mass media like newspapers, T.V, radio and conducting symposia and seminars on the bad effects of drugs will have a great influence on the young minds. As far as possible seminars etc., should be conducted in schools and colleges because it is here the youngsters are subjected to evil influences by bad company.

Thirty one

Developmental biology

General account of Gametogenesis

One of the most remarkable abilities of living beings is their capacity of reproduction. It is by this only that life has maintained itself eversince it originated on this planet.

Reproduction may be defined as the ability of the organisms to produce offspring of their own kind. Reproduction generally takes place by two methods - Asexual and Sexual. In asexual reproduction there is production of offspring without involving the fusion of cells. Only one individual is involved in reproduction. In sexual reproduction generally two individuals are involved and reproduction is brought about by the fusion of two cells which are generally called gametes. These gametes are of two types - the male gamete and the female gamete. These two fuse and the resultant products viz., the zygote develops into a new individual. This is the general pattern of sexual reproduction seen in all the organisms.

Most of the higher organisms which includes higher animals reproduce generally by sexual reproduction. The sex cells namely the gametes are produced in definite structures called gonads. In the female the gonad is known as the ovary and the gamete is called the egg or the ovum. In the male the gonad is refered to as the testis and the gametes are called the sperms or spermatozoa.

The process of formation of the gametes is known as gametegonesis (Gr., *Gamos* = marriage; *genesis* = origin). In this process the reproductive structures which are diploid produce gametes by a process known as meiosis. As a result of this the gametes will have a reduced chromosome number (haploid) as against the diploid chromosome number seen in the reproductive structures.

In the body of the animals the gonads are produced by the germinal cells. These cells eventually produce the reproductive structures wherein reduction division takes place to produce the gametes.

Gametogenesis is of two types viz., spermatogenesis and oogenesis. The developmental process leading to the formation of the male reproductive cells or the sperms is called spermatogenesis. Oogenesis deals with the development of the ovum or the egg.

SPERMATOGENESIS

The process of spermatogenesis takes place in the male gonads called the testis. In most of the vertebrates the testis consists of many fine tubules called seminiferous tubules. These tubules have a wall lined by the cells of the germinal epithelium. The sperms are formed by these cells of the germinal epithelium. The seminiferous tubules also have certain cells called the nurse cells or sertoli. These are actually sematic cells which do not undergo reduction division. There function is to provide nourishment to the developing sperms.

The development of the sperms from the germinal epithelium, although is a continuous process may be divided into two stages. These are

1. Formation of spermatids and
2. Maturation of spermatids into sperms (spermiogenesis)

Formation of spermatids

The germinal cells of the seminiferous tubules produce the spermatids. The germinal cells are also refered as primordial cells or primary germinal cells. To begin with the germ cells are diploid and they give rise to the spermatids. This process is believed to take place in three phases - multiplication phase, growth phase and the maturation phase.

Multiplication phase. At this stage the testis consists of numerous seminiferous tubules made up of undifferentiated germinal epithelium cells. These cells have large deeply staining nuclei. The primordial germ cells increase their number by undergoing repeated mitotic divisions. As a result of this a large number of cells called sperm mother cells(spermatogonia) are formed. Each sperm mother cell is diploid and is like any other body cell in its chromosome constitution.

Growth phase. This follows the multiplication phase. In this phase there is no multiplication in the number of the cells. But the cells which are already formed absorb nutrition from the surrounding cells of germinal epithelium and grow in size. These cells have a large amount of food materials and are also rich in chromatin. These enlarged cells are called *spermatocytes.* It has been observed that during the growth period the pairing or synapse of the chromatids becomes visible. Hence it may be said that meiosis (reduction division) is initiated in this stage.

Maturation phase. During this phase the process of reduction division or meiosis which has been initiated will be completed. Meiosis involves two successive cell divisions - meiosis I and meiosis II. During meiosis I there is reduction division so that the two daughter cells are formed after division will have exactly half of the number of chromosomes than present in the original cell. Meiosis II is an equational division. At the end of the division four cells are formed.

The two haploid cells that are formed after meiosis I are called secondary spermatocytes. The two secondary spermatocytes undergo meiosis II to produce four cells which are called spermatids. The spermatids do not undergo any further division but simply undergo metamorphosis to produce the sperms. Thus at the end of maturation phase each germinal cell would have produced four spermatids.

Spermiogenesis.

During this phase the spermatids metamorphose (undergo structural changes) themselves into sperms. The spermatid is a physiologically immature cell having a nucleus, mitochondira and golgi complexes. The maturation of spermatids into sperms is an extremely complex process involving many biochemical and morphological changes. These changes are basically to induce a great amount of mobility to the sperm and to make it as light as possible by shedding any superfluous material present in the spermatids. The following are the major changes that take place in the spermatid on its way to maturation into a sperm.

1. **Nuclear changes**. The nucleus gradually gets reduced in size and assumes different shapes in different animals. In human beings the nucleus which is ovoid in the beginning becomes flattened. The RNA content of the nucleus gets reduced.

2. **Formation of Acrosome**. The acrosome occurs at the anterior side of the sperm nucleus. It is derived from the golgi complex and chemically it consists of protein digesting enzymes. At the time of acrosome formation the golgi complex gets associated with one or two vacuoles. These vacuoles enlarge in size and develop a dense granule known as proacrosomal granule. This granule is rich in mucopolysaccharides. Further enlargement of the vacuole makes it come very close to the sperm nucleus. Eventually the proacrosomal granule enlarges and forms the acrosome. At this stage the rest of the materials of the golgi complex gets reduced in size and are discarded from the sperm as "golgi rest".

3. **Centrioles**. The spermatid has two centrioles. These arrange themselves behind the nucleus. The anterior one is known as the proximal centriole and the posterior is known as the distal centriole. The distal centriole functions as a basal granule and gives rise to the axial filament of the sperm. The axial filament or the flagellum of the sperm is composed of eleven fibres of which two are central and nine peripheral.

4. **Changes in mitochondria**. The mitochondrial mass becomes aggregated around the axial filament leading to the formation of the middle piece in the case of the development of the human sperm. The mitochondria form a spiral sheath around the axial filament. This spiral sheath is called the nebenkern. The cytoplasm forms a condensed layer around the middle piece.

Structurally the maturation of the sperm from a spermatid involves the formation of two stages namely, the formation of the head and the formation of the tail. The nuclear changes and the changes that take place in the golgibodies, eventually lead to the formation of the head of the sperm. The middle piece of the sperm is formed by the cytoplasm and the spiraling mitochondria. The tail is formed by the lower portion and it involves the two centrioles.

A mature sperm has three portions. namely the head, the middle piece and the tail as in the case of human beings. The tip of the head consists of the acrosome while the remainder of the head consists of the nucleus. The middle piece has the fibres surrounded by spiral mitochondria and a condensed layer of the cytoplasm. The tail is made up of the fibers.

Derivation of parts of the sperm from a spermatid.

Parts of Spermatid	Parts of the sperm
1. Nucleus	Head portion
2. Golgi complex	Acrosome
3. Distal centriole	Basal granule and the axial filament
4. Proximal centriole	Proximal centriole
5. Mitochondria	Spiral sheath around the middle sheath
6. Cytoplasm	Condensed sheath around the middle piece.

OOGENESIS

The process by which the eggs or ova develop from the primordial germ cells of the ovary is called oogenesis. The mature ovary consists of cells similar to the primordial cells in the testis. These however give rise to eggs.

The process of the development of the egg can be studied under the following stages. These are multiplication phase, growth phase and maturation phase.

1. **Multiplication phase**. The primordial germ cells divide repeatedly by means of mitosis to produce a large number of daughter cells. These are called oogonia. The oogonia are known to divide again several times mitotically to produce a large number of cells called the primary oocytes. After the formation of primary oocytes there is no more division. The cells enter into the growth phase.

2. **Growth phase**. The growth phase of oogenesis is longer and more complicated than in spermatogenesis. There is considerable enlargement in the size of the primary oocyte during the growth phase. For instance the primary oocyte of frog initially has a diameter of only about 50 microns but at the end of the growth phase the diameter of the matured egg will be as much as 2000 microns. Growth in the primary oocyte is brought about by the accumulation of a large amount of fats and proteins. These get accumulated in the form of yolk. The yolk is usually concentrated towards the lower portion of the egg called the vegetal pole. The upper side of the egg which contains cytoplasm and the egg nucleus remains often separate from the yolk and it is called the animal pole.

The cytoplasm of the oocyte is very rich in RNA,DNA, enzymes, etc., Other cytoplasmic organells like mitochondria, golgi complex and ribosomes are also found in large numbers. In Oocytes of some amphibians and birds the mitochondria accumulate at certain places forming mitochondrial clouds.

During this period, the nucleolar genes show increase activity of RNA synthesis and multiply in their number. This is known as gene amplification or redundancy.

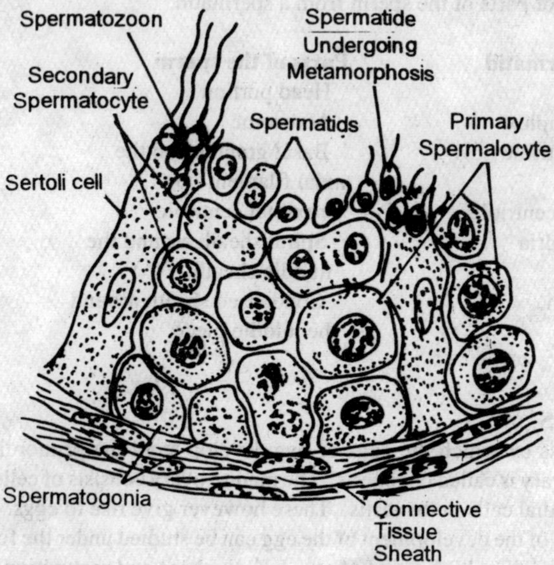

Fig. 31.1 A part of testis enlarged to show different stages of spermatogenesis and spermiogenesis

The nucleus enlarges in size due to the increased amount of nucleoplasm. The nucleolous also becomes large due to the synthesis of ribosomal RNA. At this stage in some of the amphibians the chromosomes change their shape and become giant lampbrush chromosomes.

Maturation phase: It is during this phase that the oocyte undergoes reduction division (meiosis) eventually to produce the egg or the ovum.

The maturation division of the primary oocyte differs greatly from that of the spermatocytes. While in spermatogenesis at the end of meiosis four sperms are formed, here at the end of meiosis only one large haploid egg is formed, the remaining three cells forming three small polar bodies. This unequal division results in the formation of a single egg having a large quantity of stored food that is necessary for the development of the embryo.

The first maturation division (meiosis I) reduces the chromosome into half. After the homologus chromosomes undergo chiasma formation and crossing over the nuclear membrane breaks and the chromosomes move towards the opposite poles. The nuclear division is followed by the cytoplasmic division.

This division (Cytokinesis) is unequal and results in the formation of a small polar body and a large secondary oocyte or ootid.

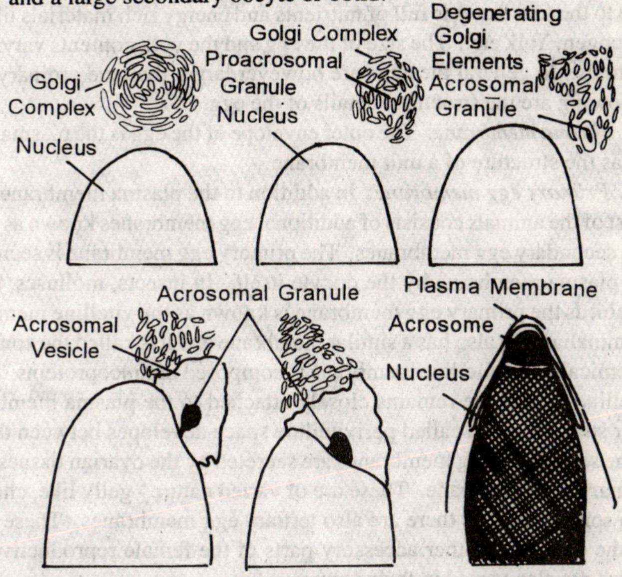

Fig. 31.2 Formation of acrosome from Golgi elements during spermiogenesis

During the second maturation division (meiosis II) the haploid secondary oocyte as well as the polar body undergo a mitotic division. As a result of this the polar body forms two polar bodies, while in the secondary oocytes the division results in the formation of a mature egg and a second polar body. Thus at the end of second maturation division there will be a large single egg and three polar bodies. These polar bodies eventually ooze out and degenerate, while the egg is ready for fertilization.

Difference between spermatogenesis and Oogenesis.

Spermatogenesis	**Oogenesis**
1. Occurs in male gonads(Testis)	Occurs in female gonads(Ovary)
2. Four spermatids are formed	Only one egg is formed along with three polar bodies.
3. Development involves three phases	Same

Structure of a mature egg:

A mature egg is several times larger when compared with a sperm. This is due to the fact that it is full of nutrients and energy rich materials like proteins, glycogen, Yolk etc., The size of the egg and the yolk contents vary in different animals. In general the eggs are however larger than the ordinary cells. The following are the structural details of the egg.

1. *Plasma membrane*: The outer envelope of the egg is the plasma membrane. It has the structure of a unit membrane.

2. *Primary egg membrane*: In addition to the plasma membrane the eggs of most of the animals consists of additional egg membranes known as the primary and secondary egg membranes. The primary egg membrane is secreted around the plasma membrane by the oocyte itself. In insects, molluscs, amphibians and birds the primary egg membrane is known as the vitelline membrane. The mammalian egg also has a similar membrane and it is called the zona pellucida. Chemically the vitelline membrane is composed of mucoproteins. Initially the vitelline membrane remains closely attached to the plasma membrane but at later stages a space called perivitelline space developes between them.

The secondary egg membranes are secreted by the ovarian tissues around the primary egg membrane. These are of varied nature - gelly like, chitinous, etc.

In some instances there are also tertiary egg membranes. These are formed by the oviduct or other accessory parts of the female reproductive structure. These are also varied in their nature.

3. *Cytoplasm*: The cytoplasm of the egg is called the ooplasm. It consists of a large amount of reserve food material stored in the form of yolk. In addition to yolk, pigment granules, water, RNA ribosomes, mitochondria and various other inclusions are also found in the cytoplasm. The outer layer of the ooplasm is refered to as the cortex and it consists of many microvilli and cortical granules. The microvilli are formed as outgrowths of the plasma membrane and they help in the transportation of the substances into the egg during its development. The cortical granules are spherical bodies of various sizes. These granules are believed to originate from the golgi complex.

Nucleus: There is a large single nucleus distributed in the cytoplasm. It has a very prominent nucleolus.

Types of eggs: The yolk content of the egg and its distribution varies in different eggs. This is used as a criterion to classify the eggs into several types.

A. Classification based on amount of yolk.

1. *Alecithal*: Here there is no yolk in the egg. Eg., eutherian mammals.

2. *Microllcithal*: Eggs have a small amount of yolks. Eg., Amphioxus.

3. *Mesolecithal*: Eggs have a moderate amount of reserve food material. Eg. Amphibians like salamander frog toads etc.

4. *Macrollcithal or polylecithal*: These eggs have a large amount of yolk. and also a little quantity of yolk free cytoplasm. Eg. reptiles, birds and prototherian mammals.

B. Classification based on distribution of yolk.

1. *Homolecithal or Isolecithal*: In this type, the quantity of yolk is very less and it is uniformly distributed all over the egg cytoplasm. eg. eggs of echinoderms.

2. *Hetrollcithal*: In these types of eggs, yolk is not evenly distributed in cytoplasm. Based on the location and area it occupies, heterollcithal eggs may be of following types.

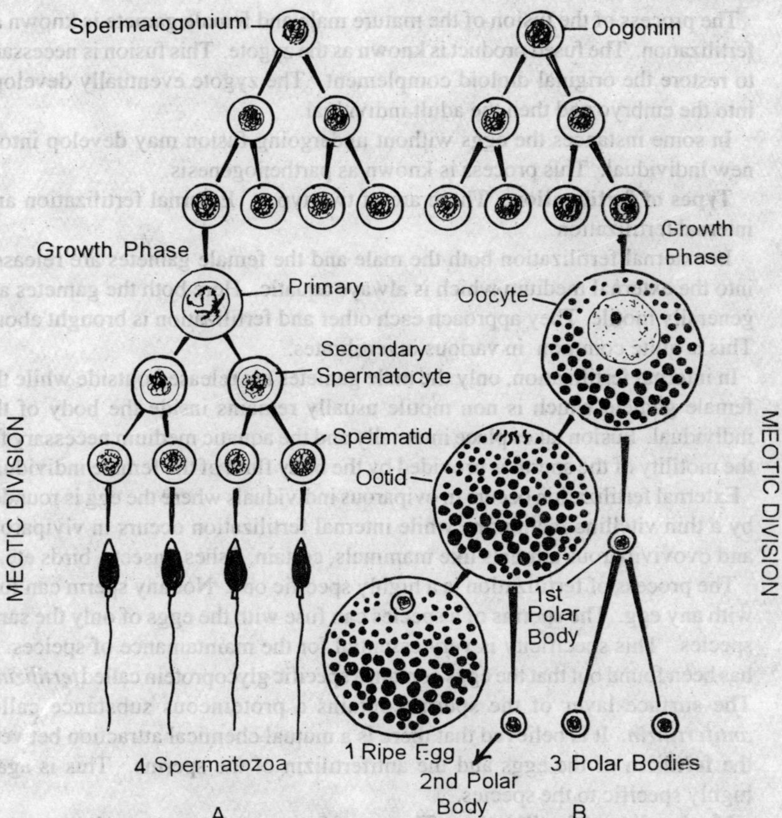

Fig. 31.3 A- spermatogenesis, B- oogenesis (after Balinshy, 1970)

a. *Telolecithal*: Here the yolk is concentrated in one half of the egg to form the vegetal pole and the yolk free portion is known as the animal pole. Eg. amphibians, reptiles, birds and prototherian mammals.

b. *Meiolecithal*: The quantity of the yolk is very large and it occupies almost the entire interior of the egg except for a small disk shaped portion of the cytoplasm. The cytoplasm contains the nucleus. Eg. Fishes, reptiles etc.,

c. *Centrollcithal*: Here the yolk occupies exactly the central portion of the egg while the cytoplasm is peripherial. Such eggs are found in insects.

Fertilization

The process of the fusion of the mature male and female gamete is known as fertilization. The fused product is known as the zygote. This fusion is necessary to restore the original diploid complement. The zygote eventually develops into the embryo and then the adult individual.

In some instances the eggs without undergoing fusion may develop into a new individual. This process is known as parthenogenesis.

Types of fertilization: These are of two types. External fertilization and internal fertilization.

In external fertilization both the male and the female gametes are released into the external medium which is always aquatic. Here both the gametes are generally motile. They approach each other and fertilization is brought about. This is quite common in various invertibrates.

In internal fertilization, only the male gametes are released outside while the female gamete which is non motile usually remains inside the body of the individual. Fusion takes place internally and the aquatic medium necessary for the motility of the sperm is provided by the body fluid of the female individual.

External fertilization occurs in oviparous individuals where the egg is rounded by a thin vitelline membrane while internal fertilization occurs in viviparous and ovoviviparous animals like mammals, certain, fishes, insects, birds etc.

The process of fertilization is a highly specific one. Not any sperm can fuse with any egg. The sperms of a species can fuse with the eggs of only the same species. This specificity is very essential for the maintainance of speices. It has been found out that the eggs contain a specific glycoprotein called *fertilizing*. The surface layer of the sperm contains a proteineous substance called *antifertilizin*. It is believed that there is a mutual chemical attraction between the fertilizin of the eggs and the antifertilizin of the sperms. This is again highly specific to the species.

Mechanism of fertilisation: The act of fusion of the sperm and the method of approach, entry, and eventual fusion of the male and female nuclei constitutes the mechamism of fertilisation. This is believed to take place in the following stages.

1. Movement of the sperm towards the egg.
2. Capacitation and contact.
3. Penetration of sperm into ovum.

4. Cortical reaction.
5. Activation of the ovum.
6. Fusion of male and female pronuclei (amphimixis)

1. **Movementof the sperm towards the egg**: This is the first step which brings the sperm in physical contact with the ovum. As the male and female gametes are produced in different individuals there are various mechanisms to bring them nearer. The initial attraction of the sperms towards the egg is supposed to be chemotactic. The sperms swim towards the egg and collide with it. Usually several sperms attach themselves to the egg.

In individuals with external fertilziation large number of eggs and sperms are released outside so that they have a favourable chance of meeting each other. In individuals with internal fertilization the sperms are discharged into the genital tract of the females by the male individual. From here the sperms move upward and reach the egg present in the uterus.

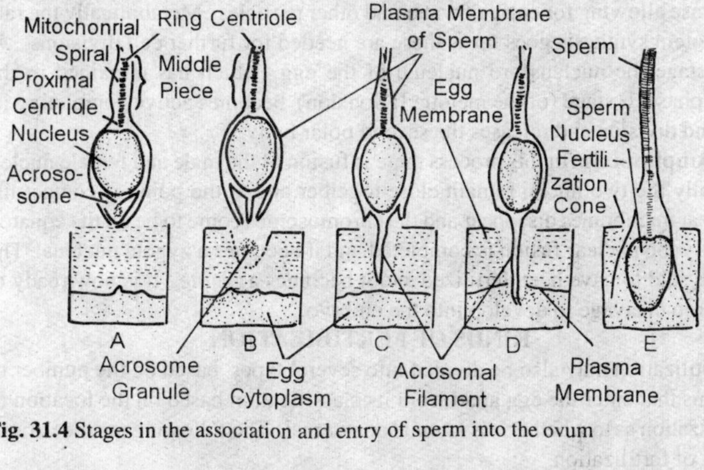

Fig. 31.4 Stages in the association and entry of sperm into the ovum

2. **Capacitation and contact**: When a large of number of sperms approach the egg for contact the fertilize in and antifertiliz in reaction ensures that only a few spermatozoa are allowed to reach the ovum. The initial attachment of the sperm to the egg is believed to be due to the chemical bonding of fertilizin and antifertilizin.

3. **Penetration of sperm into ovum**: The acrosome portions of the sperm produces some lytic enzymes called sperm lycines which have the ability to dissolve the egg membrane allowing the entry of the sperm into the egg cytoplasm. Usually only the head and the middle piece of the sperm enter the egg while the tail is left outside.

4. **Cortical reaction**: The entry of the spermhead into the egg brings about several changes in the egg surface. These are the cortical changes and the development of fertilization membrane. In some echinoderm eggs some fine granules are visible in the ooplasm after the entry of the sperm head. The vitelline membrane starts lifting itself up from the point of sperm entry and a perivitelline space is formed between it and the egg surface.

The cortical granules from the egg cortex release their contents into the perivitelline space. These contents attach themselves to the inner surface of the vitelline membrane forming what is known as a fertilization memberane. The development of the fertilization membrane prevents the entry of other sperms into the egg.

5. **Activation of the ovum**: The mature egg will be generally in a hibernating condition with very low rates of metabolism and inactive nucleus. The penetration of the sperm triggers the egg into activity. The metabolic rates increase allowing for entry of water and other particles. Metabolically the rate of protein synthesis goes up as these are needed for further cell divisions. At this stage the nucleus(pro_nucleus) of the egg which has remained in the metaphase II stage (of the meiotic II division) becomes active completing its second division and releases the second polar body.

6. **Amphimixis**: In this process there is fusion of the male and female nuclei. Initially the two nuclei remain close together and at the point of contact the nuclear membranes disappear and the chromosomes come to lie on the equator. Finally the nuclear fusion is completed and it becomes a zygote nucleus. The egg is said to have been fertilized and it becomes a *zygote*. It is now ready to undergo cleavage to develop into the embryo.

KINDS OF FERTILIZATION

Fertilization may also be divided into several types based on the number of sperms that enter the egg apart from its classification based on the location of fertilization as to whether it is internal or external. The following are the various types of fertilization.

1. *External fertilization*: Fusion in the external media.
2. *Internal fertilization*: Fusion in the internal media.
3. *Monospermic fertilization*: Only one sperm enters the egg. This is seen in most of the animals.
4. *Polyspermic fertilization*: Here many sperms enter the eggs. This is of two types -

a. *Pathological polyspermy*: This takes place under abnormal circumstances when a monospermic egg allows the entry of many sperms. In this type the egg does not develop further.

b. *Physiological polyspermy*: In animals with large yolky eggs many sperms enter the egg but eventually only one sperm nucleus fuses with egg nucleus and the rest degenerate. In this case the egg undergoes normal further development.

5. *Polyandry*: When two male pronuclei unite with a female pronucleus the condition is known as polyandry.eg. Rats.

6. *Polygamy*: Here two egg pronuclei unite with a single male pronucleus. eg. Sea urchins, rabbits etc.

7. *Gynogenesis*: In this only one sperm activates the egg but its pronucleus does not unite with the pronucleus of the egg. Eg. plannarians, nematodes etc.

DEVELOPMENT OF FROG

The egg of frog is said to be mesolecithal ie. the yolk is located in the vegetal pole of the egg, while the animal pole consists of cytoplasm and nucleus.

The ooplasm of the egg consists of many pigmented granules present in the upper region of the egg (animal pole). The vegetal pole is without any granules hence it is creamy in colour.

The egg of frog before fertilization will have a radial symmetry. After fertilization however, this changes into bilateral symmetry due to the penetration by the sperm. The sperm as it enters the egg carries along with it certain granules. As a result of this many pigment granules are carried away from the animal pole. Due to this a less pigmented area is created at the region opposite to the point of entry of the sperm and this is called the *grey crescent*. The presence of the grey crescent forms the posterior side and the opposite side forms the anterior side of the future embryo. At this stage there will be two planes in the egg. One is the *sperm enterance point plane* and the other is *grey crescent plane.*

At the initiation of the first division (cleavage) another plane is produced known as the cleavage plane which usually passes through both the animal and the vegetal poles.

The early development of the zygote consists of three main stages- cleavage, blastula and gastrula. At the end of this stage the primary germ layers are formed and from this develop all the further organs of the frog.

Cleavage: The first division of the zygote is said to be the cleavage or segmentation. This division is mitotic. The cleavage is said to be of the holoblastic type (the entire zygote divides). The first cleavage results in the formation of two cells of unequal size. Hence the cleavage is said to be of the unequal type. The pigmented half contains cells that are smaller in size than the half containing the yolk.

The first cleavage in the zygote of frog is meridional, that is it passes through both the animal and vegetal poles. The clevage begins near the animal pole and extends downwards to the vegetal poles. It appears like a shallow groove on the zygote and bisects the grey crescent area.

The two cells that are formed as a result of the first cleavage are called *blastomeres*. The formation of the first two blastomeres is completed in about 3 to 3.5 hours after fertilization.

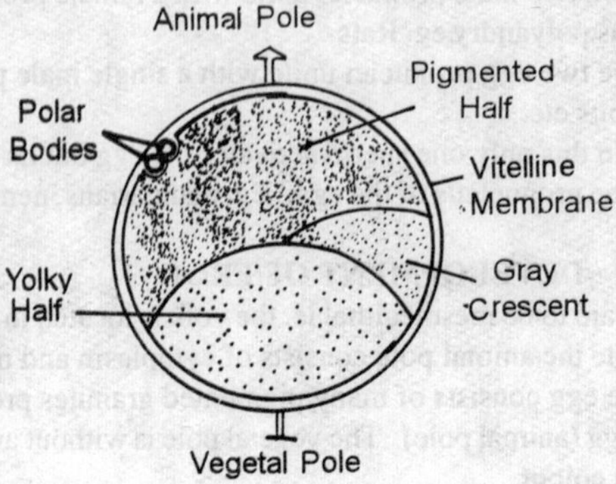

Fig. 31.5 A fertilized egg of frog (after Huetner, 1949)

The second cleavage takes place about sixty to seventy minutes after the first cleavage. This cleavage is also meridional but it takes place at right angles to the first plane of division. As a result of this division four cells or blastomeres are formed. All the four blastomeres are not identical. Two of them contain parts of the grey crescent while the other two are without it.

The third cleavage which begins about eighty minutes after the second cleavage is latitudinal(horizontal) and is at right angles to the second cleavage and passes slightly above the equator. The third cleavage results in the formation of eight cells. Of these eight cells the four cells towards the vegetal pole are larger in size and have yolk content. These are called *megameres* or *macromeres*. The four upper cells towards the animal pole are smaller in size, pigmented and are called *micromeres*. The eight cell stage is completed approximately about 5.5 hours after fertilization.

The fourth cleavage is meridionoal. And it consists of two cleavage planes passing between the first and the second cleavage. This takes place in about 20 minutes time after the third cleavage. As a result of this cleavage sixteen cells are formed of which eight are pigmented micromeres (towards the animal pole) and the remaining eight are yolk filled megameres (towards the vegetal pole). The fourth cleavage is completed approximately about six and half hours after fertilization.

The fifth cleavage is double and horizontal. It consists of two latitudinal cleavage furrows. One of these cleavages is above and the other below the third cleavage furrow. As a result of this thirty two cells are formed of which sixteen are pigmented micromeres and the other sixteen yolky megameres.

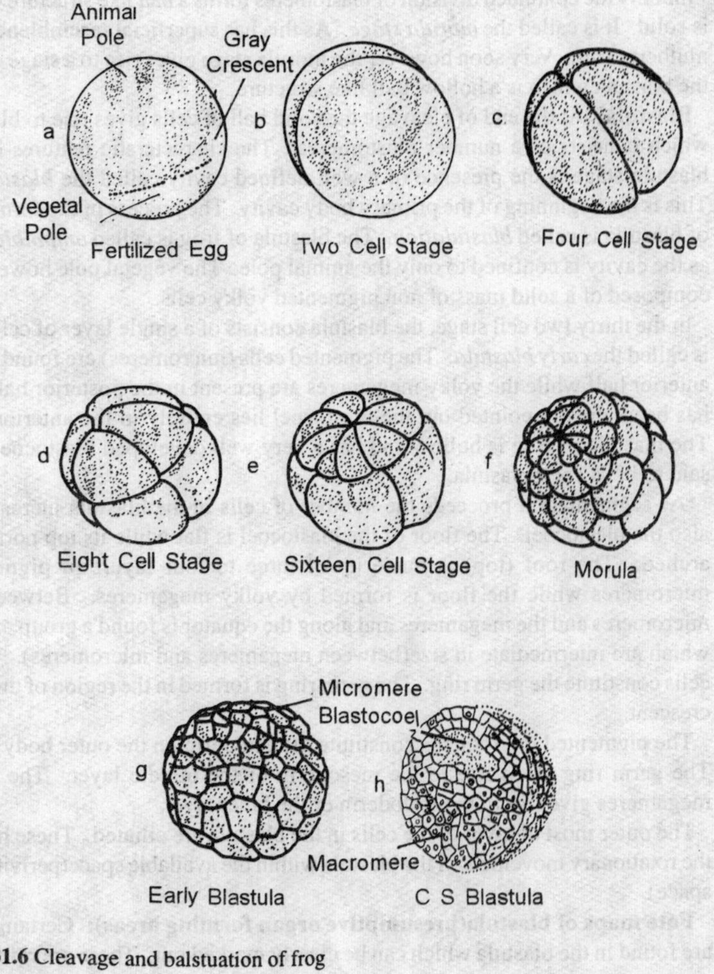

Fig. 31.6 Cleavage and balstuation of frog

These thirty two cells are arranged in four tiers of eight each. The thirty two cell stage is formed approximately about seven and half of hours after fertilization.

From this stage onwards the division becomes rather irregular (infact the unequal division begins from the third cleavage itself due to the unequal distribution of yolk which seems to determine the cleavage pattern). The rate of division also varies between the micromeres and megameres. It has been seen that the micromeres divide at a faster rate than the megameres.

Initially the continued division of blastomeres forms a ball like structure which is solid. It is called the *morula stage*. As this has superficial resemblence to a mulberry fruit. Very soon however the morula stage gives rise to a stage called the blastula which is a hollow ball like structure.

Blastula: At the end of cleavage the solid ball of cells gives rise to blastula which consists of a number blastomeres. The characterstic features of the blastula stage is the presence of a well defined cavity called the *blastocoel*. This is the beginning of the primary body cavity. The process of the formation of blastula is called *blastulation*. The blastula of frog is called *amphiblastian* as the cavity is confined to only the animal pole. The vegetal pole however is composed of a solid mass of non pigmented yolky cells.

In the thirty two cell stage, the blastula consists of a single layer of cells and is called the *early blastula*. The pigmented cells (micromeres) are found in the anterior half while the yolky megameres are present in the posterior half. As has been already pointed out the blastocoel lies entirely in the anterior half. The blastula of frog is hollow and has a very well developed blastocoel. It is said to be a coeloblastula.

As segmentation proceeds the number of cells in the blastula increase; so also the blastocoel. The floor of the blastocoel is flat while its top portion is arched. The roof (top) is made up of three to four layers of pigmented micromeres while the floor is formed by yolky megameres. Between the micromeres and the megameres and along the equator is found a group of cells which are intermediate in size(between megameres and micromeres). These cells constitute the germ ring. The germ ring is formed in the region of the grey crescent.

The pigmented micromeres constitute the ectoderm on the outer body layer. The germ ring gives rise to the mesoderm or the middle layer. The yolky megameres give rise to the endoderm or the inner layer.

The outer most of layer of the cells in the blastula are ciliated. These help in the rotationary movement of the blastula within the available space(perivitelline space).

Fate maps of blastula(presumptive organ forming areas): Certain areas are found in the blastula which can be clearly marked out. These are refered to as the presmptive organ forming areas. This nameis given because of presumption that these areas give rise to particular organs or tissues in the adult frog. The arrangement of the presumptive organ forming areas have been confirmed by marking experiments. In these experiments certain dyes are used in the early stages of division to ultimately find out which part of the future embryo consists of the known dye. By this method the presumptive organ forming areas of the blastula have been mapped and these maps are called the *fate maps.* A fate map may be defined as a topographical surface of the blastula with respect to the ultimate fate of that area in the adult frog.

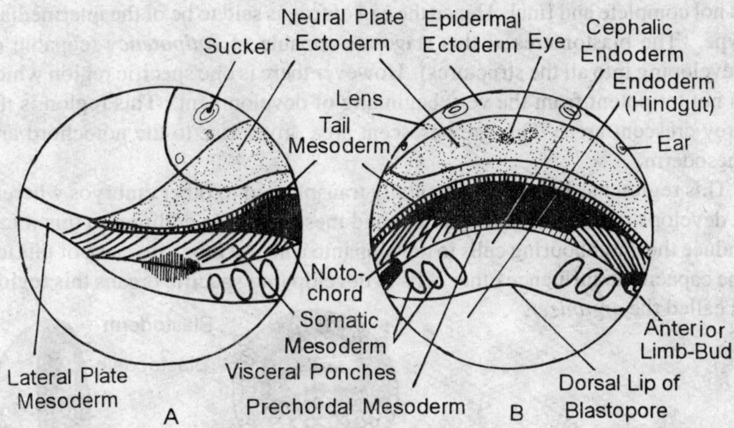

Fig. 31.7 Fate maps of blastula of frog. A-viewed from the lateral side, B-viewed from the dorsal side (after Huettner, 1949)

The significance of blastulation lies in the fact that presumptive organ forming areas are arranged in a specific plane in order to facilitate their rearrangement during the next stage of development.

The ectoderm region consists of two presumptive regions viz., the presumptive epidermis and the presumptive neural plate region. While the presumptive epidermis is present on the anterior ventral and lateral parts of the animal hemisphere, the presumptive neural plate occupies the dorsal and lateral part of the vegetal hemisphere. The other organ forming areas are

1. Head develops from the anterior parts of neural plate and epidermal region.
2. In the epidermal region of the head are found the presumptive areas of ears.
3. The presumptive eye region occupies the mid dorsal part of the neural plate.

The mesoderm forms a ring between ectoderm and endoderm. This is divided into presumptive notochordal and somite region.

The endoderm occupies the posterior half of vegetal hemisphere of the blastula.

Though the presumptive organ forming regions are marked off in blastula it is not complete and final. Hence the egg of frog is said to be of the intermediate type. The blastomeres at this stage are capable of *totipotency* (capable of developing into all the structures). However there is one specific region which is not totipotent from the very beginning of development. This region is the grey crescent area. This grey crescent area gives rise to the notochord and mesoderm.

This region has been experimentally transplanted to other embryos wherein it develops into the same notochord and mesoderm. In addition to this it can induce the neighbouring cells to develop into a neural tube. Because of this ie., the capacity to influence other cells to develop into specific organs this region is called the *organizer*.

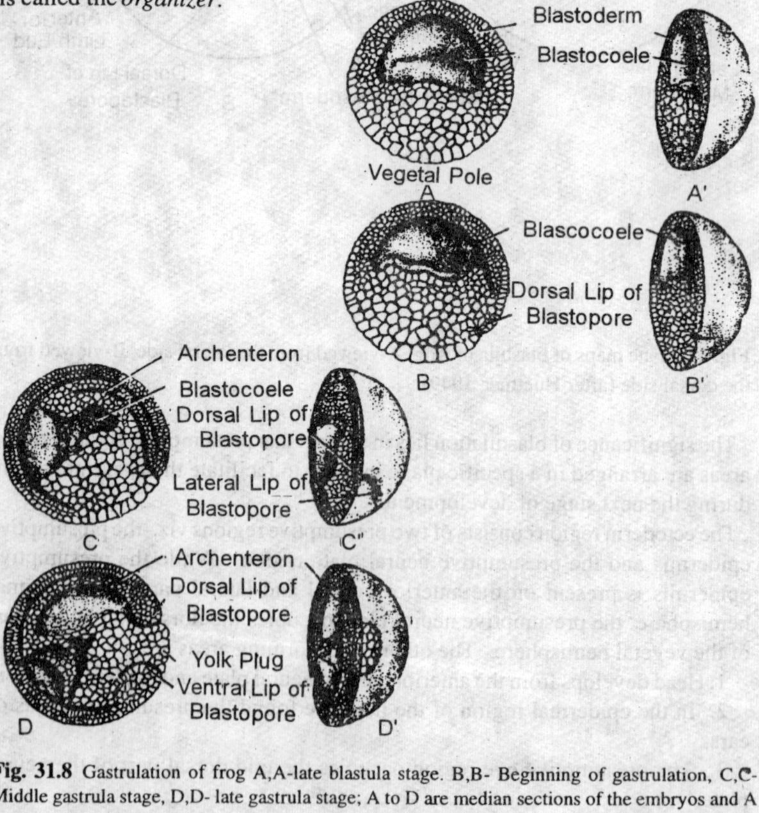

Fig. 31.8 Gastrulation of frog A,A-late blastula stage. B,B- Beginning of gastrulation, C,C- Middle gastrula stage, D,D- late gastrula stage; A to D are median sections of the embryos and A to D are dorsal views of respective sections (after Balinsky, 1970)

GASTRULA

The process of formation of gastrula is called gastrulation. In the beginning, the gastrula is a two layered stage with a cavity called *archenteron*. This archenteron opens to the exterior by means of an opening called *blastopore*.

As a result of gastrulation three primary layers are formed. These are the ectoderm, the endoderm and the mesoderm. First to develop are the ectoderm and the endoderm and eventually the mesoderm is also formed. These three layers are called the primary germ layers from which arise all the organs of the frog.

Gastrulation is a very significance and important stage in the embryogeny of frog. Because it is ' ere that all the future organ forming areas are differentiated.

The early gastrula consists of the formation of the archenteron. During the process of gastrulation there are mass movement of subcells which are necessary for rearrangement of the presumptive organ forming areas of the blastula. These movements are called formative movements. The process of gastrulation takes place at the following stages.

Epiboly: In the late blastula, the anterior half consists of micromeres which constitutes the ectoderm while the posterior megameres constitute the endoderm. The germ ring forms the mesoderm. During epiboly the ectoderm overgrows backwards on the endoderm; ultimately the entire embryo (except for the small area called the yolk plug) is covered by the ectoderm. In other words the pigmented micromeres (animal half) grow over the megameres(vegetative half). The reason for overgrowth is the rapid rate of division of micromeres.

Invagination: A small depression is formed in the region occupied by the grey crescent area. This depression grows inwards and forms the archenteron or gastrocoel or secondary body cavity. The outer opening of the gastrocoel is called the gastropole. As the gastrocoel increases in size the blastocoel gets reduced. Ultimately only a slit like semicircular cavity indicates the remnants of the blastocoel. The blastopore meanwhile becomes expanded and becomes ring shaped.

Involution: During this process the cells which have grown backward during epiboly now roll inside at the margin of the blastopore. The endoderm is the first to roll inside. The cells of the notochord and mesoderm which were formed outside now migrate over the lip of the blastopore and become internal and arrange themselves on the roof, sides and the floor of the archenteron. The notochord cells are found on the roof along the midline. While the endoderm forms the anterior, lateral and ventral walls, the mesoderm forms wing like extentions in the archenter.

Closure of the blastopore: The archenter at its beginning is marked bv a groove. The anterior margin of this groove is called the dorsal lip of the blastopore. This groove also extends laterally to form the lateral and ventral lips. At this stage the blastopore is completely ring like, surrounding the endoderm. The circular area of the endoderm is represented by a mass of cells called the yolk plug. The margins of the lips of the blastopore contract making the opening very small. At this stage the yolk plug has an oval outline and eventually becomes a small spec and is completely covered over by the ectoderm cells.

Rotation: At the end of gastrulation the yolk filled cells get rearranged. This rearrangement shifts the centre of gravity of the gastrula and results in a rotation of eighty to ninety degrees. The blastopore moves backwards and is finally located slightly above its original location at the beginning of the gastrula.

Convergence: During this process there are two kinds of cell movements. First, there is shifting of the cells of the notochord and mesoderm towards the dorsal lips. Secondly, there is a dorsal convergence of notochord and mesodermal cells towards the middorsal area of the blastopore.

Extension or elongation: This includes the growth in length of the embryo. As a result of extension the spherical gastrula becomes elongated in shape.

At the end of gastrulation three germ layers are formed. These are the ectoderm, the endoderm and the mesoderm. All the future organs of the frog are developed from this. This theory which states that organs of similar type arise from each of the three germ layers in different animals is called the *germ layer theory*.

Fate of germ layers: The three germ layers give rise to the different parts of the adult frog. The ectoderm at the end of gastrulation covers the entire embryo. It is divided into the neural plate and the epidermis. The neural plate gives rise to the neural tube and the process is known as neurulation. The epidermis forms the outer layers of the skin, the sense receptors and the lining of stomodaeum and proctodaeum. The mesoderm initially forms three regions namely epimere, mesomere and hypomere. The endoderm is derived from the cells surrounding the archenter. Eventually it forms the inner lining of the alimentary channal.

The following table gives a detailed listing of the various derivatives of the germlayers.

Ectoderm -	*Epidermis*	1. Skin epidermis 2. Lining of Stomodaeum 3. Lining of Proctodaeum 4. Lense of eye
	Neural tube	1. Central nervous system 2. Optic vesicle of eye
	Neural Crest	1. Peripheral nervous system 2. Gangilia of central and symphathetic nervous systems. 3. Pigment cells

4. Medulla of adrenal
5. Skeletal elements of visceral arches

Endoderm - 1. Lining of alimentary canal (except stomodaeum and proctodaeum)
2. Pharyngeal pouches and derivatives like thymus, parathyroid, middle ear cavity etc.
3. Thyroid gland
4. Respiratory system - Lungs and gills
5. Liver and Pancreas
6. Reproductive cells

Mesoderm - *Epimere* 1. Dermis
2. Muscles of limbs
3. Straited muscles of body wall
4. Certain muscles of the head including tongue muscle
5. Axial skeleton

Mesomere 1. Excretory system
2. Part of genital system

Hypomere 1. Muscles and skeleton of limbs
2. Parietal peritonium
3. Dermis of skin including ventral part of trunk and apendages
4. Muscles of heart, viscera and blood
5. Skull and visceral skeleton

Mesenchyme 1. Muscles of head
2. Connective tissue inclusive of blood
3. Innerlining of the heart and the blood vessles
4. A portion of the dermis

Notochord 1. Early axial skeleton which eventualy forms the centra of the vertebrae

Thirty two
Human reproduction

In human beings the sexes are separate i.e. the male and female individuals can be distinguished both by their primary sexual characters as well as by secondary sexual characters. The primary sexual characters include the sex organs namely - a pair of testis in the male and a pair of ovaries in the female. These are the basic sexual organs which produce the sperm as well as the ovum or the egg. In this chapter we will study the structure of the sperm and the ovum, the process of fertilization and the initial development upto the stage of implantation in the uterus.

Structure of the sperm

The sperm in human beings is a highly specialised structure. Also called male gamete, several thousands of them are produced and contained in a fluid called semen before it is transferred to the female for the process of fertilization. A mature sperm (also called spermatozoa) consists of the following four parts

1. Head - The head is almost conical in shape and is formed of *acrosome* and *nucleus.*

Acrosome- This is found at the anterior tip of the sperm (Gr. *akron*= extremity ; *soma* = body). The acrosome forms a cap like structure called the head cap. This occupies the space between anterior half of the nucleus and the plasma membrane of the sperm tip. In its origin (during spermatogenesis), the acrosome is formed from the golgi complex. The acrosome itself is bounded by a unit membrane. It consists of a number of hydrolytic enzymes such as acid phosphatase, hyaluronidase and others. These enzymes help in tissue lysis (dissolving) and this facilitates the penetration of the sperm into the egg membrane. The enzymes are proteolytic and help in dissolving the egg membrane.

Sperm nucleus- The nucleus occupies most of the available space of the sperm head. It is the shape of the nucleus that ultimately decides the shape of the sperm head. Structurally it is enveloped by a nuclear membrane. Some times how ever the posterior part of the nuclear membrane (towards the body of the sperm) is somewhat depressed to accommodate the proximal centriole. The nucleus consists of DNA as well as basic proteins. There is no nucleolus or any fluid contents.

2. Neck - The head is followed by a short neck to separate the middle piece of the sperm. The neck consists of just two granules (centrioles). These granules are called the *proximal centriole* and the *distal centriole.* Both these granules are situated very close and lie in the posterior depression of the sperm neck. The two centrioles enter the egg at the time of fertilization along with the nucleus.

These two centrioles are necessary to initiate division in the zygote. It is known that the centrioles help the zygotic division by forming the first mitotic spindle. The posterior or the distal centriole is responsible for the formation of the microtubules of the sperm tail.

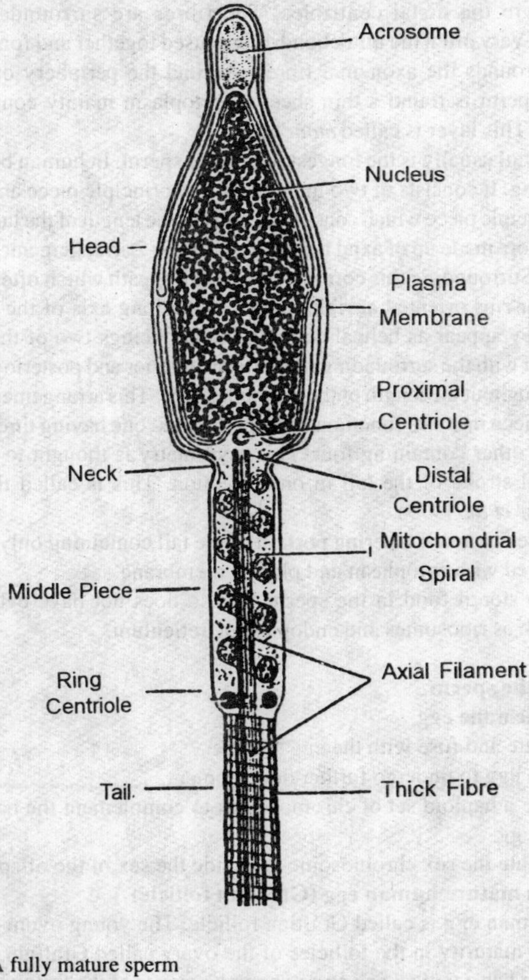

Fig. 32.1 A fully mature sperm

3. Middle piece - The middle piece of the sperm consists of the upper portion of the axial filament and in its structure it has the same organisation as the axial filament of any flagellum. It has a pair of longitudinal fibres called beta fibres surrounded by a ring of nine pairs of longitudinal fibres called alpha fibres. In human sperms, the alpha fibres of axial filament are accompanied on the outside by 9 much thicker fibres called gamma fibres or coarse fibres. The alpha, beta and gamma fibres are the sites of various enzymes. For instance alpha fibres have ATPase enzyme, while beta fibres have acetylcholinesterase enzyme. The gamma fibres have ATPase, spermosin and succinic dehydrogenese. These fibres are anchored to the distal centrioles. The fibres are surrounded by the mitochondria. Very often the mitochondria are fused together and form a spiral sheet that surrounds the axonemal fibres. Around the periphery of the mid piece of the sperm is found a thin sheet of cytoplasm mainly composed of microtubules. This layer is called *manchette.*

4. Tail - The tail usually is the longest part in the sperm. In human beings it is about 55 m long. It consists of two main parts- the principle piece and the end piece. The principle piece which constitutes most of the length of the tail consists of the central core made up of axial filaments with a 9+2 arrangement(2 central, 9 peripheral). Surrounding this core is a fibrous tail sheath which often appears as semicircular ribs oriented at right angles to the long axis of the filament. Sometimes they appear as helical coils. In human beings two of the gamma fibres are fused with the surrounding ribs to form anterior and posterior columns extending throughout the length of the principle piece. This arrangement divides the principle piece into two functional compartments- one having three gamma fibres and the other containing four. This asymmetry is thought to help in a more powerful stroke of the tail in one direction. This is called the power stroke(Wheater *et al,* 1979).

The end piece is a small tapering portion of the tail containing only the axial filament covered with cytoplasm and plasma membrane.

There is no stored food in the sperm. It also does not have cytoplasmic organelles such as ribosomes and endoplasmic reticulum.

Functions of the sperm

1. To swim near the egg.
2. To penetrate and fuse with the egg surface.
3. Induce the egg to undergo further development.
4. Contribute a haploid set of chromosomes to complement the haploid set present in the egg.
5. To contribute the sex chromosome to decide the sex of the offspring.

Structure of a mature human egg (Graffian follicle)

A mature human egg is called Graffian follicle. The young ovum develops and undergoes maturity in the follicles of the ovary called Graffian follicles. The mature egg has a typical cell structure and consists of the following parts- The outermost layer is a double layered connecting tissue sheet and it is differentiated into two layers namely *theca externa* and *theca interna.* The theca extrena is the outer fibrous layer. It is made up of spindle shaped cells and fibres. The theca interna is composed of the inner vascular layer and it is supplied by a number of blood capillaries. In a mature follicle there is a fluid filled cavity which is eccentric and it is called antrum. The fluid itself is called *liquor follicle.* The antrum is lined by a membrane called membrane granulosa. Between the theca interna and membrane granulosa is a homogenous basement membrane called glassy membrane.

Internal to the theca interna is a membrane called zona pellucida. This is comparable to the vitelline membrane. Surrounding the zona pellucida is a region called corona radiata. This is formed from the other follicle cells.

The cytoplasm of the egg is known as the ooplasm. The ooplasm consists of a large amount of reserve food materials. There is very little yolk however in the cytoplasm. The peripheral part of the ooplasm is known as the cortex. It consists of many microvilli. The microvilli are formed as protruberances of the plasmalemma (plasma membrane). The microvilli help in the transportation of the substances from follicle cells to the egg during its development. The cytoplasm consists of a number of mitochondria, lipoproteins, pigment granules, ribosomes, golgi complex and a number of other cell inclusions. The egg consists of a single nucleus which is usually vesicular in shape.

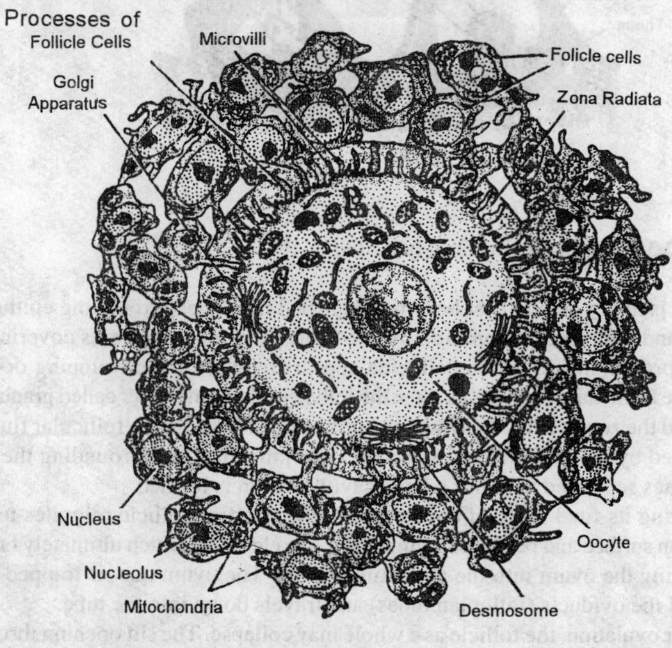

Fig. 32.2 Young oocyte of a mammal surrounded by follicle cells (after Balnisky, 1970)

Ovulation

Ovulation refers to the liberation of the ovum from the follicle. In human beings ovulation takes place in the middle of the menstrual cycle and it occurs only once in a month. Very rarely there might be two ovulations or multiples which result in the formation of twins and triplets.

In ovulation the primary follicle develops into a Graffian follicle and releases the egg in the follicle. The follicle stimulating hormone(FSH) secreted from the anterior pituitary, initiates ovulation along with several other factors including one more hormone luteinizing hormone (LH).

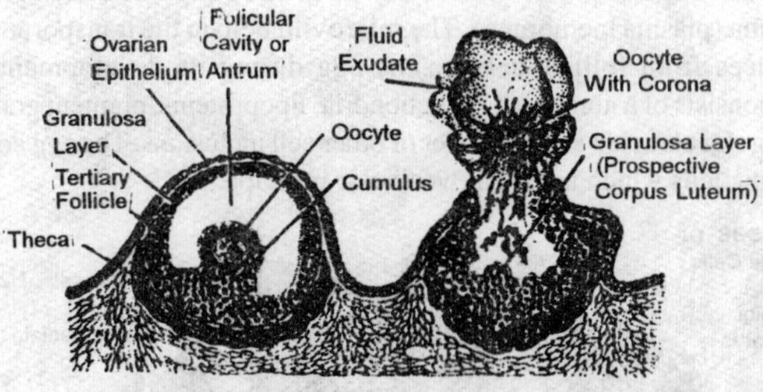

Fig. 32.3 Ovulation in a mammal (after Torrey, 1971)

The primary oocyte first becomes detached from the surrounding epithelial cells and undergoes maturation. At the same time, a homogenous covering of mucoprotein namely zona pellucida appears between the developing oocyte and the follicular cells. At this stage the follicle cells form a layer called granulosa around the zona pellucida. Under the influence of hormones a follicular fluid is secreted by the follicular cells and the cells immediately surrounding the egg becomes separated and thus a large cavity antrum is formed.

During its final stage of development, the Graffian follicle migrates to the ovarian surface and produces a thin blister like elevation which ultimately bursts releasing the ovum into the abdominal cavity. The ovum now is trapped into one of the oviducts (fallopian tubes) and travels down into the tube.

After ovulation, the follicle as a whole may collapse. The slit opening through which the egg and follicular fluid passed during ovulation very soon gets filled with a mass of coagulated blood and follicular fluid. This collapsed ovarian follicle shrinks and becomes filled with yellow coloured cells to form a solid body called the *corpus luteum.* This later secrets the hormone progesterone which helps in the development of endometrium in the uterus.

The ovum which does not get fertilised in the fallopian tube within 48 hours of ovulation will get degenerated. The corpus luteum which is meant to support the fertilized egg also degenerates and becomes *corpus albicans* which appears as disorganised globules. Simultaneously the hormone production necessary for pregnancy will also stop and the whole of the structure is discharged during menstruation and a new cycle starts.

Fertilization

In human beings the egg or the ovum should be fertilised as it is moving along the fallopian tube. During copulation about 100 to 200 million sperms are introduced into the vagina of the human female along with the discharge of the semen from the glans penis. This process is called insemination. The sperms become active and move in all directions but most of them move up, reach the cervix and then into the uterus. The contraction of the muscles during intercourse helps in sucking the sperms into the uterine opening. From the uterus the sperms find their way into the fallopian tubes where they meet the ovum. The meeting place between the egg and the sperms is usually the upper part of the fallopian tube. Many sperms reach the ovum but only one ultimately gains entry into the egg fuses with it and forms the zygote.

The process of penetration of the sperm into the egg membrane is helped by the secretion of membrane lytic enzymes by the acrosome of the sperm. After the dissolution of the membranes the male pronucleus enters the egg and reaches the female pronucleus. When the two nuclei come into close contact the nuclear membranes of the two pronuclei disappear and the male and female set of chromosomes arrange themselves together preparing for the zygotic division. The fusion of the male and female pronuclei is called amphimixis.

After its fertilization a zygote moves down reaches the uterus and becomes implanted in the endometrium.

Early development and Implantation

The human egg is very small and has practically very little yolk. Hence it is called Microlecithal. Whatever little yolk is present is distributed uniformly throughout the egg. Hence the egg is called Isolecithal or Homolecithal.

Early development includes three stages mainly clevage or segmentation, morula stage and blastula stage. At the blastula stage there is implantation.

1. *Clevage*- The human zygote undergoes clevage during its movement along the fallopian tube. The clevages in zygote are said to be holoblastic as the clevage is equal. The first clevage occurs usually within 30 hours after fertilization. The clevage furrow is vertical passing through the animal pole to vegetal pole. The two daughter cells so formed are called *blastomeres.* The second clevage is also vertical but at right angles to the first one forming four blastomeres of equal size. Subsequent divisions occur in a rapid but orderly fashion producing a number of blastomeres.

2. *Morula* - After repeated clevages, the embryo takes the form of a solid sphere of cells looking like the fruit of the mulberry hence this stage is called morula (a little mulberry) stage. The morula has 16 to 32 cells and is larger than the zygote. It has two types of cells. In about 4 to 6 days after fertilization the morula reaches the uterus. It is still surrounded by the zona pellucida which prevent its sticking to the uterine wall.

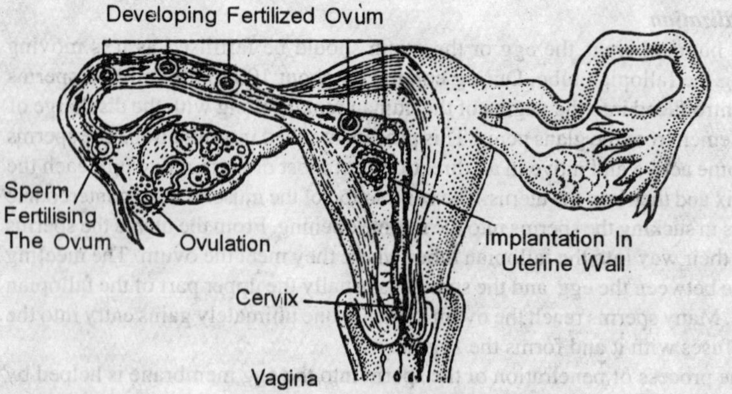

Fig. 32.4 Site of fertilisation of egg and implantation of embryo

3. *Blastula* - The outer layer of cells of morula absorb nutrition from the uterine wall and is now referred to as trophoblast. The cells of trophoblast are flattened and the fluid absorbed by this trophoblast gets collected in the cavity called the *blastocoel.* At this stage the trophoblast becomes the blastocyst and this stage of development is known as the blastula. The inner cell mass present in the blastocyst looks like a small knob at one pole and it is called the embryonal knob. The trophoblast (outer layer of blastomeres) does not take part in the formation of the embryo proper. It helps in the formation of the extraembryonic membranes. The cells of the trophoblast that remain in contact with the embryonal knob are called *cells of Rauber.*

Implantation

Implantation refers to the attachment of the developing balstocyst into the uterine wall. This is necessary for two reasons- 1. to fix the blastocyst to the uterine wall so that it does not slip down and 2. to absorb nutrition from the uterine wall which is necessary for its growth until the time of parturition (birth). Implantation usually takes place about 5 to 7 days after ovulation. The part of the blastocyst which has the embryonal knob lies against the endometrium (inner wall) of the uterus. By now the zona pellucida around the blastocyst disappears enabling the blastocyst to come into direct contact with the endometrium. This contact triggers rapid division of the trophoblast cells which destroy the endometrium cells. The destruction is partly physical and partly enzymatic. As a result of this the blastocyst sinks into a small depression formed in the endometrium. Subsequently the blastocyst will be completely surrounded by the endometrium which grows over the blastocyst. The blastocyst produces a number of projections all around to obtain the nourishment.

The endometrium which overgrows the blastocyst becomes rich in vascular supply and appears spongy. It is now called *decidua(*to shed) and refers to the pregnant endometrium. The different regions of the decidua are assigned different names and these are called -

1. Decidua basalis(decidua serotina) - This refers to that part of endometrium which is directly underlying the embryo.

2. *Decidua capsularis*(decidua reflexa) - This is the thin portion of the endometrium that covers the implantation region.

3. Decidua parietalis (decidua vera) - This refers to the endometrium lining the uterus other than the implantation region.

The pattern of implantation of the blastocyst varies in different mammals. In human beings it is said to be *interstitial* as the blastocyst burrows into the endometrium . The other two types of implantation are eccentric and superficial.

Foetal membranes (extra embryonic membranes)

These are the membranes that develop outside the embryo but in close association with it and they carry out certain specific functions. In human beings foetal membranes are amnion, chorion, yolk sac and allantois.

1. *Amnion-* This is formed above the embryo. This consists of a cavity(amniotic cavity) and encloses a fluid called amniotic fluid. The embryo is suspended into the amniotic cavity by the umbilical cord. The aminotic fluid provides a shock absorbing effect to the embryo against bumps, infections etc. The watery fluid around the embryo helps in maintaining constant temperature and pressure and protects the embryo in case the mother has a fall. The aminotic fluid is derived from the mothers blood and contains foetal cells. This is made use of in the prenatal sex test for the foetus known as *amniocentesis*. In amniocentesis the amniotic fluid is drawn out with a syringe and the cells are tested for the presence of the sex chromosomes.

2. *Chorion* - The chorion completely surrounds the embryo and has small projections all around it during early stages of development. The chorion is composed of trophoblast on the outside and mesoderm on the inside. Chorion protects the embryo and forms placenta for metabolic exchange between the mother and the foetus.

3. Yolk sac - This is formed below the embryo. In human beings this contains a fluid but no yolk. It is a vestigial organ. Its wall is made up of trophoblast and endoderm. The yolk sac functions as the region of formation of blood cells upto about 6th week of development when the liver of the foetus takes up this function.

4. Allantios- This is a small bag like structure that develops from the gut of the embryo and near the yolk sac. This membrane develops around the third week of development. Gradually the allantios shrinks in size and gets enclosed in the umbilical cord. Allantios helps in the formation of umbilical arteries and veins. The allantios also forms blood cells.

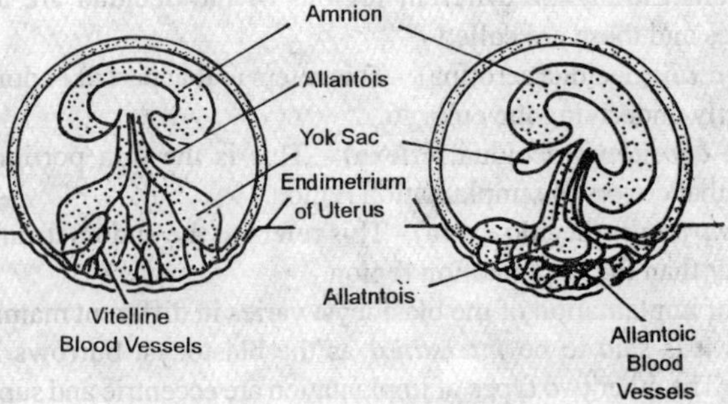

Fig. 32.5 Comparison of two types of mammalian placentae, A-chorio-vitelline placenta, B-chorio-allantoic placenta (after Philips, 1975)

PLACENTA - In viviparous animals the embryo develops inside the uterus of the mother and only a fully formed offspring comes out of the mother. This is because the amount of nutrition (yolk) stored in the egg is insufficient for the development of the embryo. Therefore the developing embryo has to depend on the maternal tissues for its nourishment. The embryo's which develop inside the uterus of the mother will be attached to the uterine wall to obtain the necessary nourishment. This attachment of the embryo to the uterus will be through an organ called placenta.

Basically the placenta has two parts from the point of view of its origin. These are foetal placenta provided by the extra embryonic membranes of the embryo and maternal placenta provided by the endometrium of the uterus.

Placenta can be classified into various categories based on their origin, on their morphological structure, on the distribution of villi and also on their histological structure.

The placenta of human beings is called allantoic or chorio allantoic placenta . Here the yolk sac is rudimentary. The allantois is very well developed, becomes vascularised and fuses with the chorion . Here the chorion bares a number of outgrowths called villi which project into the maternal tissue to help the absorption. Each villus has an outer ectoderm formed from chorion and the endoderm formed by the allantois. Between the two is a highly vascularised mesoderm consisting of the connecting tissue. The mesoderm is dual in origin. It is made up of both chorionic as well as allantoic parts.

The maternal part of placenta consists of the outer layer of the endomertirum and also the stroma with its associated glands and the uterine blood vessels.

The chorio-allantoic placenta has been classified into various types on criteria such as morphological structure, anatomical structure etc. Based on the morphological structure the chorio-allantoic placenta of human beings may be classified as deciduous placenta or placenta vera. Here the association between the foetal and maternal tissues is very intimate. The uterine wall gets eroded due to the action of the penetration of embryonic tissues to facilitate two way transportation of materials (from mother to foetus and *vice versa*). Here the chorionic villi fuse with the uterine mucosa to bring out better communication. At the end of pregnancy the entire placenta comes out causing extensive haemorrhage from the uterine wall during child birth. This haemorrhage is due to the intimate association of foetal and maternal tissues.

Based on the anatomical structure the human placenta is designated as haemo-chorial placenta. Here there are only three layers from the side of the foetus. The three layers from the side of the mother disappear. In such cases the endothelial walls of the uterine blood vessel disappear and the chorionic villi directly come into contact with the maternal blood. The chorionic villi are surrounded by sinuses (spaces) without any endothelial covering and into this the maternal blood enters through the arteries of the uterus and the blood from this flows back into the uterine vein.

Functions of Placenta

1. Nutrition- Placenta acts as a link between the foetus and the mother helping in the transportation of nutrients from the latter to the former.

2. Respiration - Placenta helps in the exchange of gases such as carbon dioxide and oxygen between the mother and the foetus thus playing an important role in foetal respiration. Oxygen from the maternal blood diffuses into the foetal blood while the carbon dioxide from the foetal blood is taken into the maternal blood.

3. Excretion - The waste products produced due to the metabolic activities of the foetus are transferred to the maternal blood with the help of the placenta.

4. Immunity - The maternal blood provides antibodies against several diseases to the foetus to keep it healthy.

5. Transport of drugs - Some of the drugs taken by the mother during pregnancy can pass on into the blood of the foetus through the placenta. Many a time certain drugs taken for ailments of the mother may harm the foetus. Pregnant women should be very careful in the use of drugs even for small ailments. For instance thalidomide which is used as a sedative to prevent morning sickness is known to cause many deformities in the foetus.

6. Storage of nutrition - Placenta can store materials like fat, glycogen, iron etc.

7. Secretion of hormones - The placenta can secrete a number of hormones such as progesterone, estrogen, gonadotropin etc., which are helpful in maintaining pregnancy.

8. Transport of pathogens - Microbial pathogens such as viruses can enter the foetus through the placenta. Particularly those causing syphilis, measles, small pox are known to enter the blood stream of the foetus if the mother gets infected during pregnancy.

Role of sex hormones

A number of hormones are secreted by the human body to initiate, maintain and complete the process of sexual reproduction.

Basically two hormones secreted by the adenohypophysis of the pituitary gland to control the process of sexual reproduction. Generally called gonadotropins there are two types of hormones which promote the sex organs to produce sex hormones. The two gonadotropins are follicle stimulating hormone (FSH) in females and Luteinizing hormone (LH). The LH in the case of females stimulates ovulation while in the case of males it stimulates the production of male hormones called androgens.

Female Hormones (secreted by the ovary)

Under the influence of FSH the ovary secrets two hormones namely estrogen and progesterone. Estrogen is necessary to initiate the menstrual cycle. It also helps in the development of secondary sexual characters. The progesterone is a hormone that is necessary to maintain pregnancy. It helps in the formation of endometrium to allow for the implantation of the egg.

Male Hormones (secreted by the testis)

Under the influence of LH, the testis in human males secrets hormones generally called androgens. Testosterone is a hormone secreted by the interstitial cells of the testis and helps in the development of the secondary sexual characters. Besides it also helps in the maturation of male sex organs.

The placenta is also known to produce several hormones which are essentially meant to suppress fresh ovulation during pregnancy and continue the pregnancy. Some of these hormones are progesterone, estradion, human chorionic gonadotropin (HCG) and human chorionic growth hormone prolactin(CGP). It is known that CGP induces prolactin in the development of mammary glands in females. The corpus lutuem is helped by HCG.

Need for Birth control

The history of human beings is about 50,000 years old. Ever since his origin, man because of his intellectual superiority has dominated both the environment as well as other organisms and has tried to use both of them for his benefit. This has resulted not only in altering of the environment but gradual increase in the number of human beings. In the remote past when man had lived the life of nomadic, the increase in his population was gradual. But as man started conquering the environment, population increase has been geometric.

Statistical study shows that during the last 40 years there has been a steep increase in world population but there has been no corresponding increase in resources. With the birth of every new child the area of the earth does not increase even by one square inch. The following table gives the population size during the last 150 years.

Table World population

year	Population (in millions)
1850	1000
1925	2000
1965	3000
1980	4000
1990	5000
2000	6500-7000

As one can observe from the above table it took nearly 75 years (1850-1925) for the human population to double from one billion to two billion, but in the next 65 years it has increased to 5 billion and the projected rate of growth will have more than 6.5 billion by the turn of the 20th century.

The reasons for such rapid increase of population is the tremendous advancement in the medical field which has almost conquered the infectious diseases and has brought the death rate down. Life expectancy has increased in all countries.

The problem of population increase : The problem of uncontrolled increase in population, particularly in developing countries is indeed enormous. Overpopulation brings into naught all efforts to increase the standard of living. Medicare facilities, drinking water supply, dwelling houses, and all the basic necessities of life become totally inadequate. Pauncity of housing and poverty leads to slum dwelling. With no sanitation facitlities and too many people in too small a place leads to the emergence of epidemics. Sociologically the problems caused by population increase are many- education and employment will be wanting, unemployment and illiteracy leads to frustration and ultimately people turn to criminal activities. Further, even within a family, if there are many children, they will not get adequate attention, nourishment and education.

When such a dissatisfied child grows into an adult he will be a problem in society. On the other hand a small family will be a nucleus of happiness. Hence population control is a must for any country and more so for a developing country like India.

Indian population has increased enormously and while it is the 5th largest country in the world in terms of geographic area, in terms of population it is second. *It has been estimated that every year India adds on to the population as many human beings as there are in Australia.* This increase in population if remains unchecked will bring (it has already brought) untold miseries. Hence planned growth of population at the society level and planned growth of family at the individual level is a must. This is called family planning or birth control.

A survey of family planning methods:

Family planning refers to planned growth of the family i.e., it is *children by choice and not chance.* In order to have children only when the parents wish, a number of methods are followed. These are usually refered to as contraceptives as they prevent conception or pregnancy. An ideal contraceptive should be 100% effective, simple, safe, acceptable and inexpensive. Various contraceptive methods can be discussed under the following heads.

1. Control of sexual intercourse
2. Mechanical barriers to prevent the entry of sperms
3. Use of spermicidal chemicals
4. Intra-uterine devices
5. Surgical methods
6. Pharmacological approach

1. Control of sexual intercourse :

*a. Withdrawal method :*Here the male organ is withdrawn from the vagina just before the discharge of the semen. This however is not only fool proof but may lead to psychic disturbances.

b. Rhythm method (for females) : This method depends on the avoiding of intercourse and non availability of sperms during the fertile period (ovulation period). In this method where the normal menstrual cycle is of 28 days, ovulation occurs on the 14th day starting from the beginning of the menstrual period. Since the ovum can be fertilized only when it is traveling in the fallopian tube it will be comparatively safe to have intercourse a week after the period and a week before the period. The middle 10 days can be regarded as fertile period when conception can take place.

Rhythm method is a natural method as it does not involve any artificial devices. However it is not 100% fool proof as the menstrual cycle may vary in women. As the cycle is under hormonal control, stress, strain and other psychological factor may alter the duration. Hence it is difficult to calculate the safe period.

c. Rhythm method based on body temperature : In this method women have to monitor their body temperature daily. The body temperature falls before ovulation and rises after ovulation. Women have to avoid intercourse from 5th day after menstruation is over to until 3 days after the temperature has risen. The body temperature is recorded using a special thermometer.

Another variation of the temperature method is to avoid intercourse every month until 3 days after the rise in body temperature. The disadvantages in this method are- body temperature has to be carefully measured every day and it many not be totally safe.

2. Mechanical barriers to prevent the entry of sperms

*a. Condom :*This is of three types- latex rubber, membrane (skin) and rubber cap (American type). The latex rubber type is available in several brand names and is a rubber sheath. It is rolled over and has to be put over the male sex organ in order to prevent the entry of sperms during intercourse. Some condoms have a teat at the tip to allow for the collection of the sperms. In addition to physically preventing the sperms from entering the uterus, condom also gives protection to the individual from contacting communicable diseases such as syphilis, gonorrhoea and AIDS.

b. Diaphragm : It is a shallow cup of thin rubber stretched over a flexible ring. This is inserted into the vagina and placed at the mouth of the cervix. Initially it requires the help of a doctor to place the diaphragm in proper position. Additionally the diaphragm may be coated with spermidial jelly for extra safety. The diaphragm is to be removed after every intercourse.

3. Chemical methods

These include a number of methods such as - a. Douching, b. Sponges, tampoons and foam tablets, c. spermicidal jelly and d. oral contraceptives.

*a. Douching :*This refers to washing the vaginal canal immediately after the intercourse with fluids such as salt water, lime juice, detergent water etc. A douching nozzle is to be used for this purpose.

b. Spermicidal jelly : Contraceptive jelly is to be smeared all over the interior of the vagina as deep as possible. For this purpose special applicators are available.

Foam tablets or pessaries which are to be kept in vagina before the intercourse dissolve and form a spermicidal foam. A number of brands of spermicidal pessaries are available in the market.

4. Intra uterine contraceptive device (IUD)

These are devices that are placed inside the uterus. The basic principle behind the use of IUD is that any foreign material placed inside the uterus prevents the implantation of fertilized ovum. Common IUDs are- loop and copper 'T'.

a. Loop : this consists of a polythene 'S' shaped loop with an attached string and can be inserted into the uterus with the help of a plunger (insertion device).

An experienced gynaceologist only can do this. When once inside the uterus the loop uncoils. The string is necessary to pull out the loop when it is not necessary (ie. when conception is needed).

Loop has mertis as well as some demerits. While it is 100% fool proof, it may cause pain or increase menstrual flow.

b. Copper 'T' : This is just like loop, but 'T' shaped and emits small quantities of copper which acts as a spermicidal. This has to be inserted into the uterus with the help of a gynaecologist.

5. Surgical methods

a. Vasectomy (in men) : Vasectomy refers to cutting and rejoining the vas deferens through which the sperms move. As a result a block is created and sperms do not reach urethra even though semen is produced. Vasectomy is a minor surgical procedure for men and does not require hospitalization.

b. Tubectomy (in women) : In this method which is a surgical procedure requiring general anesthesia, the abdomen of the woman is opened and the fallopian tubes of both ovaries are cut and ligated (tied) again. This prevents the movement of the ovum beyond the tied end of the fallopian tube.

As tubectomy requires hospitalization, it is best done after delivery when the women well be in hospital any way. Tubectomy should not be done on pregnant women.

Sterlization surgery is not irreversible. In male and female (when necessary) it is possible to join the severed vas deferens and fallopian tube to allow conception.

c. Laproscopy (in women) : In this method an instrument called laproscope is introduced through the abdomen with the help of a small cut near the naval. The instrument is taken into the fallopian tubes (it is possible to see the internal organs through the laproscope) and the tubes are tied or their lumen is sealed.

Laproscopy is a very simple and effective procedure when compared to tubectomy as it does not involve opening of the abdomen. It can be done under local anesthesia and needs one hour of hospitalization and can be done any time. Usually laproscopy is done after the 5th day of menstruation and not during the middle period the cycle.

6. Pharmacological methods (Hormone pills for women)

In this method the hormone pills inhibit the secretion of hormones that help in maturation of ovum and ovulation. The hormonal pills have progesterone and a little of estrogen. The pills (oral contraceptives) have to be taken every day starting from the 5th day of the menstrual cycle up to the 25th day of the cycle. The pills are to be stopped on the 26th day when menstrual cycle is initiated and ovulation begins and menstruation starts at the appointed date. Oral contraceptive pills are of various types- combination pills, sequential pills, mini pills and morning after pills etc.

Side effects of oral contraceptive : Since the pills interfere with the normal hormone production in the body they have several side effects. Prolonged use results in painful breasts, nausea, increased weight, increased sugar in the blood, vaginal bleeding etc. A relationship between the use of pill and blood clots (in legs, lungs etc) has been observed. Women above the age of 40 should not use pills.

Male contraceptive vaccine : In this an injection (Antiandrogen injection) given to the male prohibits production of sperms for a long time. However this method is still under experimental stage. Researches on this are being carried out at the Indian Institute of science Bangalore and results are encouraging.

Infertility control

Infertility or sterility refers to the inability of sperms to fertilize the egg. This may be due to various reasons. Male fertility requires, the production of adequate quantity of viable sperms, their satisfactory unobstructed passage through the urethra and their deposition in the vagina. In female, fertility depends on proper production of ova, their unhindered passage in the fallopian tube and finally proper implantation in the uterus. Infertility is the result of abnormality in any one of the functions of the reproduction by steps both in male and female. There are a variety of methods to treat infertility. In males, if there is any blockage in the vas deferens, it can be set right by surgery; if the sperm count (each 2.5 to 6 ml of semen must have 50 to 100 million/ml; if it is less than 20 million/ml the individual is regarded as sterile) is less it can be brought to the normal level by hormonal treatment.

In females if any structural deformity is causing infertility it can be surgically corrected. If there is any blockage in the fallopian tubes, inflation by carbon dioxide can be tried. In case there is failure of ovulation, it can be induced by a synthetic drug (clomiphine citrate). Women who have a totally blocked fallopian tube can now have children by resorting to *in vitro* fertilization or what is popularly called *test tube babies*.

In cases where women are normal but husbands have a very low sperm count or no sperms at all are artificially inseminated with either husbands or donor's sperms. In the latter case, for various sociological and legal reasons, the identity of the donor is kept secret.

Artificial insemination in human beings

This may be defined as deposition of the sperms in the vagina, cervical canal or the uterus of the female by artificial means so that it brings about pregnancy in an otherwise healthy women who is unable to conceive through intercourse with her husband.

In our country artificial insemination generally is not practised due to the sociological conditions of our society. For introduction of the sperms usually sperms are taken from *sperm bank* or *semen bank*. In sperm banks, semen with healthy sperms are collected from normal men and are preserved under very low temperature (upto -1960 C). The semen gets frozen and is then stored in liquid nitrogen until it is required for use. Artificial insemination is of the following types -

(i) Artificial insemination homologus (AIH) - In this method semen from the husband of the woman is artificially introduced. This is resorted to when the husband produces semen less than the normal quantity. Semen from the husband is collected several times and is stored in the sperm bank. After sufficient quantity is obtained it is deposited into the womb of the wife.

(ii) Artificial insemination donor (AID)- Also called heterologus artificial insemination, here sperms from an anonymous healthy donor are obtained from the sperm bank and are deposited into the womb of the women.

Surrogate birth

In this technique artificial insemination is done to another women with the sperms of the husband of a barren women. In this case after the birth of the child the surrogate mother should hand over the child to its biological father and his wife. In another type of the same technique if the ovaries of the otherwise barren women are producing normal ova they are taken out and fertilised *in vitro* (in special laboratory equipments outside the body) and then reimplanted into the womb of another women (surrogate mother). This may be called prenatal baby sitting to help in the birth of the child. This method is also widely practised in USA.

Test tube baby

Researches in the reproductive biology of human beings have yielded wonderful results both for the control of the population and for production of human beings in cases where the parents are unable to produce children. Literally test tube baby means production of babies in a test tube. But in practice test tube baby means artificial fertilization of egg and sperm outside the body and implantation of the zygote into the womb of the mother. So far no one has succeeded in growing a human embryo completely into a chold on culture media.

Test tube baby production involves two main steps- *in vitro* fertilization (IVF) and embryo transfer (ET).

Historically the first attempt to produce the test tube baby was by an Italian scientist Petrucci (1959). In his experiment he removed an ovam from a patient and put in the middle of spermatozoa in a glass dome. One of the sperms fertilized the egg and a small embryo formed, but this survived only for 29 days. However Petrucci's experiment opened up a new horizon and a new hope for infertile parents. In recent years two British scientists Edwards and Steptoe have done pioneering work in the field of test tube babies. The first scientifically recorded test tube baby in the world was born on July 25 1978 and the baby girl was named Louise Joy Brown. In our country the first recorded instance of a test tube baby was under the supervision of Dr. Indira Hinduja in KEM Hospital Bombay in 1986. Subsequently test tube babies have been reported by Dr. Baidyanath Charkravarthi of Calcutta.

According to Dr. Charkavarthi the production of the test tube baby involves the following 5 steps-

1. Ovulation (by natural cycle or hormonal stimulation)
2. Follicular aspiration(either by laproscopy or ultrasonographic technique)
3. In vitro fertilization (IVF)
4. Clevage
5. Embryo transfer (ET)

The following procedure is adopted in the production of a test tube baby.

A suitable ovum is obtained from the mother or any other donor women. First an inert gas like carbon dioxide is pumped into a small cut made just below the naval of th woman. This will help the abdomen to droop which will allow the proper manipulation of instruments. Next, a laproscope is inserted through the cut and brought near the ovaries. The laproscope has a provision for viewing the ovaries. When the ovary is spotted a syringe is inserted into a follicle on the surface of the ovary and eggs are sucked into the syringe. This procedure is performed on the woman under general anesthesia. (If there is any problem of ovulation, before laproscopy the woman is injected with gonadotropin hormones which will stimulate the production of eggs).

In vitro fertilization(IVF)

After the egg has been removed it is carefully placed in culture medium containing calf serum or human serum along with other necessary nutrients. The whole thing should be kept in completely sterile conditions. The spermatozoa of the husband (or of the donor) are introduced into the culture medium. Normally the egg is fertilised between 12 to 24 hours. Clevage begins within 38 hours and 2 cells are formed. The clevage continues and within 72 hours after sperm entry, the zygote reaches the 8 celled stage. Within 5 days the embryo reaches the 16 celled blastocyst stage. The blastocyst is now ready for implantation.

Embryo transfer (ET)

The blastocyst stage of the embryo is now ready for transfer into the womb of the mother. Prior to the implantation of the blastocystt the uterus of the recipient mother should be made ready to receive the embryo. This is usually done with the hormonal treatment. The blastocyst is introduced into the uterus by one of the two methods- by using a catheter through the vagina and cervical canal to the womb or directly into the uterus through a cut made in the wall of the uterus.

The womb of the mother under the influence of the hormones develops the endometrium and the blastocystt gets implanted and the normal process of pregnancy continues.

Test tube baby production is a highly complicated task requiring a high degree of skill and advanced facilities in the hospital. The following precautions are to be taken for a successful delivery of a test tube baby.

a. The ovum that is taken out should be fully mature

b. Culture conditions should be ideal providing normal body temperature

c. Strict aseptic conditions must be maintained

d. The uterine wall of the recipient mother should be prepared to allow for implantation

e. After implantation there should be regular check up throughout the period of pregnancy to prevent miscarriage.

Advantages and Disadvantages of test tube baby

Advantages

In vitro fertilization enables sterile woman with defective fallopian tubes to give birth to her own child. This method also helps in situations where the sperm count of her husband is very low or there are some problems where the uterus of the woman does not allow the movement of the sperms.

Disadvantages - The whole process is highly cost intensive and possible only for rich individuals. This technique cannot be performed on women whose uterine wall is damaged. In a country like ours there could be religious and moral objections to follow this method of conception.

***Gamete-Intra-fallopian Transfer* (GIFT)**

This is a method in which sperms are artifically introduced directly into the fallopian tube via a minor surgical procedure. This can be performed on women who have atleast one healthy fallopian tube and for certain reasons the fimbriae of the infundibelum of the fallopian tube cannot capture the eggs. This is also useful in instances where cervical problems prevent conception. In this method sperms which are thoroughly washed and ova obtained from the ovary are transfered using a laproscope into the ampulla of the fallopian tube. Within the fallopian tube which is the normal site of fertilization the two gametes fuse and clevage occurs according to the normal process. The blastocyst travels down the fallopian tube and gets implanted in the endomentrium of the uterus as a normal process. When compared with IVF and ET, GIFT is a less cost intensive procedure and does not require costly equipments. Here except for the artificial introduction of ova and sperms into the fallopian tube everything else (fertilization and implantation) is a normal process.

***Sexually transmitted diseases* (STD)**

Also called venereal diseases these diseases are transmitted from one individual to another through intercourse with an infected individual. The following are the major STD's-

1. Syphilis, 2. Chancroid, 3. Gonorrhoea, 4. Granuloma inguinale, 5. Lymphogranuloma venereum, and 6. AIDS (Acquired immunity deficiency syndrome). Of the above diseases syphilis, gonnorhoea and AIDS are the most important. Strictly speaking AIDS is not transmitted only through the sexual means. It is transmitted through blood transfusion also. Since it is one of the most dreaded disease of human beings also transmitted through sexual contact it is included here.

SYPHILIS

This disease is a specific chronic infection caused by the *Microbe Treponema pallidum*. The microbe was first discovered by Shaudinn and Hoffmann (1905). The microbe is a bacterium ranging in length from 5 to 24 m. The bacterium is sensitive to cold, dryness and detergents. Transmission of disease is through sexual intercourse. Disease is transmitted only in early stages. Sometimes accidental wounding of the skin and introduction of the bacteria from the primary sores can also transmit the pathogen.

Clinical features - The disease is manifested in three or four stages with a latent period (non disease producing) between each phase.

The primary stage has an incubation period which ranges between 9 to 90 days with an average of 20 days when the symptoms start appearing. Lesions or sores appear at the site of contact on the genital organs including prepuce, labia, cervix, uteri etc. Extra genital regions where sores appear are lips, neck, cheek, shoulder and breasts. To start with the sores are painless but gradually they swell leading to ulceration. In men a painless hard sore called *chancre* develops on the penis and in women chancre develops in the vagina or in the uterine cervix.

Second stage starts with the appearance of rose coloured rashes after the primary sores. This may start appearing any time between 6 weeks to 6 months after the sexual contact with an infected individual. General symptoms include discomfort, headache, sorethroat, fever and general uneasiness. Eyes also may be affected some times. In the second stage the lesions or sores may be of the following types-

1. Cutaneous lesion : found on chest, back, abdomen etc.
2. Lymphadenopathy : lymph glands become rubbery, discrete and nontender (adenitis)
3. Snail tract ulcers : found on the mucous membrane of the mouth ; highly infectious.
4. Condylomata : moist papilate outgrowths found on the labia majora in women.
5. Eye infection : acute iridocyclitis of one or both the eyes may be seen.
6. Miscellaneous symptoms : hepatitis, meningitis, epididymitis.

The third stage occurs between two to 10 years of infection. The leisons known as gummatous lesions are very destructive in character and involve visceral organs. Liver, heart, larynx, bone, tongue and testis may be affected. Fever is one of the important symptoms. Sometimes central nervous system may be affected.

The fourth stage or the final stage may start even after 20 to 30 years after the initial infection and at this stage there is involvement of almost all parts of the body. Complications include paralysis, blindness and even death.

Syphilis may be congenital or acquired. Congenital syphilis is very rare. In this, the foetus gets the infection before birth. The child may be still born (dead at birth). In case the child survives it will have several abnormalities which includes papulo(eruptions on buttocks, palms and soles), bullous eruptions on the legs, palms and face. Acquired syphilis is through sexual contact.

Treatment - Early diagnosis of the disease is very essential for treatment and it yields good results when treated during the first two stages. The bacteria are extremely sensitive to procaine penicillin which should be taken for a longer duration. Some times benzyl penicillin in a single mega dose may be enough to cure the disease. For patients who are sensitive to penicillin, tetracycline, erythromycin etc may be given in suitable dosages.

GONORRHOEA

This disease is caused by a gram negative bacterium called *Neisseria gonorrhoeae,* a gonococcus discovered for the first time by Meisser. The bacterial cells occur in groups of two hence called diplococcus. The disease is caused through sexual contact and the primary site of infection in both men and women is the lower urinogential tract.

Aetiology and *Epidemiology* - Infection follows intercourse with an infected individual with prostitution playing an important role in transmission of the disease. Homosexuals also transmit the disease. Occasionally infection may be picked up by young girls from infected lavatory seats or infected towels.

Clinical features and symptoms in male - The symptoms usually appear, between three days to two weeks after the sexual contact. The severity of the symptoms depends on the extent of the infection and the resistance of the individual. In man dysuria is the first symptom. Patient complains of itching and burning at the tip of the penis. This is usually accompanied by frequent urination. The discharge from the urethra is first mucoid and yellowish in colour (gleet). The discharge from the urethra becomes purulent. As a result of infection there may be other complications also like the infection of prostatic glands, damage to the heart valves, arthritis etc.

Clinical features and symptoms in female - In about 50% of the affected individuals symptoms may not appear. When symptoms appear, they include burning sensation on urination, yellow vaginal discharge, dysuria, frequency of micturition etc. Secondary symptoms include backache. In a few patients vulvitis may develop due to profuse discharge. The complications due to infection include endometritis, pelvic peritonitis and pelvic abscess.

In some instances when the seminal vesicles (in males) and uterus (in females) are involved, the gonococci may invade the blood stream carrying the infection to the other parts of the body. These are polyarthritis, conjunctivitis etc.

Diagnosis - Microscópic examinations of the smears taken from the urethra or cervix will reveal the presence of gram negative cocci and is an indication of the disease. Further confirmation may be done by cultures of bacteria on chocolate haemolysed horse blood.

Treatment - Antibiotics give satisfactory results if the disease is detected at an early stage before the infection enters the blood stream. Among the antibiotics pencillin is a drug of choice. Other antibiotics like ampicillin and chloromycetin are also very useful. Other palliative treatment includes avoiding alcohol and excessive physical exertion. Patients should also take lots of liquid.

Preventive measures - STDs have a social dimension also besides a medical angle which comes into play only after an individual contacts the disease. Promiscuous behavior which leads to free sex with unknown individuals, contact with prostitutes etc., are the leading sources for infection. While prostitution cannot be totally avoided, prostitutes should be made to undergo regular medical checkup. Young men should be warned of the dangerous consequences of sex with unknown individuals. The disease when detected should be treated at the earliest stage.

AIDS

AIDS or Acquired immunity deficiency syndrome is caused by virus belonging to a Retrovirus group. It has been given the name HIV (Human immunodeficiency virus). In this disease the viruses attack the lymphocytes (T4) and destroy them. Lymphocytes are necessary to provide immunity against any disease whenever our body gets any kind of infection. Lymphocytes are produced in large numbers and they kill the infecting microbes. With the destruction of the lymphocytes the entire defence mechanism (immunity) is disturbed and the body will not be able to combat even ordinary infections. While AIDS does not kill the individual directly it renders the body weak and susceptible to all kinds of infections.

The first few AIDS cases were reported in 1981 in Atlanta USA; since then it has spread to all countries including India.

Transmission of AIDS: The following are the methods of viral transmission.

a. Through sexual intercourse with an infected person (Homosexuals are more susceptible)

b. Through Contaminated injecting needles

c. Through blood transfusion

d. Infants of infected mothers also get the disease through maternal blood.

Symptoms : All HIV infected persons need not show visible symptoms. The virus after gaining entry may remain dormant for a period ranging between 9-300 months. In blood transfusion cases this may be between 4-14 months. Persons with AIDS infection can be classified into (3) categories.

1. Healthy Carriers : This is the first stage; there are no symptoms. But they are capable of infecting others. They may develop full blown symptoms or may remain healthy for a long period.

2. Prodromal Phase : This is called AIDS related complex (ARC). This group has a milder form of the disease with fever, diarrhoea, weight loss and enlarged lymph glands. They exhibit depression of immune system.

3. The End Stage : Here AIDS is fully developed; the patient dies of Incontrollable infections of lung, exertion, chest pain, fever, loss of weight, severe headache, loss of memory, convulsions, development of skin malignancy (Kaposis sacroma) etc.

Identification of AIDS : Serological test of the blood by elisa or western blot will indicate the presence of antibodies against HIV in infected patients.

Cure : Till now no cure has been found for AIDS. Drugs Azidothymidine, suramin etc., may give a temporary relief. Attempts are being made to develop a vaccine against the virus. However chances of contacting AIDS may be avoided by taking the following precautions.

a. Sterilizing the needles properly before injection or use disposable syringes.

b. Utmost care during blood transfusion; only HIV screened blood to be used.

c. This is a moral step. Avoid sexual contact with several partners. Sexual contact to be had with only one, known partner. Homosexuality should be strictly avoided.

Thirty three

Oirgin of life

One of the most basic problems which as yet has not found a satisfactory answer is the problem of origin of life. Right from the time man started thinking about his surroundings he has pondered over many problems concerning life. In the beginning when the knowledge of man was limited, problems that could not be solved satisfactorily were simply attributed to super natural powers. So the early theories regarding the origin of life are mostly philosophical. Subsequently as knowledge progressed, man began to think more rationally and tried to find answers based on materialistic evidences. In this chapter we shall study a brief account of all the theories concering the origin of life and try to analyse the state of our present knowledge concerning the origin of life.

Life could not have existed on this planet since eternity. In fact earth itself has not existed since eternity. Then it is obvious that life on earth is more recent than earth itself.

Many people believe that life arose in some remote corner of the universe and then it has been transferred to earth. Many astronomers opine that there are chances of life being present on other planets. Even then one cannot say with certainity whether life really exists on other planets and even if it exists in what form it is existing. But even if you believe that life arose in some other planet the problem still persists as to how life arose even on a different planet. Before we start the discussion on the origin of life it is but necessary for us to understand as to what constitutes life. However it is very difficult to define life. Generally the characteristics of living beings are explained and then defined that life possesses all these characteristic features. Possession of cells, development, growth, irritability, reproduction and organisation are regarded as the most basic features of a living being.

The above description of life is a very generalised one, because there are certain life forms which do not fit into the kind of description that we give for life. For instance viruses are the smallest living entities. They are capable of reproduction, have nucleic acids and proteins. But like an ordinary chemical they can be crystallised and kept in a test tube for any number of years. When they are introduced into a living system they start reproducing like any other living being. Viruses seem to possess the characters of both living and non living. Hence they are often regarded as chemicals in a test tube but living beings inside a host. When we are considering the problems of origin of life the discussion on viruses is very important because they are on the border line between living and non living.

There have been several theories concerning the origin of life. These may be regarded as religious, semireligious and scientific. The following are some of the theories.

1. Theory of special creation.
2. Theory of spontaneous generation (Abiogenesis theory)
3. Theory of Eternity of life.
4. Theory of Panspermia or cosmozoic theory.
5. Theory of Vital forces
6. Theory of Chemical evolution.

1. **Theory of special creation**: According to this theory, all the life forms have been created specially and particularly by some supernatural force. Many religions of the world hold this opinion. The Christian theory of the origin of life as explained in the Bible believes that God created different forms of life in the beginning and as the final act of creation man came into existence. As this theory is purely based on faith no evidence is available.

2. **Theory of spontaneous generation**: This theory is also known as the Abiogenesis theory, meaning life coming from a non living source. According to this theory, life is being continuously generated from non living sources. Many of the early European scientists notably - Anaximander, Anaxagoras, Needham and others believed that the small forms of life are continuously generated from the non living material. In the 18th century, people believed that maggots, flies and worms suddenly arise from decaying vegetable matter. They also believed that frogs come to the earth through rain water. Even the great Aristotle believed in the theory of Abiogenetic origin of life.

It is true that some time in the remote past life should have come from a non living source. That must have occurred only once. Since then life is not being generated from a non living source. But it has maintained itself through millions of years because of its capacity of reproduction.

Many European scientists notably Fransesco Rheedi, Lazzaro Spallan zani and Louis Pasteur have clearly proved through their experimental evidences that the theory of spontaneous generation is not correct.

Franscesco Rheedi was perhaps the first person to provide experimental evidence against the theory of spontaneous generation. He showed that the white maggots that appear in the meat were the larvae of flies and nothing else. He showed this by setting up an experiment whereby the meat was protected by a muslin cloth. When the meat was not exposed to air, worms did not appear in the meat. He therefore concluded that the decaying meat was not generating the maggots but it only offered a suitable environment for the laying of the eggs by the flies and their subsequent development into flies.

Lazzaro Spallanzani an Italian priest and scientist disproved the theory of spontaneous generation in the year 1765. He took a piece of meat kept it in a long necked flask, boiled it and then sealed the flask. Even after several months the broth remained clear and there was no development of maggots or worms.

John Needham who was a supporter of Abiogenesis theory argued that by boiling the broth the vital force necessary for the origin of life was destroyed. That is why worms did not develop in the sealed broth. Spallanzani could not satisfactorily answer this criticism.

It was left to the great microbiologist Louis Pasteur (1822 - 1895) to give a onclusive evidence against the theory of spontaneous generation. For his experiments he used a goose necked or swan necked flask. He boiled a solution of sugar and yeast for several hours. Then the flask was left unsealed for the free exchange of air with the outside environment. Even after several weeks there was no development of microbes in it. Because the neck of the flask was shaped in such a way as to trap the outside dust particles or microbes in the neck allowing only the air to reach the solution. In this way the solution inside the flask remained germ free. By breaking the neck of the flask Pasteur reported the development of microbes once again.

The experiment of Louis Pastuer clearly showed that life can arise only from a preexisting life and the abiogenetic theory of life is not correct. This means that the present life forms can only come from preexisting life. This is called the biogenetic theory.

Theory of Eternity of life: This theory believes that life has been in existence indefinitely. Hence the question of origin of life does not arise.

Theory of Panspermia or Cosmozoic theory: According to this theory, life did not originate on earth. It must have originated in another part of the universe from where it has been transferred to earth in the form of resistant spores or seeds called cosmozoans. These cosmozoans during their space travel have reached the earth accidentally and have evolved on this planet. To date there is no tangible evidence for this theory.

Theory of vital forces: This theory is some what similar to the theory of special creation. According to this theory the living beings and all the chemical compounds present in them require a vital force and they also believed that this vital force is responsible for all the unique properties of living beings. They argued that the compounds present in living beings - organic compounds are present only in the living beings and are not present outside of the body of living organisms. Many attempts by chemists to synthesise organic compounds in the laboratory had failed. And this seemed to support the vital force theory. The supporters of vital theory argued that chemicals of living beings have a vital force, and can be created only by a supernatural force, and man being a product of this force can neither control nor is in a position to repeat this.

In the year 1828, Frederick Wohler for the first time synthesised urea (chemical compound present in the urine of animals) in the laboratory using inorganic compounds. This was a great blow to the supporters vital force theory, as they believed no organic compound can exist much less be synthesised outside the body of living organisms. Since then a number of organic compounds have been synthesised in the laboratory and today synthesis of organic compounds is a very normal affair in the chemistry laboratory.

Theory of chemical evolution : This is the most modern and widely accepted theory. According to this theory right at the beginning, organic compounds must have been produced from inorganic sources. And sometime later under certain special circumstances, the chance combination of these organic compounds must have produced some forms similar to life. Those who believe in this theory are Haldane, Oparin, Urey, Stanley Miller, Sydney Fox etc. Ernst von Haeckle, a German biologist was actually the first person who proposed that the first form of life must have been due to a chance combination of organic compounds under certain special environmental conditions such as electrical charges, excessive heat, moisture, Ultraviolet rays etc.,

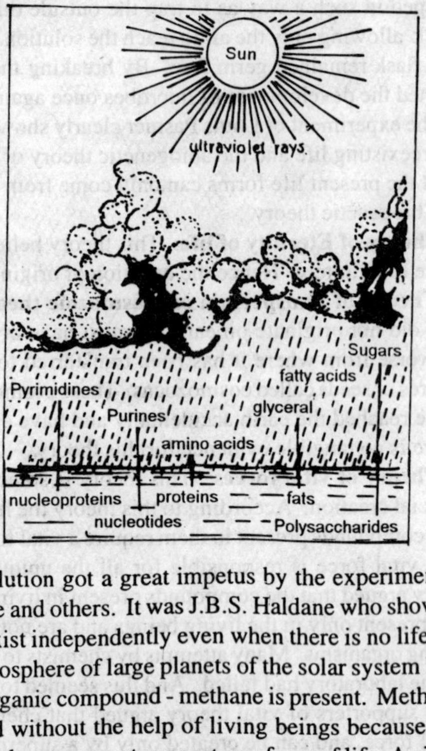

Fig. 33.1 Formation of complex organs molecules in primitive sea

The theory of chemical evolution got a great impetus by the experimental evidences provided by Haldane and others. It was J.B.S. Haldane who showed that organic compounds can exist independently even when there is no life. It has been shown that in the atmosphere of large planets of the solar system like Jupiter, Saturn and Neptune, organic compound - methane is present. Methane here must have been produced without the help of living beings because the freezing temperature of these planets cannot support any form of life. A similar condition must have existed on the earth also in the remote past. The molecules of water, must have reacted with many chemicals to produce Hydrocarbons. On the nascent earth the atmosphere perhaps did not have oxygen nitrogen and carbon dioxide. But the atmosphere was filled with hydrogen sulphide, Methane, Ammonia and molecules of water. All these must have undergone different types of permutations and combinations to produce methyl alcohol, Aldehydes and amino compounds. Many of the molecules of methane react with water and ammonia to produce amino acids. These amino acids are very important as they are the building blocks of proteins, which constitute the basis of life along with nucleic acids.

Experiment of Stanley Miller

The experiment of Stanley Miller was most significant as it provides a conclusive evidence for the chemical synthesis of life. The experiment conducted by Miller in the year 1953 consisted of a specifically prepared flask in which was kept a mixture of gases, methane, ammonia and Hydrogen. The flask was sealed after sterilization. A high frequency electric spark was discharged in a constantly circulating manner so that it keeps continuous contact with the gases. In order to provide water vapour a small quantity of water was boiled and supplied through a tube to the mixture. In short, the entire environment was simulated as it must have existed on the nascent earth.

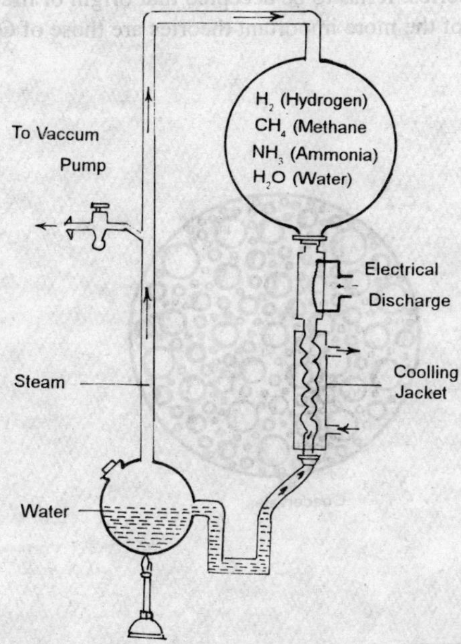

Fig. 33.2 Miller's and Urey's apparatus to create atmospheric conditions of primitive earth

After about a week, the analysis of the mixture in the flask (which was in the form of a solution due to the condensation of water vapour) was analysed as to its chemical constituents. It showed that the mixture had amminoacids like gylcine, alanine, aspartic acid, glutamic acid etc., In addition to this, many aldehydes and HCN were also present.

Subsequent experiments by many scientists clearly showed that amino acids can be synthesised by inorganic compounds. On the primitive earth with a reducing atmosphere moderate energy sources were sufficient to produce major groups of organic compounds necessary for living systems on earth. Under the influence of reducing atmosphere dilute alkaline aqueous systems can produce amino acids, sugars, fatty acids, purines, peptides and polypeptides. The energy for the formation of these organic compounds must have come from ultra violet radiation, intermittent lightning and ionizing radiations etc.

The experiment of Stanley Miller shows that organic compounds can come from inorganic compounds. But the question is how these organic compounds have reached the stage of the first life form even though there are several theories which try to explain this stage many doubts also have been expressed regarding these theories. It has to be accepted that origin of life is still an open question. Some of the more important theories are those of Oparin and S.W. Fox.

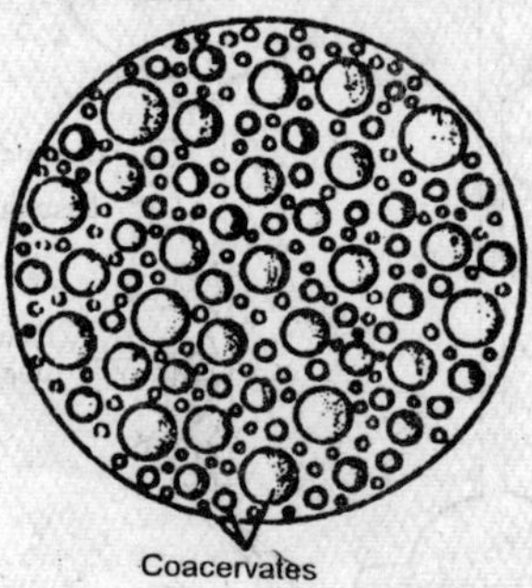

Fig. 33.3 The coacervates

Oparin's Theory: According to Oparin, the organic molecules after having come into existence became organised into supramolecular structres of some types. These supramoleculer structures are called coecervates. A coecervate may be defined as a solution made up of high molecular weight chemicals. According to Oparin, under certain special circumstances the coecervate seperates into two phases - The sol phase and the gel phase. The gel phase in contact with acqueous solution shows great osmotic activity and brings about oxidation, reduction and polymerisation. Oparin called these systems as protobiont. It is these protobiont which subsequently must have given rise to protoplasm.

Fox's theory: S.W.Fox in his experiments simulated the conditions of volcanic eruptions that he belived were existing on the primitive earth. In these experiments he showed the polymerisation of aminoacids into molecules called proteinoids as they were unlike true proteins. These proteinoids group together and form what are known as microspheres. These proteinoid microspheres according to Fox can metabolize glucose. Thus according to Fox and Oparin the microspheres and coecervates are the first structures that constituted the ancestery of all living beings. It is these that subsequently gave rise to the first life forms.

Thirty four
Organic evolution

The living world consists of immense variety of forms of life occuring in every possible region of land, air and water. They are found in fascinating varieties of size, shape and structure. When and how did they appear on earth was a mystery man tried to solve through the ages in the past. Two explanations have been offered by the thinkers of the world, 'special creation' and 'Evolution'. Besides, a number of relevant references are seen in the eastern and western philosophies. Specially the ancient Indian scriptures throw revealing light on the mystery of creation of earth and life.

According to the 'Christian theory of Special Creation' all types of plant and animal life that live today were there since the beginning of the Universe and will continue till eternity. It further states that living organisms are 'unchangeable' and that the present day forms are not modified forms of the ones that lived in remote past.

Eventually with the progress of biological sciences, the doctrine of evolution came to be accepted by the scientific world. The term 'evolution' means a gradual unfolding and orderly change from one state to the next. The two aspects of evolution are inorganic evolution and organic evolution. Inorganic evolution deals with the origin of planets, stars, topography of earth and the changes of chemical compounds of the Universe. Organic evolution concerns the origin of life. According to it, life first originated as very simple form , the unicellular organisms. From these evolved the simple multicellular organisms. The simple multicellular organisms through time and space gave rise to myriads of plants and animals as found today.

The process of evolution through thousands of million years, from inert matter through plant and animal life to man can be compared to the hop, step and jump of the athletics!

The three important theories which support Organic evolution are 1. Lamarckism 2. Darwinism and 3. Mutation theory of De Vries.

In this book the theories of Darwin and Mutation theory are explained besides an account on NeoDarwinism.

Darwinism : Any discussion on evolution without Charles Darwin will be absolutely meaningless as he more than any other biologist has contributed a great lot for the understanding of the theory of evolution. In fact the concept of evolution and the name of Darwin are almost regarded as synonymous. In the foregoing paragraphs we will briefly discuss Charles Darwin's biography and his contribution towards the mechanism of evolution.

Charles Robert Darwin (1809 - 1882) was born on Feb 12, 1809 in England. His father Robert Waring Darwin practised medicine at Shrewsbury, England and Darwin had his early education there itself. Darwin showed his love towards nature at a very early age and generally showed little interest in formal studies. After his early school education, Charles Darwin was sent to a medical school at Edinburgh when he was 16 years old. Being a great lover of nature, Darwin did not find any interest in medical studies and gave up medicine after two years of futile attempt. Quite dissatisfied with Darwin's disinterest in studies his father thought that he might do better in religious studies, so advised him to be a clergyman. Initially Darwin was attracted towards this and got himself admitted to the Christ's College, Cambridge for a degree course as the clergyman's position then required a graduate degree. Even though Darwin was not interested in the formal studies he graduated in the year 1831. But even during this period Darwin spent most of his time in observing the nature and studying the natural History. During this period Darwin came into contact with Dr.Henslow an eminent Botanist.

Fig. 34.1 Charles Darwin

The personal contact with Dr.Henslow provided a great opportunity to young Charles Darwin when he was offered the post of a naturalist on the recommendation of Dr.Henslow in the ship H.M.S. Beagle which was planning to undertake a voyage of global exploration. Darwin's father initially protested but ultimately agreed on the persuation of Dr.Henslow. The ship set sail from Plymouth on December 27th, 1831 on a five year voyage exploring South America, the Galapagos Islands and various other Islands of the south pacific.

During this trip, Darwin took extensive notes and collected a lot of material and published several scholarly articles on Barnacles, Mammalian Fossils, Geology and coral reefs. He made extensive collection of both animal and plant specimens and also did a critical study on the comparison between living and fossil specimens. The ship finally returned on October 2nd 1836 after, a trip of five years.

In the year 1837, Darwin had an opportunity of reading an essay on "population" by the eminent economist Malthus in which he had explained as to how populations increased in geometrical ratios untill they were naturally controlled by various limiting factors. It was this essay and the wide array of collections of specimens he had made during the voyage , influenced Charles Darwin to speculate on the struggle for existence among the individuals and the resultant effect of such a struggle. Darwin came to the conclusion that in such a struggle only organisms possessing advantageous or favourable variations would be preserved and the rest would be eliminated. The agency to preserve or eliminate these organisms was nature itself and Darwin called this action of nature as natural selection. Darwin first wrote his theory in 1842 and expanded it after collection of some more data in the year 1844. He went on collecting more and more data and in the year 1858 when he was about to write his classical treatise on the origin of species encompassing all his ideas, he received a short essay from a young English naturalist, Alfred Russel Wallace. Wallace's essay was entitled "On the tendencies of varieties to depart indefinitely from the original type". Darwin was greatly surprised to see that the ideas of Wallace closely resembled his own. But it is to the credit of Darwin that instead of ignoring the essay of Wallace, Darwin included his name also as a joint author in the paper entitled "on the origin of species by means of natural selection or the preservation of favoured races in the struggle for life". This work was published in the proceedings of the Linnaean Society in 1859. This work of Charles Darwin has been acclaimed as a monumental work on organic evolution.

Apart from this paper Charles Darwin has written several papers and monographs. Some of these are - Origin of Species, A Monograph on Cirripedia (four volumes), the fertilization of Orchids, The variation of plants and Animals under Domestication, The Descent of Man, The insctivorous Plants, etc., Charles Darwin died on April 19th, 1882 when he was 73 years of age.

The main aspects of the theory of natural selection are as follows.

1. Rapid multiplication.
2. Limited food and space causing struggle for existence.
3. Natural selection or survivalof the fit and elimination of the unfit.
4. Origin of new species.

1. **Rapid Multiplication**: Every living organism in this world reproduces itself. The rate of increase of the organisms is always in a geometric proportion. Darwin illustrated this point with the example of the slowest breeder among animals, the elephant. An elephant starts breeding at its 30th year continues to do so until it is 90 years. It lives up to 100 years. So in its life time it produces 6 offsprings. Darwin calculated that if all these survive, within about 750 years there would be about 19 million elephants. Oysters produce about 80 million eggs in a season. A pair of sparrows would produce 275 billion individuals in about 10 years if there is no death.

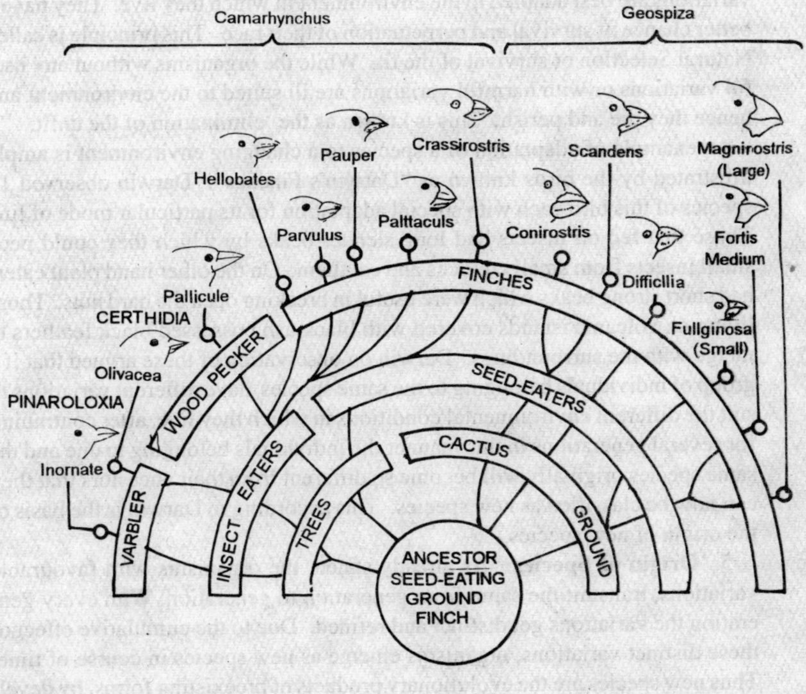

Fig. 34.2 Daarwin's finches

2. **Struggle for existence**: The rate of multiplication and existence of organisms are more or less common for all living things. But the food supply and space remain unchanged, in other words they are limited. Under these conditions, there starts a competition among the organisms to fulfil their needs of food and space. This is known as struggle for existence. It may be interspecific, that is between the members of different species or intraspecific between members of the same species or between them and the environmental factors.

3. **Variations**: Every organism has its own specific characterstics and relationship with its environment, the variability caused by 'variations' may be large or small. Large variations are known as 'macrovariations'. Small variations are termed as 'microvariations'. Variations may be favourable or harmful. Those with useful variations survive while the others perish.

4. **Natural selection or Survival of the Fit**: The organisms with favourable variations are best adapted to the environment in which they live. They have a better chance of survival and perpetuation of their race. This principle is called Natural Selection or survival of the fit. While the organisms without any useful variations or with harmful variations are ill suited to the environment and hence they die and perish. This is known as the 'elimination of the unfit.

An example of adaptation of a species to a changing environment is amply illustrated by the birds known as "Darwin's Finches". Darwin observed 13 species of this bird each with special adaptation for its particular mode of life. Those that fed on insects had long slender beaks by which they could peck small insects from small crevices and eat them. On the other hand plant eaters had short strong beaks which were useful in breaking open the hard nuts. Those living on volcanic islands covered with black ash possessed black feathers to merge with the surroundings. Darwin on observation of these argued that if a group of individuals belonging to the same species have different variations to suit the different environmental conditions in which they live, after continuing for several generations in this manner the individuals belonging to one and the same species originally will become so different from their ancestors that they can now be classified as new species. This according to Darwin is the basis of the origin of new species.

5. **Origin of Species**: As already stated, the organisms with favourable variations, transmit the same from generation to generation. With every generation the variations get distinct and refined. Due to the cumulative effect of these distinct variations, organisms emerge as new species in course of time. Thus new species are the evolutionary products of preexisting forms, by developing useful variations.

Objections to theory of natural selection.

The role of natural selection has no doubt been regarded as a chief factor in evolution. But it cannot be the sole factor to bring about evolution. Darwin himself accepted the insufficiency of natural selection being the sole arbiter for evolution. Some of the objections raised against the theory of natural se-

1. Darwin belived evolution by the accumulation of slow and small variations. But in many instances the occurence of such variations when they are not fully developed are of no use to the individuals in the struggle for existence. For instance, the eye in the case of animals is useful only in a fully formed condition and surely all the features of the eye according to Darwin could not have come into existence at one go. It should have been by accumulation of several mutations. But then each change by itself is of no advantage.

2. In many instances the adaptations of individuals seem to be more of a disadvantage rather than any use. For instance the antlers in deer are mor obstructive because of their huge size than of any use, when the animal has to run to avoid from predators. In this instance natural selection indeed should have limited the structures not to grow beyond a point of usefulness.

3. The main drawback in Darwin's theory is that he assumed that the organisms are having favourable and unfavourable variations. The main aspect in the mechanism of evolution is the origin of variations. Only then one can explain the role of natural selection. Natural selection according to Darwin is only an arbiter to decide which variation is favourable or unfavourable. But the crux of evolutionary theory is to explain how a variation can arise. Further, Darwin also did not distinguish between heritable and non heritable variations.

4. Natural selection theory does not satisfactorily attribute the reasons for the presence of vestigial organs. For example appendix in human beings. These organs should have been eliminated by natural selection during the course of evolution.

5. If all the species were to be evolved by the accumulation of small variations, there should have been transitional forms in almost all the species. But this is not the case. Darwin however explained this away by attributing it to the faulty preservation of fossils. In areas where no interlinking forms were found, Darwin used the term "missing links". For instance between man and monkey.

6. Natural selection as belived by Darwin was only a selective force and not a creative force. Many of the current observations have indeed shown that the environment not only chooses favourable variations but also is a causative agent for variations.

Theory of Pangenesis: In order to explain the lacuna in the theory of natural selection that it did not account for the fact of origin of variations, Darwin proposed the theory of Pangenesis. Darwin belived that the understanding of the mechanism of heredity was indeed essential to understand the origin of variations, Darwin did not agree with the idea of Lamarck but unfortunately had not known the epoch making work of Mendle on pea plants, even though it was published long before Darwin's death. This is the reason why in the book Origin of species Darwin states that the fundamental principles of Heredity are still unknown.

In order to explain the origin of variations, Darwin proposed the theory of Pangenesis. According to this theory all the organs of an individual produce certain discrete particles called pangenes which carry the information regarding the organ from which they come. The pangenes travel through the blood stream, reach the reproductive structures and are included in the gametes. Each gamete will have different pangenes for different organs. After the formations of zygote in the offspring each pangene develops the organ from which it has come. This in essence means that if there is any change in the organ of an individual during its life time the pangene will also be changed and the changed pangene will produce a changed organ in the next individual. This theory is in no way different from the Lamarckin theory of inheritence of acquired characters.

Mutation theory : Hugo de Vries (1840 - 1935) was a Dutch Botanist. In his book- The Mutation theory written originally in German and later translated to English, de Vries described a number of changes and new types in many species of plants that he had studied. In particular he made a detailed study of the genus *Oenothera* (evening primrose plant) belonging to the family Onagraceae. In the species *Oenothera lamarckiana,* he noticed several changes. In 1887 de Vries noted two other species *O. brevistylis* and *O. laevifolia.* He collected the seeds from all these and sowed them to find out if they bred true. In the case of *O.lamarckiana,* some of the offspring showed some exceptional characters that were not present in the parents. These changes which occured suddenly were called Mutations by de Vries. A new type which appeared much larger than the parent was described by de Vries as a new species *O.gigas.*

On the basis of these observations de Vries proposed the theory of mutation to account for evolution. According to this, mutations throw up suddenly fresh variations which are then acted upon by natural selection. Since mutations were known to be random evolution was also considered a random process. Many biologists thought that all variations are chance occurences and that there is no question of adaptation.

de Vries continued the breeding experiments and discovered 15 different types of mutations and grouped them into 15 different species. As these were markedly different from the original plants he concluded that mutations constitute the basis of evolution. He stated that origin of species or evolution takes place by sudden mutations and not by microvariations as proposed by Darwin. He announced this theory in 1901.

However it should be pointed out that the so called mutations observed by de Vries were actually chromosomal rearrangements rather than the gene mutations or point mutations as it was belived by de Vries. But the theory of mutation proposed by de Vries as providing the basic raw material for evolution is essentially correct. Further, the mutation theory explains some objections raised against the natural selection theory. The sudden changes or sports to which Darwin did not seem to attach much importance was regarded as very significant by de Vries. It is on the basis of mutations that one can explain how heriditary variations arise in individuals as against the superficial variations.

NEO DARWINISM.

As the theory of natural selection proposed by Darwin had many inaccuracies, by the year 1890 doubts began to arise as to its usefulness. Around the year 1910 many critics opined that Darwin theory of selection is purely speculative. With the discovery of mutations by de Vries it proposed that evolution proceeded by was large jumps and not by accumulation of minor variations.

The contribution of Mendel (even though he was unaware) provided the necessary proof to Darwin's concept regarding the origin of heritable variations not necessarily by means of mutation. Mendle's work on recombination of characters showed that existing characters of individuals might produce a number of new combinations offering new variations for natural selection to act upon. Many of the supporters of Darwin worked in various fields and tried to interpret Darwinism in terms of modern developments. The essential synthesis of all these theories along with the Darwinian theory of natural selection has come to be known as Neo Darwinism. The names of Huxley, Haldane, Goldschmidt, Dobzhansky, Fischer and others are important in the NeoDarwinian theory of evolution. The Neo Darwinism is also regarded as the modern synthetic theory of evolution. This theory combines the good aspects of mutation theory of de Vries, the natural selection theory of Darwin,the principles of inheritence of Mendlc and the germplasam theory of Weismann. According to the NeoDarwinian theory of evolution there are five basic factors involved in the process of organic evolution. These are

1. Gene Mutations
2. Changes in chromosomes structure and number
3. Genetic recombination
4. Natural selection and
5. Reproductive isolation

While the the first three factors are responsible for providing genetic variability, the last two provide a direction to the evolutionary process.

The NeoDarwinian theory states that "evolution is the product of genetic changes occuring under the influence of natural selection." Another factor which is basically essential is the isolation of genetically changed forms. Genetic changes are caused by gene mutations. Natural selection acts upon the new genotypes in the population and picks up the favoured ones and allows them to reproduce. when populations with useful genotypes are isolated for a long time they tend to originate into a distinct species from the one they belonged to originally.

Thus isolation is one of the fundamental factors for evolution. Isolation which is brought about by migration may be geographical, climatic, genetic, reproductive or ecological. Isolation checks free inbreeding of organisms. This explains as to how new species originate when free inbreeding among the individuals or species is controlled by isolation.

Thirty five

Dairy

India is basically an agricultural country since time immemorial. Inspite of industrialisation the major percentage of national income is from agriculture. In addition to this, a large majority of Indians live in villages practising agriculture. In agricultural operations animals are of great help to man. Domestication of animals perhaps dates back to prehistorical times. Ever since man formed agricultural settlements he needed animals to help him in ploughing the field and to provide him with milk besides also providing meat. Among the livestock, cattle and buffaloes are the backbone of Indian agriculture. Bullocks and even cows are used for tilling the lands. Cattle provide milk and milk products which provide the much required animal protein in the human diet. With all these reasons no wonder then a farmer considers cattle to be his prized possession next to his land holdings.

The Indian cattle belongs to two species *Bos indicus* commonly known as the Bull and *Bos bubalus* the Buffalo.

Utility of cattle (livestock) in agriculture, transport and food

India can boast of the largest population of cattle in the world. It is belived that for more than 20% of the world's cattle population, India is the home. The cattle have become a part and parcel of an Indian farmer's family. They help in numerous ways. Inspite of the introduction of modern techniques of agriculture like tilling of the land through mechanised methods such as use of tractors, tillers etc., even today a majority of Indian farmers still use cattle (bull) for ploughing their lands. When land holdings are small it will be uneconomical to use the tractor as it is costly and has recurring expenditure. Majority of the Indian farmers have very small land holdings hence there is no wonder that even today major ploughing activity in Indian fields is by the cattle. Besides ploughing the land the cattle also help the farmer in several other agricultural operations. These are sowing of the seeds, deweeding operations etc. Cattle are very much necessary during the threshing of the grains after the harvest. In the traditional method of the threshing of the grains, a portion of the field is made even and smooth; the harvested crop is spread on the ground. A huge rolling stone is tied to a yolk which is then pulled by the cattle in a circle over the crop. This will help in the separation of the grain from the plant. The grains along with the chaff are then winnowed.

Cattle also help the farmer in irrigation purposes. In Karnataka in a method called *Kapile,* the cattle help to lift the water from the well. In this method a series of small buckets are tied on to a pully which is turned around with the help of cattle. The buckets collect the water as they move round in a circle and pour the water on to a small canal.

In villages, tiny oil mills use cattle to extract oil from oil seeds. In this a rotating weight is placed inside a big wooden vessel in which the oil seeds are kept. Cattle tied to a yolk help in rotation of the weight. Oil is extracted from the crushed seeds.

Inspite of numerous modern methods of transportation the Indian farmer still depends on his bullock cart for transportation. This is because the bullock cart does not demand properly laid out roads which in any case are absent in villages. The bullock cart are used by the farmer to transport his agricultural produce to the market. It is also a mode of family transport for him. Considering the importance of bullock cart, many researches have been carried out (particularly in IIM, Bangalore) to improve the design of the bullock cart mainly to reduce the weight and the burden to the cattle and to increase the mobility by using ball bearings, lubricants etc.

The cattle is useful to the farmer in providing the much required organic manure. The cow dung perhaps is the best organic manure. Now a days cow dung has been used in the production of Gobar gas which provides the much needed fuel for cooking, lighting etc. This will go a long way in the conservation of forests as otherwise fire wood is used for cooking. Another special practise in Indian villages is to fashion the cow dung in the form of cakes. These dried cakes are used as fuel. From the point of view of food there is no other animal perhaps which can parallel the cow and the buffalo. Cow's milk which has the best balanced diet is regarded as next best to mother's milk. Milk is the source of dairy industry. A number of milk based products are produced in dairy. Butter, ghee, curds (yoghurt), cheese are all milk based products. Most of the sweet items used by man are also milk based (pedha and others).

Dairy Industry

This industry is based on milk and milk products. Considering the importance of cattle in the economy and the rate of production of milk many attempts have been made to improve the yield of milk. Dairy industry is involved in all activities that are concerned with the production of milk. These include selection and rearing of good breeds of cattle, breeding of these animals to obtain a better quality animal etc. Dairy industry also deals with production of milk, procurement of milk, supply of milk and also produciton of other products from milk such as milk powder, butter, ghee, sweets etc.

Importance of milk in human diet

Milk is the most important product in dairy industry. It (milk) is described as nature's most perfect food and it is the only food for all new born mammals. Infact even for human beings, mother's milk is the only source of food. In the absence of which cow's milk is the best. Milk is important in the diet because of three of its ingredients protein, calcium and riboflavin. Proteins provide many of the essential ammino acids that are absent in the cereal foods. One glass of milk per day can supply all the protein requirement of children upto 6 years of age and more than 60% of the requirements of growing children upto

14 years of age. From 14-20 years a glass of milk per day supplies approximately 50% of the daily protein requirement. Another important element in milk is calcium. The daily calcium requirement (from 1.2-1.4 gms) is effectively provided by milk. For the elderly, milk is very essential because it provides the necessary calcium for the protection of their bones.

Composition of milk

Milk is a white liquid produced from the mammary glands of the cow. The mammary glands are called *udder* and it has four nipples through which the milk is secreted out.

Physical properties - The unadultered milk is a thick fluid with viscosity ranging between 1-2. The boiling point of milk is about 100ºC.

Chemical composition of milk

The following components are found in milk-water,lipids, carbohydrates, proteins, vitamins and minerals.

Water - This constitutes about 85-87% of milk. This is the main medium for the suspension of all other components.

Lipids - Milk is an emulsion and the lipids are found in a globular form. The main lipids present in milk are triglycerides, phospholipids and cholesterol. The triglycerides are formed of a number of fatty acids such as palmitic acid, stearic acid, lauric acid, linolic acid etc. Lecithin, Cholin and many cerebrosides are other forms of fats present in milk. The percentage of fats in milk varies in different breeds; it may vary from 3.5-5%.

The fatty substances in milk can be separated in a solid form by applying centrifugal force after it is allowed to get curdled. Butter, the concentrated fat of milk is an important food ingredient in human diet. Ghee is another product obtained by melting the butter.

Carbohydrates - The most important carbohydrate present in milk is lactose. It is commonly called milk sugar. Lactose is a disaccharide formed of two monomeres of monosaccharides-glucose and galactose.

Protein- Among the proteins, casine commonly called the milk protein is the most important constituent. In milk, casine combines with calcium forming calcium caseinate. For growing children casine is a very essential protein. Other proteins of milk include lactoalbumins and lactoglobulins.

Vitamins and minerals - Milk contains a number of essential mineral elements such as sodium, potasium, calcium, magnesium, iron, copper, iodine etc. Among the vitamins, milk has B complex and vitamin C and A. Even vitamin D and E are present in milk.

With all these essential ingredients present no wonder the milk is an important part in the balanced diet. This is especially so for growing children and convalceng people.

Regarding the composition of milk of cows and buffaloes, the latter yield generally more quantity of milk and with a higher fat content. Generally however buffaloe milk is not given to infants as it is not easily digestible.

Breeds of cattle (livestock)

The breeds of Indian cattle are distributed all over the country though a majority of them are present in the dry regions. It has been found out that there are 26 breeds of cattle and 7 breeds of buffaloes in India.. The criteria for differentiation being the body colour, horns and forehead. These breeds are classified into three categories- milch breeds, draught breeds and general utility breeds. The milch breeds (dairy breeds) are generally meant to provide milk. While the cows yield very good milk the bullocks are not good draught animals. Some of the milch breeds include Gir, Sindhi, Sahiwal etc. The draught breeds are not good milk yielders but the bullocks are very hardy and are capable of helping the farmer in farm work such as lifting water, pulling carts and ploughing the fields. Some of the draught breeds are Amrithamahal, Hallikar, Bargur etc. The general utility breeds are useful for both the purposes in the sense they also yield satisfactory quantity of milk and the bullocks are also capable of hard work. The important general utility breeds are Haryana, Ongole etc.

Breeds of Buffaloes

There are 7 Indian breeds such as Murrah, Bhadawari, Jaffrabadi, Surthi, Mehsana, Nagpuri or Ellichpuri and Niliravi. These are distributed in several Indian states like Punjab, Uttarpradesh, Haryana, Gujarath, Rajasthan, Central and South India etc.

Imported breeds (exotic breeds)

These breeds are imported from Western countries like USA and countries of Europe. Some of these breeds are Ayrshire, Brown swiss, Holstein-Friesian, Jersey, Red dane etc. These breeds are imported for two purposes. 1. to use them as milch cows 2. to use them for breeding with native breeds to obtain hybrids having a high yielding capacity. Many of the imported breeds have a very high milk yielding capacity.

In this book the Hallikar breed of cattle and the Nagpuri breed of buffaloe is explained in some detail are -

Hallikar breed - This breed of cattle is a native of Karnataka. It is mainly distributed in the districts of Hassan and Tumkur. The characteristic features of the animal include its medium size and a compact muscular appearance. The head is some what longer with a bulging forehead having a median furrow. The horns are longer and are placed close to each other and are carried back on each side of the neck. The horns gradually curve up ending in sharp points. The face is also long but the ears are small. The colour of the skin varies between gray to dark gray. Deep shades of this colour are found on the front and hind parts of the body. The Hallikar breeds do not have a pronounced hump.

The Hallikar breeds are very hardy animals and help the farmer in both transportation and field work. The cows however cannot be regarded as good milch breeds as the quantity of the milk yield is poor.

Nagpuri breed - This breed of buffaloe is distributed mainly in centeral and south India. The characteristic features of the animal include a longish head and long curved horns. The colour of the skin is black while white patches may be seen on the face, tail and legs. The yield of milk in female is about 2,000 litres during the lactation period. The males are very hardy and are helpful in agricultural operations.

Cross breeding

The purpose of breeding cattle is to combine desirable traits that are spread in different breeds. In the breeding programme two animals belonging to two different stocks are chosen and are bred together hoping that the progeny would contain the desirable traits. The first important task in cross breeding is the selection of the parents. The male and female are carefully chosen after they have proved to posses the desirable characters. The next important step is to find out whether the character in question can be inherited properly. Other important aspects in selecting traits for breeding purposes are the economic value of the trait, the range of variation of expression of each trait, the correlation among the traits and the cost of such a selection programme.

In cattle, breeding programme is conducted mainly for two purposes- dairy purpose and for the purpose of obtaining beef (beef cattle). For the purpose of dairy the important traits for selection are yield of milk, fat content in milk, protein content in milk, yield of milk solids etc. For the purpose of beef the traits that are looked into are the quality of the beef, the weight of the calf at birth, the mature weight and the quality of carcass.

The following table gives the average milk yield per cow per annum in different countries

USA	4250 kg
Netherlands	4220 kg
Denmark	3710 kg
Pakistan	420 kg
India	220 kg

In our country however the main aim of breeding is for dairy purpose. Generally, Indian cows even though quite hardy, have very less milk yield. The average annual yield is about 150-200 kgs/cow which is far less when compared to the annual yield of imported breeds which works out to more than 3000kgs/annum. For the purpose of crossing, exotic breeds like Red dane, Jersey and Holstein are chosen as their milk yield is very high. The European breeds are crossed with Indian breeds like Sindhi, Sahiwal, Haryana etc. It has been found that the progenies of the cross breed have a higher milk yield than their native parents. Sometimes the cross breed progenies give a better yield even when compared with their European parents (Jersey). In breeding programme, both the males and females of the exotic breeds are utilized.

Breeding programmes are being conducted with the help of research carried out in several agricultural universities including Indian Veterinary Research Institute (IVRI).

The Australian milking zebui (AMZ) has been developed as a new breed mainly for use in tropical areas. This animal is a cross between the Sahiwal and European Jersey and it has three unique qualities of heat tolerance, resistance to ticks and a high level of milk and butter fat.

The National Dairy Research Institute in Karnal has developed two cross breed progenies after crossing the Indian cows with the bulls of the European breeds such as Holestin, Jersey, Red dane etc. The two new cross breeds are Karanswis and Sunandini.

There are many advantages in the cross breeding programme between the Indian breeds and the exotic breeds. There has been a phenomenal increase in the yield of milk giving the farmer a better source of income.

Artificial methods of breeding

In the natural methods of cross breeding explained above the two parents are chosen for selective mating. Other than this there is no human interference in the process of reproduction. But in artificial methods of breeding after careful selection of parents there is manipulation in the reproductive process so as to yield better results.

Artificial breeding methods are also followed in such instances where there could be a failure in the natural methods of breeding. In the paragraphs below the methods of artificial breeding are explained

1. Artificial insemination- In this process, semen from a high quality pedigreed bull is selected and the cows are artificially inseminated. For this purposes there are several mechanical devices to introduce the sperm in the vaginal tracts of the cows. Semen collected from bulls is usually kept frozen at temperatures upto -196°C in liquid nitrogen. At the appropriate time it is brought to normal temperature and used for insemination.

2. Superovulation and Embryo transplantation - Just as in human beings, in cattle also the reproductive process is controlled by hormones. These hormones regulate the production of ova at a predesignated time and in a predesignated quantity. Superovulation refers to increasing the quantity of production of ova in cows by injecting hormones. For this purpose a good quality cow with a high yield of milk is selected and is given hormonal injections to induce superovulation. The semen from a good quality bull is obtained and is artificially inseminated to induce fertilization. When embryos are formed (more than the normal number), 4-10 embryos are collected at a time. The embryos would be in the blastocyst stage because it is at this stage they can undergo implantation. Each embryo is then transplanted to another cow which is generally refered to as *carrier cow* or *Surrogate mother.* The pregnancy of the cow continues in a normal manner and the birth of the calf takes place after the usual period of gestation.

The embryos obtained from the cow after superovulation are usually preserved by deep freezing techniques by storing them at -196° C. It has been found out that the 7 days old embroys can be preserved for several years by deep freezing techniques and can be used whenever needed. Experiments have also been conducted by cutting the embryo into two at very early stages of clevage and the two pieces so obtained can be made to develop into individual embryos. The progeny produced by this technique will be homozygous twins. The advantages of superovulation and embryo transplantation are many. This technique has helped to increase the number of quality breed cattle which can be later used for breeding purposes.

In *vitro* fertilization and embryo transfer (IVF and ET) can also be practised in cattle as is done in human beings. The Veterinary college of Agricultural universtiy at Hebbal, Bangalore is using this technique to produce quality breeds of cattle.

Importance of cattle in leather industry, biogas, fertilizers and gelatin

There is a proverb in Kannada which states that the cow is useful to the farmer when it is alive as well as when it is dead. This is very true because apart from the main produce the milk, the cow is useful to the man in a number of ways. Agricultural field work, transportation, fertilizers, tanning industry, food industry are some of the fields where cattle are extremely useful.

Leather industry- Leather finds a number of uses in various fields. Footwear, clothing, purses, belts are all made of leather. Leather is obtained from animals and is then processed before being fashioned into various kinds of goods. Leather industry is also called tanning industry because it is here that the raw skin obtained from animals is "prepared " so that it could be used for various purposes.

Among animals which provide hide(skin) for leather industry livestock are most important. Cattle and buffaloe skin is the principle source of leather supply. The skin of cow is called kips and that of the buffaloe is known as buff.

In tannaries, the raw skin obtained from animals is processed following various steps such as flaying, curing, salting and tanning. The tanning process is a very complicated process and involves three main steps. These are the pre-tanning, tanning and post-tanning. The pre-tanning process itself involves a number of steps like soaking, liming, unhairing, fleshing, deliming and pickling. The tanning process involves treating the skin with organic, inorganic and synthetic tanning agents. Basically there are three types in tanning. These are vegetable tanning, chrome tanning and oil tanning. In the post tanning step the skin is subjected to a number of mechanical, physical and chemical treatments. The population of cattle in India is one of the highest in the world and hence no wonder India is one of the leading suppliers of leather goods to the world market. In order to imporve the quality of the leather researches are going on and there is an Institute for this purpose- Central Leather Research Institute (CLRI) at Madras. Tannaries are also present in Karnataka - Mysore Chrome

Tannaries, Bangalore. In our state Athani and towns nearby are very famous for producing quality footwear. Kanpur in North India is also very famous for the manufacture of leather goods.

Biogas- The Indian livestock produce an enormous amount of dung which can be put into proper use for the purpose of harnessing energy. The dung of cattle undergoes putrefaction in the normal course by decaying bacteria and releases a gas. Researches have shown that this gas can be used for the same purposes as natural gas . Popularly called gobar gas or biogas, the gas obtained from the dung of cattle can virtually create an energy revolution in rural India. An estimate shows that the average annual production of cattle dung in our country exceeds 1,500 million tonnes. Even if a small quantity of this is used for the purpose of biogas production it will solve the entire energy problem of rural India.

In the production of biogas, the cattle dung, urine and some of the agricultrual wastes are put into huge tanks built for this purpose along with some quantity of water. Bacteria normally present in these wastes bring about putrefaction and release the gas which can be supplied through pipes. The biogas can be used as a cooking fuel and it also can be used for the purpose of lighting. Biogas plants can be built on different sizes depending on the requirement and the availability of animal dung and plant wastes. Even a single family with 40-50 kgs of daily production of cow dung can have a small gobar gas plant which will supply all their daily fuel and lighting requirements.

Use of biogas also has an indirect benefit in that it protects the forests which should have otherwise been cut to provide fuel wood.

Fertilizers- Fertilizers are of two types chemical fertilizers and organic fertilizers. Eventhough chemical fertilizers have a targetting effect such as providing a particular kind of nutrient, they are very costly and many a time they may have some bad effects on the plants also. On the other hand organic fertilizers commonly called organic manure or farm yard manure absolutely have no bad effects on plants. Besides, they are very cheap in the sense every farmer can produce manure himself in his backyard. In the production of farmyard manure the role of cattle, buffaloes, sheep, goat and fowls play a very important role. The excrement of these animals is a very rich source of nitrogenous materials. Usually in the farm yard a pit is dug into which all the animal dung and plant wastes are dumped. In course of time due to the activity of bacteria and also earth worms the whole thing gets converted into a very rich manure called compost. The cattle dung, even after the production of biogas can be used as a manure.

Gelatin - This is a protein derived by the degradation of collagen which is a fibrous protein present in the connective tissues of animals. In cattle, many parts of the body contain collagen which can be used for the production of gelatin. Gelatin from animal sources is mostly produced in Western countries. Another source of gelain are marine algae belonging to the classes Phaeophyceae and Rhodophyceae.

Gelatin is insoluble in cold water but increases in volume after taking in water. There are a number of uses for gelatin including in food and chemical industries. In the food industry ice creams, jelly, candies and confectionary items contain gelatin. In chemical industries gelatin is useful in the manufacture of photographic emulsions. In the pharmaceutical industries gelatin finds extensive use in the manufacture of the capsules. The capsule covering is made of gelatin and on being swallowed, the gelatin covering dissolves, releasing the medicine.

Thirty six

Poultry

Human beings in their never ending search for nutritious food have identified various sources both from the animal kingdom and the plant kingdom. Both vegetarian diet and non vegetarian diet have their own merits and demerits. Early man used to gather food from plants and used to hunt animals and birds to obtain food. When agricultural settlements evolved, man started domesticating animals both for farm labour as well as for his food. Cattle, buffaloe, sheep, goat etc were meant to provide meat as well as milk and some of them even helped him in his farm work. Other kinds of domesticated animals include birds which were mainly meant to provide him nutritious food both as meat as well as eggs.

The practice of rearing the birds for the purpose of meat as well as egg is known as poultry or poultry keeping. Poultry includes keeping of several kinds of birds like chicken, duck, geese, turkey and guinea fowl. However major activity in poultry is involved in rearing of chickens.

The domestication of chicken in India perhaps dates back to 3000 BC. Ancient Indians were attracted by fowls both for food as well as for sports (cock fighting). It is widely believed that India and its neighboring countries are the original home of chicken.

Modern poultry in India started only in recent times when compared to its antiquity in our country. Before the starting of 5 year plans (1951) poultry keeping was largely a backyard venture without much attention being paid to breeding, feeding, disease control and proper management on a commercial scale. The first step towards scientific poultry management in our country was taken in 1949 with the establishment of a division for poultry research at the Indian Veterinary Research Institute (IVRI) at Izatnagar in Uttar pradesh. A team of scientists working there developed a vaccine against the disease (Raniket) which provided the birds lifelong immunity.

Importance of eggs in combating protein malnutrition

Eggs are a very rich source of a number of nutrients in an easily digestible form. Besides, it has a number of vitamins and minerals. As the nutritional value of egg is very high particularly from the point of view of proteins its daily use is highly recommended. Even vegetarians can safely take the egg as the cereal foods lack in adequate quantity of proteins. Egg either in raw condition or in semi cooked or cooked condition can be easily given for growing children to combat malnutrition.

Nutritive value of the egg

Egg is a rich source of all nutrients except vitamin C. The egg proteins are regarded as of high biological value as they can be easily digested, absorbed and used. The following is a list of different nutrients in 100gms of egg.

water 74.5 gms; protein 12.1 gms; carbohydrates 0.7 gms; lipids 11.5 gms; The proteins contain almost all valuable aminoacids- Arginine 0.75 gms; Histidine 0.31 gms; Isoleucine 0.80 gms; Leucine 0.95 gms; Lysine 0.78 gms; Methionine 0.38 gms; Phenylalanine 0.62 gms; Threonine 0.63 gms; Tryptophan 0.22 gms; Valine 0.93 gms.

The lipids which constitute about 11.5 gms have both monounsaturated fatty acids and polyunsaturated fatty acids to the extent of 3.97 gms. Besides, there are also linoleic acid (2.22 gms), linolenic acid (0.3gms) and arachidonic acid (0.24gms).

Egg is also a rich source of vitamins except C. Vitamin A(1015 I.U), D(925 I.U)and E (1.8mg) are present in the egg besides vitamin B. The vitamin B composition is as follows - Thiamine B1 0.09 mg, Riboflavin B2 0.26 mg, Pantothenic acid 1.48 mg, Choline 5.32 mg, Niacine 0.09 mg, Pyridoxine B6 111 mg, Folic acid 5.54 mg, Biotin 9.25 mg, Vitamin B12 0.93 mg, Inositol 20.3 mg.

The mineral elements present in the egg include calcium, iron, phosphorus, sodium, potassium, magnesium, sulphur, zinc, chloride, manganese, iodine, copper,fluorine and selenium. Eggs also have trace elements like chromium, aluminum, molybdenum etc.

Some of the birds are mainly bred for the purpose of meat. The chicken meat is easily digestible and even recommended for persons suffering from heart ailments as it has very little cholesterol.

The annual egg production in India exceeds 25,000 million and is ranked as 5th in the World.

Poultry as a subsidiary source of income

Even before the advent of modern methods of poultry farmers in India and elsewhere have been keeping the fowls for their use as well as for marketing purposes. Both the eggs as well as the chicken are sold and they provide an additional source of income for farmers.

With the introduction of modern methods of poultry and the interest taken by national agencies like Intensive poultry development project (IPDP) which was setup during 3rd 5 year plant poultry keeping has taken a big boost. During the 4th 5 year plan the Indian Council of Agriculture Research developed many research projects to improve suitable strains both for egg laying as well as for meat production.

The other national level agencies which have given a big boost to poultry are National Agricultural Cooperative Marketing Federation (NAFED) , National Egg Coordination Committee (NECC) etc. In the state of Karnataka also near Hesaraghatta in Bangalore there is a Government poultry farm engaged in the production of new breeds. As a result of all these activities farmers in Karnataka have taken to poultry keeping on a large scale. Apart from small attempts by farmers maintaining 100 to 1000 birds, large hatcheries have been established where scientific system of rearing the birds is practised on a very large scale. Some of these hatcheries have 50,000 to 1,00,000 birds. In large hatcheries special breeds are maintained for egg production while broiler breeds are maintained mainly for meat.

Apart from the egg and the chicken meat the fowls are also helpful to farmers in providing a very good manure. The faecal matter of birds is a rich source of nitrogen and a very good organic manure.

Breeds of Poultry

The original habitat of the modern breeds of chicken is south and central India, the Himalayan region, Assam. Burma and Ceylon. There are four species of the wild fowl and all of them belong to the same genus-*Gallus* which means a cock. The four species are -

1. *Gallus gallus or Gallus bankiva*(red jungle fowl)
2. *Gallus lafayetti* (Ceylon jungle fowl)
3. *Gallus sonneratti* (Grey jungle fowl)
4. *Gallus varius* (Jawa jungle fowl)

All the four species can cross with one another and the hybrids are more or less fertile. The domestic fowl can be classified into two categories. These are the indigenous or Desi breed and the exotic or improved breed. The Desi breed are indigenous to India but generally they are poor layers and the eggs are small in size. There are four important varieties of Desi chicken. These are Aseel, Chittagong, Ghagus and Basra.

The exotic or improved breeds are modern breeds of chicken. They include all those breeds which are produced outside India and later imported to our country. Most of the breeds however are successfully bred in our country and have also acclimatised to our country.

The exotic breeds are basically of four types -

American class - This includes birds which were developed in USA. They serve the dual purpose of both meat and egg. The American class is produced by crossing the Mediterranean and the Asian classes. There are eleven different breeds belonging to the American class. Of these however the important ones are Rhode Island red, Rhode Island white, Wyandotte, New Hampshire, Plymouth Rock and Delawares.

Asiatic class- This includes breeds that are developed in the Asian continent. These are usually for food and they are poor layers of egg. The important varieties in this class are Brahma, Cochin, Langshan, Giriraj etc.

English class -This group of chicken are popular for their good quality meat. The breeds have white skin and red ear lobes. They lay brown shelled eggs. The six known varieties are Sussex, Australorp, Orpington, Dorking, Cornish and Red cap.

Mediterranean class- This includes birds which originated in Europe near the Mediterranean sea. These are generally small sized and light bodied. They mature early and start laying eggs when they are just 5 or 6 months old. They are good layers and consume less quantity of food. There are several varieties in this class but the important ones are Leghorn, Minorca and Ancola.

In this book White Leghorn, Rhode Island, Plymouth Rock, New Hampshire and Giriraj are discussed in some detail.

White Leghorn

Barred Plymouth

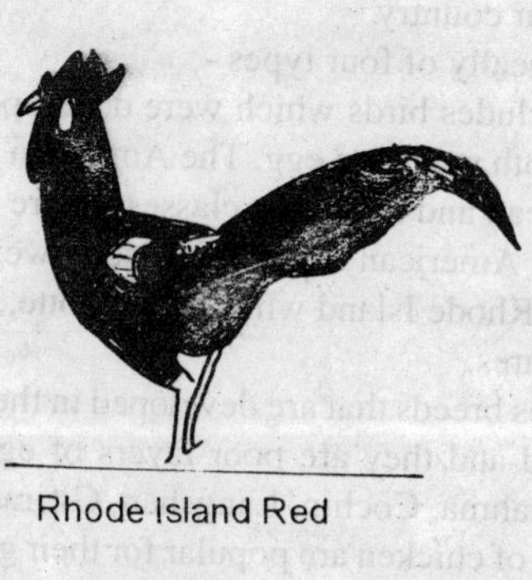

Rhode Island Red

Fig. 36.1 Some exotic varieties of chicken

White Leghorn- The leghorn birds belong to the Mediterranean class. They are the most popular breed for poultry farming as they are excellent layers. However the flesh is not as delicious as in other birds. The body is short with a long back and a protruding breast. The ear lobes are creamy with rose comb. The White Leghorns were first introduced to out country by foreign missionaries about 75 yrs ago. This variety is good only for egg laying as the meat is not delicious. The birds can be kept very well in dry regions as they do not prefer wet areas. White Leghorns are very popular as layers as they produce the eggs within about 6 months of their birth. The average weight of a leghorn cock is about 2.75kg while that of the hen is 1.75kg.

Other varieties of leghorn include the brown leghorn, the black leghorn and the buff.

Rhode Island- These birds belong to the American class. There are two varieties in Rhode Island. These are Rhode Island Red and Rhode Island White. The Rhode Island Red is the most popular of the American class. It has a hardy nature and can survive even in damp rainy regions. The body is somewhat rectangular and the plumage is dark red or brownish red. It has a good quality of meat and a good layer of eggs. The eggs are brown shelled , the chicks grow very quickly and easy to rear. There are two varieties of Rhode Island Red - the single comb type and the rose comb type. The single comb variety is more popular than the other . The average weight of the cock is 3.8 kgs and of the hen 3kgs.

The Rhode Island White is an uncommon breed. It has rose comb and developed from white Wyandottes and white leghorns. The plumage is pure white and does not have any tint.

Plymouthrock- This bird belongs to the American class. It has a good size and rather long bodied with a prominent breast and fat body. It is the most popular breed in America and the cocks have a single comb. The average weight of the bird - cock 4.5 kg and hen 2.2 kg. This breed is not very popular in India but the males are used for breeding with the Desi varieties. The Plymouthrock has several varieties such as barred, Columbian, silver pencilled, blue and partridge etc.

Newhampshire- The Newhampshire is an American class bird and has developed in Newhampshire from light coloured strains of Rhode Island Reds. The Newhampshire has a single comb and the body is less rectangular than the Rhode Island Red. In the Newhampshire males the back is deep chestnut red while the females have the parts of the body having medium chestnut red. The birds are good layers and the eggs are brown shelled. It is also a good meat producer. The average weight of the cock is 3.5kg and that of the hen 2.5kg.

Giriraj -The bird belongs to Asiatic class. This is a hardy breed obtained in Karnataka. It is a good broiler used for meat. The meat is soft and tasty. The average weight of the bird is about 3-4 kg.

Thirty seven

Sericulture

Sericulture refers to the production of silk fabric by rearing silk worms. The production of silk has a very ancient past dating back to more than 3,000 years. While it is generally believed that silk industry. Originated in china there is another opinion that states silk worm rearing was known to the Indians living at the foot of Himalayas. It is from here it spread to China. According to the Chinese origin it is believed that a Chinese princess Siling Chi was the first one who discovered the art of reeling an unbroken thread of Silk from the cocoon. This art of silk rearing was kept a secret and any one who tried to take the eggs of silk worm out of the country was given capital punishment. In the year 555 A D the technique was smuggled out of the country by two monks. Later sericulture was spread to a number of countries including Europe and other countries of Asia. Presently Sericulture is an Agro based cottage Industry in a number of countries like Japan, China, India, Korea, Brazil, Italy and France. In India Investigation into Silk worm rearing was first started at the Pusa Institute at New Delhi in the year 1905.

Sericulture as an Agro based Industry:

India is one of the important silk producing countries. Karnataka State also is playing a great role in production of silk. There are separate Government organizations which include Central Silk Board as well as the State Silk Board which are engaged in the various activities controlling the sericltural industry. Sericulture is practised by our farmers as a subsidiary activity along with the main agricultural activity. Basically there are 3 phases in sericlutre. These are cultivation of mulberry, production and rearing of silk worms including production of cocoons and reeling of silk threads from the cocoons. The last activity in sericulture provides raw material for textile industry where fabrics are woven using the silk threads.

For the silk worms (*Bombyx mori*), the mulberry plant happens to be the only natural food, hence the first step in sericulture is cultivation of the mulberry plant the leaves of which constitute the food for silk worm. The next step is obtaining the eggs and feeding the silk worms with the leaves of mulberry. The silk worms feed on the leaves and develop into cocoons, which are then boiled in water and thread from it is obtained as silk fibre.

Sericulture in the state of Karnataka is not only an important agricltural activity but also a profitable cottage industry providing them an additional source of income. Using modern methods of agriculture which includes improved varieties of mulberry, quality breeds of silk worm, sericulture has become an important activity for farmers.

Importance of Mulberry Cultivation :

It has been mentioned above that for the silk worm-*Bombyx mori* the mulberry leaves happen to be the only food . Hence cultivation and proper yield of the mulberry plants occupies an important place in sericulture. Extensive research is done on the improvement of mulberry plant which includes disease resistance and high yield form the plants.

Taxonomically the genus Morus has more than 12 species distributed very extensively in India and other East Asian countries like Japan, Indonesia etc., Only about 3 or 4 species of the genus-*Morus alba, M.indica, M.rubra* and *M.nigra* are used in silk industry.

The mulberry plant is a medium sized tree belonging to the family Moraceae. This deciduous tree has a soft white wood which is in great demand in the manufacture of sports goods while the leaves constitute the food for silk worm. The cultivated mulberry plants however generally do not reach the size of a tree as they are pruned regularly.

Soil: Mulberry plants grow very well in a loam soil. The soil should have proper drainage. A pH rang of 7.0 is ideal for the growth of plants

Planting: Two methods are followed

(i)*Seeds:* Seeds collected from ripe fruits of mulberry plant are sown in the soil generally during the months of December and May. The seedlings spring up very fast and ready for transplanting in June, July or November. Seed beds should have good manure like lime and ash. In order to prevent soil borne infections fungicides may be added to the soil.

(ii)*Cutting:* Stem cuttings offer a rapid method of propagation of plants. Cuttings are obtained from improved varieties of plants in the months of July - August and are planted in the field. Two methods are followed in Karnataka in planting the cuttings.

(1)*Pit planting:* Cuttings are planted in individual pits with the space between pits being 3'x3'.

(2) *Row planting:* This is of two types again- single row planting and multiple row planting. In single row planting, Mulberry is a subsidiary crop. Plants are raised along the boundary of the field. About 400 plants are required to cover the boundary of one hectare of area. In the multiple row system, the entire plot is used for mulberry cultivation. About 10,000 plants can be cultivated per hectare with the distance between plants being one square meter. Between the rows of plants the soil is made into furrows for easy drainage of water.

Fertilizers and Irrigation: Mulberry plants require adequate irrigation and fertilizers for proper growth and yield. Weekly irrigation is a necessity for a good yield. There should be proper chanelling of the soil for the easy flow of water. Good drainage that does not allow water logging near the base of the plants is required. Agricultural practises like mulching will help in the preservation of moisture. Deweeding should be done regularly.

Organic as well as chemical fertilizers are required for the healthy growth of mulberry. Quantity and frequency of manuring the soil depends on the type of soil, fertility etc. Generally however farmyard manure of 10 kg per plant per year (for one year old plant) and super phosphate 250 gms per plant per year (for one year old plant) are required. The doses of fertilizer should be increased with the increase in age of the plant.

Purning and Thinning: Pruning of the plant is very essential to give a proper orientation to the growth of the plant. Thinning also helps in avoiding overgrowing and results in a better yield. The months of December and January are best suited for pruning. In mulberry plantations, plants will be ready for harvest in 4-5 years time for the feeding of silk worms.

There are two types of pruning- bottom pruning and middle pruning. In the former, pruning is done at the base and in the latter it is done at the middle region.

Harvesting : The main crop in mulberry plant are the leaves. Leaves are plucked from the plant in the following three ways.

(i) Branch cutting: leaves along with short branches are cut from the plant.

(ii) Leaf picking: Individual leaves are plucked from the plants and are fed to the worms. This is however a tedious task requiring more manual labour.

(iii) Shoot leaf picking: Shoots are removed from their base for feeding the worms.

Dried and diseased leaves should not be fed to the larvae. As far as possible fresh leaves should be provided and for this purpose leaves may be collected in baskets and covered with moist cloth.

Fungal (powdery mildew and leaf spot) and insect diseases (leaf eating caterpillars, Jassids, mealy bugs etc) can be controlled by proper fungicidal and insecticidal sprays.

Life cycle of silk moth :

The most popular among the silk moths is *Bombyx mori* which is popularly called the mulberry silkworm, as its main feed is mulberry plant. The life history of mulberry silk worm consists of four stages- the adult, the egg, the larva(catripillar) and the pupa stage.

Adult stage - The adult silk moth is about 2.5 cms in length and pale creamy or white in colour. The moths are dioecious (male and female moths are separate). In the female, due to the heavy body and generally weak wings flight is impossible. The female moths are larger than the male.

The body of the moth is divided into three parts namely head, thorax and abdomen. The head has got a pair of compound eyes and there are two antennae. There are three segments in the thorax- these are the prothorax, the mesothorax and the metathorax. The prothorax is quite big. The moth has got three pairs of legs and there are two pairs of wings.

The female moth generally does not feed during its very short life of about 2-3 days.

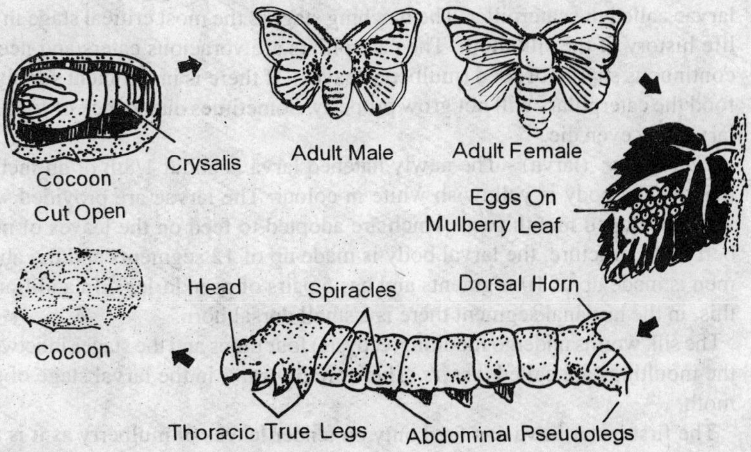

Fig. 37.1 Life history of Bombay mori

Fertilization- After copulation, fertilization takes place. Just after the emergence from the pupa, the male moth copulates with the female for about 2-3 hours and then they separate. After fertilization the female starts laying eggs and it takes about 24 hours to complete the process. On an average a single moth can lay about 400-500 eggs. The quantity of eggs depends on the climatic conditions and the supply of food material to the caterpillar from which the female moth is developed. They (eggs) are usually laid on the upper surface of the mulberry leaves. The eggs are usually in clusters and are enclosed in a gelatinous substance which is secreted by the female moth. The gelatinous substance helps in the attachment of the eggs to the leaves.

The eggs are minute in size, spherical in shape and white in colour. The weight of about 1000eggs will be less than one gram. Gradually the eggs become dark in colour. It has been obsorved that in mulberry silk worms there are two types of eggs. These are the *diapause type* and the *nondiapause* type. The silk worms of temperate regions lay the diapause type whereas the silk worms of subtropical regions (as in India) lay the nondiapause type of eggs. In diapause type of eggs there is a hibernating stage (resting stage). The nodiapause type of eggs are nonhibernating and consequently the moth will take a very short time to complete the life cycle. In fact, the mulberry silk moth in India can complete five life cycles per year.

Hatching of eggs - After 10 days of incubation, the eggs are hatched into larvae called the caterpillar. The hatching stage is the most critical stage in the life history of the silk moth. The caterpillars are voracious eaters and need a continuous supply of food (mulberry leaves). If there is insufficient supply of food the caterpillars will not grow properly. Sometimes due to lack of food the larva may even die.

Caterpillar (larva) - The newly hatched larva is about 1/8th of an inch in length. The body is yellowish white in colour. The larvae are provided with well developed mouth parts which are adopted to feed on the leaves of mulberry . In structure, the larval body is made up of 12 segments and the abdomen is made up of 10 segments and has 5 pairs of pseudo-legs. In addition to this, in the last anal segment there is a small dorsal horn.

The silk worms undergo moulting(ecdysis) four times and the stages inbetween the moultings are called *instars.* There are 5 *instars* in the larval stage of silk moth.

The first instar larva can feed only on tender leaves of mulberry as it is at a very young stage. The first instar larva feeds continuously and grows rapidly. After the first, second, third and fourth moultings the larva grows into second, third, fourth and fifth instars respectively. The whole moulting process upto the 5th instar stage takes about 21-25 days after hatching.

The 5th stage larva or the caterpillar can be regarded as the final stage of growth. The caterpillars at this stage are fully grown and measure about 7.5 cms in length and in weight it ranges between 4-6 gms. The caterpillars develop salivary glands and generally stop feeding before they reache the pupa stage.

Pupa - This is the next stage of development. Initially the fully grown caterpillars stop eating and gradually withdraw towards a corner among the leaves in the feeding trays. They secrete a sticky fluid through the silk glands. The fluid comes out through a narrow pore located on the hypopharynx of the caterpillar. The fluid gradually comes out in the form of delicate threads which hardens on exposure to air. Gradually the number of threads coming out of the caterpillar increase and they form a covering around the body called the cocoon.

Cocoon - The cocoon is a covering generally white in colour and is made up of closed weaving of the silk threads. In the cocoon the threads on the exterior are somewhat irregular while those present towards the inner region of the cocoon are regular in arrangement. It has been estimated that the length of a single continuous thread secreted by a pupa for the formation of cocoon is about 1000-2000 mts in length. The secretion of this thread requires about 3 days time. The arrangment of the thread in the cocoon is in a concentric manner. In order to arrange the thread in a concentric manner the pupa rotates its head at the rate of 65 times per minute constantly.

A fully formed cocoon has the pupa inside. Estimates have shown that 2500 cocoons can yield upto one pound of silk. In the weight of the cocoon (1.8-2.2 gms) the shell which is useful in extracting silk is only about 0.45 gms. The pupal stage takes about 10-12 days to complete the life cycle. At the end of this stage the pupa would have grown into an adult (imago) and it comes out of the cocoon by cutting it open.

Emergence of the adult moth *(imago)*- The pupa when it is inside the cocoon undergoes physiological and morphological changes called metamorphosis. The abdominal pseudo legs disappear and in its place two pairs of wings develop. The moth when it is inside the cocoon secretes an alkaline fluid which moistens one corner of the cocoon. At this point, the cocoon is soft and the moth cuts the threads to make a hole. It is through this the adult moth comes out of the cocoon.

Races of mulberry silk worm - Extensive research by sericulturists have resulted in the production of several races of mulberry silk worms. These races are mainly categorised based on the number of life cycles that they can complete within a year. This is very important from the point of view of production of silk. If the worm has more number of life cycles per year to that extent the quantity of the silk produced will also be more. Generally the number of life cylces per year varies on the temperature and other climatic conditions. In temperate countires the number of life cycles is less whereas in tropical countries like India the moth may have 2-7 life cycles per year. Based on the number of life cycles the races of mulberry silk moth are classified into the following -

Univoltine - One life cycle per year (one crop per year).

Bivoltine - Two life cycles per year.

Multivoltne - More than two life cycles per year.

Varieties of silk (Mulberry and Nonmulberry)

There are several varieties of silk and these are based on the type of silk worm from which it is produced. The following are some of the varieties of silk.

Mulberry silk - This is produced by the worm *Bombyx mori* belonging to the family Bombicidiae. This moth is belived to belong to China originally. Presently, it is the major silk moth and is found in all silk producing countires of the world like Japan, India , Korea, Italy, France, Russia etc. The silk threads produced by this moth is white or yellowish in colour.

Tasar silk - This is produced by the worm called the Tasar silk worm. The moth (zoological name- *Antheraea paphia*) belongs to the family Saturniidae. It is mostly found in India, China and Srilanka. This worm feeds on a number of plants such as oak (*Quercus*), Sal(*Shorea*), Fig (*Ficus*) etc. The cocoon of this moth is very big and almost of the size of a hen's egg. The surface of this cocoon is very hard and the silk threads are brown coloured. Originally the Tasar silk worm was not cultivated. Recently however cross breeds have been produced and the tasar silk worm has been domesticated.

Breeding of tasar silk worms is a very difficult process as the moths do not breed in captivity. In our country Bihar, Madhya pradesh, Uttar pradesh and Orissa produce tasar silk.

Muga silk - This silk is produced from the worm *Antheraea assama* also belonging to the family Saturniidae. This worm originally belongs to Assam where it is cultivated on a sufficiently large scale. Muga silk is a cottage industry in Assam. The caterpillars of this moth feed on Machilus plant (M*achilus bombycina*).

Eri silk - This is produced by the worm *Attacus ricinii* belonging to the family Saturniidae. This silk is mainly produced in East Asia. The silk worm feeds on leaves of the castor plant (*Ricinus communis*). The cocoons have the silk threads arranged in an irregular manner hence it has to be spun and cannot be reeled. The silk threads on tne cocoon are very loose and glossy. But they are not as durable as mulberry silk . In our country, Eri silk (also called Arandi silk) is produced in Bihar and Assam.

Oak silk - This silk is produced from the worm *Antheraea pernyi* also belong to the family Saturniidae. It is found in China and Japan. This worm produces good quality silk.

Uses of silk - Silk fabric is in great demand since time immemorial. The threads are woven in various methods such as laying, crepe,georgette etc. It can be made into dress materials, sarees, jackets, shawls etc. Besides this silk finds a number of non garment use. These are parachutes, parachute cords, fishing lines, insulation for electric cables and telephone cables etc.

Byproducts of silk industry

The main product in silk industry is the production of thread obtained from the cocoons. In addition the silk industry can produce a number of byproducts. Some of these are -

Silk worm oil - This is an oil which can be extracted from the dead pupae. It is used in the manufacture of soaps. After the extraction of the oil, the waste (cake) is used as poultry feed.

Silk worm faeces - The excreta of silk worms finds a number of uses. It can be used in the production of vitamins (K and E), in the manufacture of acid resistant plastic sheets etc. The faeces is also a good organic manure and can be added to the cattle dung to produce gobar gas.

Manufacture of garlands - The remains of the cocoon after the removal of silk threads can be fashioned into garlands. These silk garlands are manufactured in Channapatna of Bangalore district.

Uses of Mulberry plant

The mulberry plant which is the main stay of silk worm moth has a number of uses. Some of these are -

(i) the fruits are edible; can also be used in the manufacture of alcoholic drinks

(ii) the stem in the trees is white in colour, soft and pliable. It is in great demand in the manufacture of sports goods specially cricket bats. Many agricultural implements can also be fashioned out of the stem. The wood pulp of the stem is a good raw material for the manufacture of quality paper.

(iii) mulberry leaves constitute a good cattle feed. It is known that this feed increases the milk yield.

(iv) various parts of the plant body have a high medicianl value and are used in the Ayurvedic system of medicine.

(a) the fruit is aromatic, cooling, laxative, removes thirst and good in the treatment of fevers.

(b) bark is anthelmentic and purgative.

(c) leaves in the form of decoction are used to gorgle in the treatment of inflammation of the vocal cards.

(d) roots are anthelmentic and astringent (acts as a binding agent) and useful in the treatment of diahhroea.

There are four species of mulberry - *Morus acedosa, M.alba, M. indica and M. nigra.* All these species are of medicinal value.

Thirty eight

Pisciculture

About 5/7th of the earth by volume is occupied by water. Salt waters like oceans, seas and fresh waters like rivers, streams, ponds, lakes etc., constitute the total quantity of water. Water plays a very important role in the life of all organisms including human beings. Apart from helping in navigation, commerce etc., water, both marine as well as fresh water is the source of an immense variety of protein rich food which can be harnessed for the welfare of mankind. Among the important items of food used by man fishes play a very pivotal role. Varieties of fishes rich in nutrieants can be effectively used to overcome the food problem to certain extent.

Fishes may be obtained from two sources. These are marine and fresh water sources. In India which is a peninsula, the country is surrounded by sea water on three sides with a very long coast line. The sea waters of the Indian coast contain varieties of fish constituting a major source of revenue. The Bengal and Arabian sea and a large number of islands with their swamps occupying almost 2,59,000 Sq.Kms. are a rich source of fish population.

Apart from marine water vast streches of fresh water consisting of rivers, streams, lakes, ponds and even paddy fields offer immense scope for inland fishing.

Fish are usually caught either in the sea or in fresh water using varieties of methods. The different types of fish vary in the nutritional contents. After realising the food potential of fishes man has started culturing fish in salt water as well as fresh water in addition to catching them wild. The science of culturing fish in protected bodies of water is called *Pisiculture*. Another branch of science-acquaculture is closely associated with pisiculture. In acquaculture a variety of acquatic animals (fish,prawn,lobster,crab etc.,)which are of nutritional value are cultured.

Pisiculture has advanced very rapidly in recent days with refined techniques and scientific methods. Maximum productivity, low mortality, increased nutritional content have all been achieved by using scientific techniques in pisiculture.

Economics of Fish Forming

India has a very vast coastline and a large quantity of different types of fresh water resources which are virtually a gold mine from the point of view of fish farming. In fish farming required types of fish are cultured both in protected bodies of sea water as well as fresh water. As a source of food, fish is second to none from the point of human welfare. It has been estimated that in our country about 8.5 million tonnes of fish are required annually to meet the present day demand. However as against this the annual production of fish in our country is only about 1.7 million tonnes.

A survey conducted has shown that the inland waters of India which run into several million acres have a vast potentiality of producing required quantity of fish. Researches conducted by the Central Inland Fisheries Research Institute have introduced modern methods of pisiculture and a production rate of 85,000 Kg. per hectare per year has already been achieved.

Pisiculture provides gainful employment to a large number of people besides providing various benfits like earning foreign exchange through export of fish to various countries. In our country many coastal states export fish and fish products to various countries.

Fishing industry has several by-products in addition to the main product namely the fish. The many by-products of the fishing industry which are a potential source of revenue are body oil, liver oil, fertilizers and other food products etc.

The oil obtained from fish is of two types. Body oil and liver oil.

The body oil extracted from the whole fish is of two types. Dry oil and semi drying oil. Dry oil is obtained from sardines, salmon etc. While semidrying oil is obtained from carps. Body oil of fishes finds extensive use as an industrial lubricant.

The liver oil obtained from several fishes is of great medicinal value and is a rich source of vitamin C,D and E. Liver oil is mainly obtained from Cod, Shark etc.

Another by-product of fishing industry is the fish meal which is prepared from the wastes of fish oil.

Fish protein is also extracted from fish wastes by digesting the underisable fats with solvents. Fish protein is used in preparation of icecreams, varnishes, textile, paper, cosmetics etc.

INLAND FISHERIES

As has already been pointed out, our country has a vast and varied source of inland water which can be economically exploited for the culture of various kinds of edible fish. In addition to the normal water reservoirs, ponds, lakes, rivers etc., man made water systems like dams, irrigation canals and even paddy fields, provide immense scope for the cultured propagation of edible fish.

Inland fisheries deals with two types of fish catching. 1) direct capturing of naturally grown fish from water bodies 2) by employing artificial methods of cultivation fish eggs may be hatched and grown in artificially created bodies of water.

Inland fisheries offer an immense scope to increase the production of fish. The main aim being increase in quality as well as quantity of the product. Inland fisheries is also useful in utilising vast resources of unexploited water bodies. Many a time these water bodies are a rich source of mosquito breeding. By cultivating fish these bodies of water can be brought to good use.

Types of fish cultivated inland in India:

Carps

Catla catla
Labeo rohita — *Labeo calbasu*
Labeo bata — *Cirrhina mrigala*
Cirrhina reba
Cirrhina cirrhosa

Cat fish
Mystus seenghala
Mystus oar
Rota rota
Eutropichthys vacha
Wallago attu

Sheat- fish
Ompok bimaculatus
Notopterus chitala

Feather backs
Notopterus notopterus
Clarias batrachus
Hetropneustes fossilis

Live-fish
Anabas testudineus
Channa striatus
Channa gachua
Channa punctatus
Liza corsula

Mullet

Liza carcasia

Herring
Hilsa ilisha

Anchovies
Sctipinna phasa
Anguilla bengalensis

Eel
Mastocembalus armatus

Management of Inland Pisiculture:

Culture of fish for productive purposes is a difficult program and involves various techniques and also a thorough understanding of the life history charactaristic features and physiological requirements of the fish to be cultured. In addition the abiotic factors of the environment like the type of water, source, suspended material in water, oxygen content, sources of food for fish in water are all very important factors that one should bare in mind in a Pisiculture program.

Pisiculture involves the following steps.

Collection of eggs: Required types of fish eggs are collected during monsoon months from rivers or other bodies of water. The eggs are collected twelve to fourteen hours after fertilization and made to hatch under protected conditions. For the purpose of collection of eggs mosquito net are other types of suitable net may be used.

Collection of seeds: Fish seed means collection of young ones from the breeding ponds or from natural sources. The fish seeds are of two types *fries* and *fingerlings*. Fries are just hatched fishes while fingerlings are little older than fries. In our country a large number of collection stations have been established near the banks of Ganga, Yamuna etc.

Hatching pits: The fertilized eggs are released into small pits located very close to the breeding ground. These pits are specially made to promote the hatching of the eggs. The convenient size of a hatching pit is 8'X4'X2'. Usually a series of hatching pits are established with interconnections so as to allow the free flow of water from one pit to another.

Hatching pits are of two types. *Hatcheries* and *hatching hapas*. Hatcheries are small sized pits into which fertilized eggs are transfered. Hatcheries have some drawbacks such as vagaries of temperature, attack by fish predators and drying up of the pit etc.

Hatching hapas is an improved method when compared to hatcheries. These are rectangular troughs made of a very coarse cloth of three layers. The outermost layer is made of a cheap coarse cloth. The middle layer of the tank is somewhat smaller and is made up of a round meshed mosquito net. The inner most cloth tank is still smaller in size.

In some parts of India, Hatcheries are made of cement and consist of two walls.

Hapas have an advantage over plain hatcheries as they are generally protected from predators and are not subject to vagaries of temperatures.

Transport of fries to cultured ponds. The fish fries are collected from hatcheries and transfered to large culture ponds. The fish hatchery is not a place for the complete growth of fish as its area is small. It is mainly meant to produce fish seeds. These seeds should be transported to larger ponds. These ponds are of three types. Nursery ponds, rearing ponds and stocking ponds.

Nursery pond: The fries which are transported to nursery ponds are very fragile and they should be carefully maintained. As there is a possibility of high rate of mortality during transport it is advisable to have nursery ponds as close as possible to hatcheries.

The nursery pond is a shallow water reservior ranging in depth from 3 to 8 feet. The extent of a nursery pond may vary from 2000 Sq feet to about half an acre. A very vast nursery pond is not recommended as it will not be possible to control the conditions necessary for fish culture.

The nursery pond generally has less natural food and it should be agumented by adding food and promoting development of natural food for fish. For this purpose chemical fertilizers should be added to the water which will promote the growth of zooplanktons and phytoplanktons. By the use of these manures it has been seen that the pond water will be full of tiny acquatic life within a matter of 10 to 15 days. When once the natural food is available in nursery ponds, the fries of fish can feed on them comfortably.

One of the important precautions to be taken in a nursery pond is to prevent the presence of predators which will create havoc to the fish culture. Another care to be taken is not to allow excess growth of planktons which may decrease the oxygen content of water.

After the fish fries are grown to about 10 to 15 cms they are transfered to another type of a pond called *rearing pond* which is necessary for their further growth.

Rearing ponds: After the fish fries have grown to a sufficient size, they should be transfered to larger ponds for two reasons. 1. As the size of the fish increases they should have free mobility in water without hindrance.
2. Large bodies of water like the rearing pond will have higher quantity of food resources which is necessary for fish.

Generally the rearing ponds are much larger than nursery ponds (not necessarily in depth). They are usually narrower and longer to allow the fish to swim for long distances. Paddy fields having a depth of water, up to 2 or 3 feet may be an ideal rearing pond. The depth of the water in any case should not be more than six feet. The water of rearing pond is usually seasonal and it should retain water for atleast two or three months.

The rearing pond, even though it may be a seasonal body of water, should be prepared properly before the fries are transfered. The preparation includes removal of extra weeds, eliminating the predators, manuring the ponds to facilitate the growth of suitable planktons etc.

The fish fries are usually transfered by collecting them in nets and transporting them into metal containers. In rearing ponds the fish reach a maximum size of about 20 cms when they should be transfered to the stocking ponds for further growth.

Stocking ponds: The fingerlings that have reached the maximum size in rearing ponds are transfered to stocking ponds in containers of 1000 liter capacity. These containers are specially made to prevent injury to young fish. These containers should have facility for aeration. In some instances, in order to prevent excessive movement of the fish, mild sedatives are also used.

Before transportation, the fingerlings should be carefully washed by using medicated water (such as antibiotics, potassium permanganate) in order to prevent the spread of any possible diseases that the fish might have superficially contacted when they are in rearing ponds. One of the fungi, *Saprolegnia parasitica* is known to create havoc in fish hatcheries as it can kill the entire stock due to infection.

The stocking ponds are usually large perennial ponds covering an area ranging from 2 to 20 hectares with an average depth of about 6 feet. Some precautions have to be taken to prepare the stocking ponds before the fingerlings are released into them. The process of preparing is same as that followed for rearing ponds. Preferably all predators must be eliminated eventhough it is difficult as the stocking pond is very wide and deep also. As increased quantity of food is required because the fish are in a rapid stage of growth, measures are to be taken to promote the growth of natural food for fish. Manuring is an important step to promote the growth of natural food. Organic manuring to the extent of 20,000 to 25,000 Kgs. of cow dung per hectare per year is essential for the proper growth of planktons. Chemical fertilizers such as super phosphate, ammonium nitrate and ammonium sulphate at the rate of 1000 to 1500 Kgs. per hectare per year should also be used. In addition to this, artificial food like powdered rice, oil cakes, mustard seeds, groundnut powder, etc., may also be added to augument the food resources. The artificial food should not be added in large quantities at a time. Small quantities of this are to be added daily, preferably in the morning hours.

The carrying capacity of the stocking pond depends on several factors like the type of fish, availability of food, oxygen content etc.

It has been seen that a mixture of different species of fish(polyculture) in a stocking pond would have a higher yield when compared with a single species (monoculture). No fixed ratio of different species of fish in a polyculture has been evolved as yet. But experiments have shown that a combination of *catla, rohu* and *mrigala* in the ratio of 3:3:4 would yield best result. On an average polyculture of above species would yield between 1000 to 1500 pounds per acre.

Harvesting : Capturing of the fully grown fish after they attain their maximum body length and weight is known as harvesting. Well grown fish are marketed while the small ones that are caught are released into water for continued growth.

Different fish are harvested by using different methods. These include stranding, angling, trapping, scooping and using varieties of nets.

Stranding is a method widely employed to catch fish in shallow waters. water is thrown out from the shallow pit after separating it from the main body of water. After scooping out the water, fish are caught.

Angling is a very old and common practice of fishing. Fishing rods with baits are dipped into the water and when the fishes are caught while swallowing the bait, they are scooped up. Angling is practised both in shallow and deep waters.

Traps are special devices used for catching fish, particularly through moving water. Traps are usually woven baskets of cane and are kept in a place where the water is moving. After some time the traps are lifted and the fishes if anv are caught.

In *scooping,* a net is tied to a rectangular or circular bamboo frame and is dipped in water and lifted up. The fish any will be trapped in this net.

Varieties of nets are used to catch the fish. These are lift nets, cast nets, purse nets, gill nets, drag nets etc. The basic method of all these nets are same. To trap the fish and to allow the free flow of water.

Fishes of Karnataka

The state of Karnataka is gifted with plenty of water resources both fresh water as well as marine. It has a long costal line and a number of rivers like Cauvery, Sharavathy, Kabini, etc., flow across the state. All these are very rich in a variety of edible fishes. The following is a description of the edible fishes of Karnataka. The major types of edible fishes are *carps* and *cat* fishes.

Carps: Carps are generally soft finned fresh water fishes that possess large scales on their body. There are a number of carps that belong to the following types.

1. Catla: This fish is found distributed in almost all the rivers. At maturity it attains a maximum length of about 180cms but it is suitable as a food with high nutritive value when it is about 60 cms in length. The body of the fish is stout and has a broad snout. The mouth is large having a prominent lower lip with a large gill apparatus. The fins on the body of the fish are somewhat dark in colour. The colour of the body is greyish on the dorsal side while it is a shiny silvery on the lateral surface.

Catla is one of the dominant fishes of fresh water throughout India. It migrates to the rivers of plains during the breeding season (June to August). The larvae are about 4 to 5 mm in length. The young ones grow to 15 to 20 cms in length and feed on water fleas.

***Systematic position of Catla*:**

Class........Teleostomi
Order........Cypriniformes
Family.......Cyprinidae
Genus........*Catla*
Species......*catla*

Labeo : The genus Labeo has a number of species of which *Labeo rohita, L.calbasu*, and *L.gonius* are important.

Labeo rohita is commonly called rohu. This is a very important carp fish distributed in the plains of India. A Fully grown rohu is more than 90 cms long. It has an elongated body with a projecting snout. The colour of the body is bluish or brownish green. The lips are thick and the fins are grey or blakish in colour.

The rohu fish has feebly developed *barbles* and herbivorous in its food habit. It is a bottom feeder with the body covered by cycloid scales. This is a most suitable fish for pond culture.

The rohu fish breeds during the monsoon period and the females have an enormous output of egg which is estimated to be 2 million eggs at a time. The young fish may reach a length of 45 cms.

Systematic Position of *L.rohita.*

Class......Teleostomi
Order......Cypriniformes
Family.....Cyprinidae
Genus......*Labeo*
Species....*rohita*

Labeo calbasu: This is one of the commenest fishes in the rivers of south India also found in Punjab, West Bengal etc., This is also a bottom feeding fish feeding on algae, vegetable matter, snails, worms etc. These are found in large numbers in the ponds. The body of these fishes is stout with a somewhat long conical head. One characterstic feature with this fish is the presence of elongated barbles and special colour of the cone. A fully grown fish may reach a length of 90 cms but in fish ponds they usually grow up to 25 to 30 cms per year with a maximum length of 42 cms.

Systematic position of *L.calbasu*

Class......Teleostomi
Order......Cypriniformes
Family.....Cyprinidae
Genus......*Labeo*
Species....*calbasu*

Labeo gonius: This is generally found in North India. But sometimes also found in the South Indian rivers like Krishna. It has a long body with smaller scales provided with dark margins. The lips are some what protruded and thick. This fish is also a bottom feeder and feeds on submerged or planktonic algae, organic matter, leaves of submerged plants etc. The fish is supposed to have a very good taste among all the carps. Like other Labeo fishes this also breeds during monsoon season.

Systematic position of *L.gonius*

Class.......Teleostomi
Order.......Cypriniformes
Family......Cyprinidae
Genus.......*Labeo*
Species.....*gonius*

Cirrhina : Among the several species of *Cirrhinus,* two species are generally distributed uniformly. These are *C.mrigala* commonly called mrigala and *C.reba* commonly called reba.

Cirrhinus mrigala : It is generally a North Indian fresh water fish found in Punjab, West Bengal and Orissa, but also is a useful food fish of Karnataka. The body structure of mrigala fish is almost like rohu except for a very wide mouth and thin lips. The body has a characterstic colouration - it is orange on the ventral surface and dark grey on the dorsal surface. A fully grown fish ranges in length between 90 to 100 cms. It has been estimated that the fish can grow more than 20 cms in about 6 to 8 months. The food habits of mrigala are like those of other carps. The food includes sand, mud, decaying vegetable matter, planktonic algae etc.

The fishes attain maturity after about a year when they are ready for reproduction. The females however may mature earlier than a year. Breeding season is same as those for other carp fishes.

Systematic position of *C.mrigala*

Class.....Teleostomi
Order.....Cypriniformes
Family....Cyprinidae
Genus.....*Cirrhinus*
Species...*mrigala*

Cirrhinus reba is another important cirrhina carp fish found distributed throughout the rivers of India. The fish attains a maximum length of 30 cms with males smaller than the females. The fish feeds on planktonic algae, zoo planktons, mud, decaying vegetable matter etc. However the adult fish seem to be entirely vegetarian.

The characterstic features of the reba include hexagonal scales, dark colouration and a small narrowing head.

Systematic position of *C.reba*

Class......Teleostomi
Order......Cypriniformes
Family.....Cyprinidae
Genus......*Cirrhinus*
Species....*reba*

Cat Fishes: Apart from carps, the other important food fish of Karnataka are cat fishes. These are characterised by very small eyes with very well developed barbles. As they have a very powerful vision they easily move about rapidly and are given the name cat fish. Generally cat fish are distributed in rivuletes, streams, lakes, ponds etc. These fishes have a high nutritive value and can grow in water bodies with less amount of oxygen.

The following are some of the examples of cat fishes.

1. *Wallago attu:* Commonly called fresh water sharks, these have a slender and elongated body with a length ranging between 60 to 90 cms. Under favourable growth conditions they may even reach the length of 185 cms. Wallago is a very nutritive and tasty fish and can be a good game fish also. In fish hatcheries it is known that wallago feeds on young carps. As such one has to be careful about the presence of the wallago fish in the stocking ponds.

The bodily features include a large mouth with very sharp teeth. The wallago fish is known to be a meat eater. It can even feed on human carcasses. The breeding of this fish is like other fishes. ie. during rainly season.

Systematic position of *W.attu*

Class......Teleostomi
Order......Cypriniformes
Family.....Siluridae
Genus......*Wallago*
Species....*attu*

2. *Mystus:* The genus Mystus has several species of which the important ones are *M.seenghala*(Tengra),*M.aor*(Dariai tengar) and *M.cavasius*(Sutahawa tengara).

Mystus seenghala is found distributed in river and rivuletes. It has also been reported in the North Indian rivers like Ganga and Jamuna. This fish is a meat eater and is known to attack other small fresh water fish and prawns. The body has a brown dorsal surface and a shiny silvery lateral surface. At maturity it is known to attain a length of 46 cms. But there are instances where the fish may be as long as 180 cms and weighing nearly 10 kilograms. The breeding season of this fish is between April and July.

Systematic position of *M.seenghala*

Class......Teleostomi
Order......Cypriniformes
Family.....Bagridae
Genus......*Mystus*
Species....*seenghala*

Mystus aor has a bluish grey body with a whitish ventral surface that reaches a length of 90 cms. The fish is very widely distributed. Peculiarly in this fish is the female parent does not take part in caring for the offspring.

Systematic position of *M.aor*

Class......Teleostomi
Order......Cypriniformes
Family.....Bagridae
Genus......*Mystus*
Species....*aor*

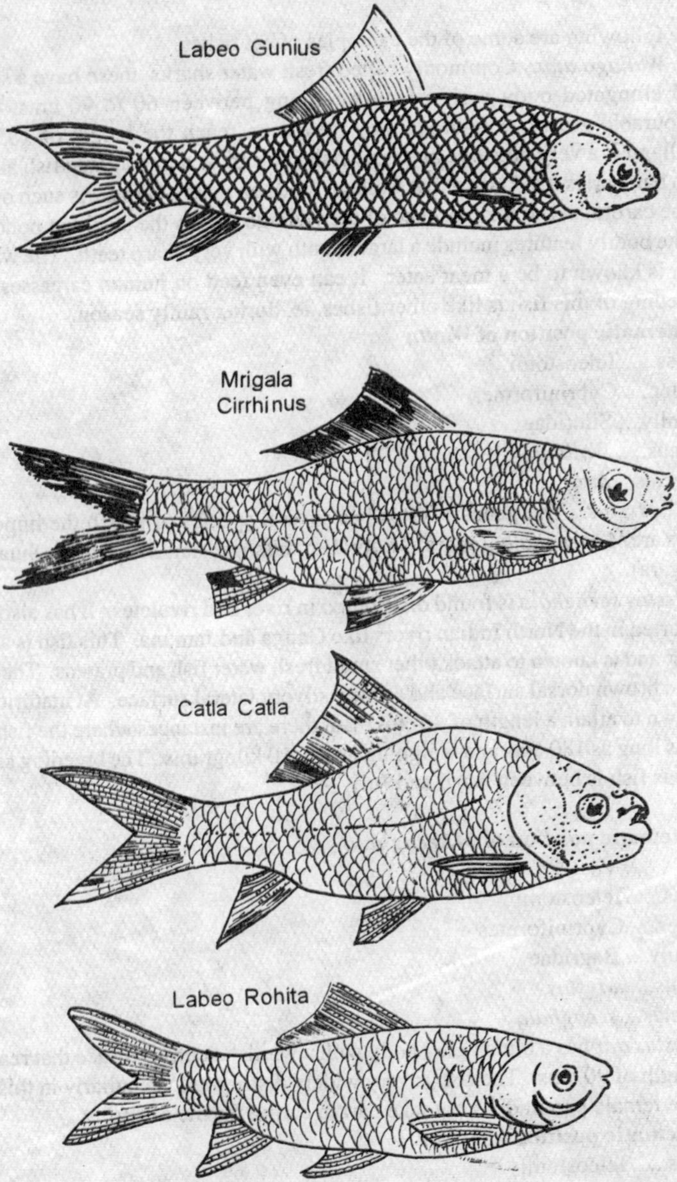

Fig. 38.1 Some edible fishes of Karnataka cat fish

Mystus cavasius is another common fresh water fish found throughout India. Adult fish has a body length up to 45 cms. It is one of the delicious food of human beings. The body has a blakish dorsal surface and slightly yellowish ventral surface.

Systematic position of *M.cavasius*

Class.......Teleostomi
Order.......Cypriniformes
Family......Bagridae
Genus.......*Mystus*
Species.....*cavasius*

3. Heteropneustes fossilis: Commonly called singhi, this cat fish is found in rivers throughout India. The body is about 30 cms long with a wide flattened head. The pectoral of this fish are poisonous. Though they are carnivorous they normally do not feed on fish. Insects, worms, and other vegetable matter constitute the food. The fish breeds between April to June.

Systematic position of *H.fossilis*

Class......Teleostomi
Order......Cypriniformes
Family.....Heteropneustidae
Genus......*Heteropneustes*
Species....*fossilis*

4. Pangasius pangasius: Found very widely distributed in the rivers this fish has an elongated body with a somewhat compressed head. The head is covered by a tender and soft skin. A fully grown fish may be as long as 120 cms. The fish is known to be carnivorous. It feeds on molluscs and also on carps. The breeding season of the fish is during monsoon.

Systematic position of *P.pangasius*

Class.....Teleostomi
Order.....Cypriniformes
Family....Schilbeidae
Genus.....*Pengasius*
Species...*pengasius*

5. Clarias batrachus: This is generally a semi acquatic fish found surviving in the mud but generally found in the fresh water ponds, lakes and rivers. Regarded as one of the delicious fish, it has an elongated body with a laterally compressed head. The colour of the body is reddish brown or greyish black.

Known to attain a length of nearly 45 cms. The fish feeds on prawns,small insects,larvae and vegetable material like algae. The breeding season of the fish is between April to June.

Systematic position of *C.batrachus*

Class.......Toleostomi
Order.......Cypriniformes
Family......Clariidae
Genus.......*Clarias*
Species.....*batrachus*